T. PIENKOWSKI

Human Health and Forests

People and Plants International Conservation Series

People and Plants International (PPI) is a non-profit organization of ethnoecologists devoted to conservation and the sustainable use of plant resources around the world. PPI follows and builds on the 12-year People and Plants Initiative, a joint project of the WWF, UNESCO, and the Royal Botanic Gardens at Kew, UK, which came to an end in December, 2004. Registered as a non-profit organization, PPI is composed of an international steering committee that develops projects in six primary programme areas in regions of high biological diversity.

Applied Ethnobotany: People, Wild Plant Use and Conservation
Anthony B. Cunningham

Biodiversity and Traditional Knowledge: Equitable Partnerships in Practice
Sarah A. Laird (ed)

Carving Out a Future:
Forests, Livelihoods and the International Woodcarving Trade
Anthony Cunningham, Brian Belcher and Bruce Campbell

Ethnobotany: A Methods Manual
Gary J. Martin

Human Health and Forests
A Global Overview of Issues, Practice and Policy
Carol J. Pierce Colfer (ed)

People, Plants and Protected Areas: A Guide to In Situ *Management*
John Tuxill and Gary Paul Nabhan

Plant Conservation: An Ecosystem Approach
Alan Hamilton and Patrick Hamilton

Plant Identification:
Creating User-Friendly Field Guides for Biodiversity Management
Anna Lawrence and William Hawthorne

Plant Invaders: The Threat to Natural Ecosystems
Quentin C. B. Cronk and Janice L. Fuller

Tapping the Green Market:
Certification and Management of Non-Timber Forest Products
Patricia Shanley, Alan R. Pierce, Sarah A. Laird and Abraham Guillén (eds)

Uncovering the Hidden Harvest:
Valuation Methods for Woodland and Forest Resources
Bruce M. Campbell and Martin K. Luckert (eds)

Human Health and Forests

A Global Overview of Issues, Practice and Policy

Edited by Carol J. Pierce Colfer

London • New York

This book is dedicated to

My aunt, **Helen Harris**, who –
along with Albert Schweitzer and the
fictional 1950s nurse, Cherry Ames –
inspired my interest in health

and

The **women of Long Segar**,
who struggle daily with minimal resources
to maintain their own health
and that of their families

First published by Earthscan in the UK and USA in 2008

ISBN: 978-1-84407-532-4 (hardback)
ISBN: 978-0-41584-887-9 (paperback)
Typeset by Domex e-Data, India

Cover design by Susanne Harris

For a full list of publications please contact:

Earthscan
2 Park Square, Milton Park, Abingdon, Oxfordshire OX14 4RN
Simultaneously published in the USA and Canada by Earthscan
711 Third Avenue, New York, NY 10017
Earthscan is an imprint of the Taylor & Francis Group, an informa business

First issued in paperback 2013

Earthscan publishes in association with the International Institute
for Environment and Development

A catalogue record for this book is available from the British Library

Library of Congress Cataloging-in-Publication Data

Human health and forests : a global overview of issues, practice, and policy / edited by
Carol J. Pierce Colfer.
p. cm.
Includes bibliographical references.
ISBN-13: 978-1-84407-532-4 (hardback)
1. Environmental health. 2. Forests and forestry–Health aspects. 3. Forest policy–Health aspects. I. Colfer, Carol J. Pierce.
[DNLM: 1. Environmental Health. 2. Trees. 3. Biological Products. 4. Conservation of Natural Resources. 5. Developing Countries. 6. Disease--etiology.
WA 30 H9168 2008]
RA566.H86 2008
614.4'2252--dc22

2007034789

Contents

PART II – THEMATIC AND REGIONAL HEALTH SLICES

PART III – HEALTH-CARE DELIVERY IN FORESTS

List of Boxes, Figures and Tables

Boxes

Figures

TABLES

Foreword

Forests Need Us – We Need Them

This book is an extraordinary achievement. It comes 20 years after Gro Harlem Brundtland tabled her report, 'Our Common Future', which defined sustainable development as 'development that meets the needs of the present without compromising the ability of future generations to meet their own needs'. The report placed human health at the centre of concerns for sustainable development.

Since then there has been a growing awareness of the relationships between health, the environment and economic development. Much of the focus has been on the rapid pace of urbanization and industrialization and the consequences of this for both health and environment. Far less attention has been given to forests. In fact over the last 20 years, and during the major United Nations conferences on environment and development in Rio de Janeiro in 1992 and in Johannesburg in 2002, there was virtually no discussion about the major topics covered in this book. Rather, they focused almost solely on the need to address deforestation in a general sense. Some attention has been given in recent years to addressing the ecological impact of extractive industries' action in forests; less attention has been given to the human consequences of such development projects in remote forests where forest people's first connection with modernity is associated with forest clearing.

Carol Colfer and the distinguished panel of authors who have contributed to this book have corrected this imbalance and provided us with evidence-based and ethical reasons to seriously address how the health of people and the health of forests are closely entwined.

They show how forest-dwellers are suffering from deforestation; how little we know about the nutritional and health benefits of many fruits, vegetables, tubers and other forest plants at exactly the time that many species are under threat of extinction; and how many recent viral infectious disease outbreaks originate in forests, and affect people only when they intrude into or disrupt the delicate ecology of forests.

I read this book en route to Brazil, where we met with groups working with forest people in the Amazon. Many of the findings in the book were echoed by people we met. They felt that forests held the promise of nutritional and health solutions for urban-dwellers. They highlighted a deep sense of the neglect of their own wellbeing by authorities at the cost of rapid urban development. They shared fears about how their culture and way of life was under threat like never before. This was obvious while flying into São Paulo. Deep and wide cuts are being made into the rich green forests that surround the mega-city, and this process is being repeated in every emerging economy in Asia, Africa and Latin America. The cumulative loss could in time threaten the health of all.

Importantly, the authors do not naïvely call for a hands-off approach to forests. Rather, they list sensible and sound recommendations that would go a long way towards understanding and preserving forests and the myriad species that live in them. They recommend improved governance of forests, engaging political, non-governmental and corporate actors, and specific ways in which the health needs of forest people could be addressed. The need for such recommendations to be acted upon has never been so urgent. New threats to forests arise daily. The latest threats include a push to find forest plants that could be used as biofuels, and proposals to clear forests to cultivate such plants.

This book is truly multidisciplinary and international. It draws upon the wisdom of people with a deep commitment to improving planetary health and is required reading for all who share this commitment and who seek ways of making it a reality.

Derek Yach MBChB MPH
Director of Global Health Policy, PepsiCo
New York
August 2007

Notes on Authors

Robbie Ali, MD, MPH, MPPM, is an American physician specializing in environmental health who has worked on several conservation health projects in Asia and Africa. He is currently Director of the Center for Healthy Environments and Communities at the University of Pittsburgh Graduate School of Public Health.

Pascale Allotey is a Ghanaian-Australian Professor of Race and Diversity in the School of Health Sciences and Social Care and the Centre for Public Health Research at Brunel University in Uxbridge, United Kingdom. She is also Director of the doctoral programme there.

Colin Butler, BMed, DTM&H, MSc, PhD, is an Australian physician and Visiting Fellow at the National Centre for Epidemiology and Population Health, Australian National University.

Manuel Cesario, MD, PhD, FLS, is the Brazilian founder and leader of OASCA – the Advanced Observatory on Health and Environmental Changes, which explores the relationships between the emergence of infectious diseases and socio-environmental changes in southwestern Amazonia. He is now Professor of Sustainable Development and Global Environmental Change at the Graduate Programme on Health Promotion at the University of Franca, São Paulo, Brazil.

Carol J. Pierce Colfer, PhD, MPH, is an American anthropologist with public health training, and Principal Scientist at the Center for International Forestry Research, Bogor, Indonesia.

Tony Cunningham, PhD, is an ethno-ecologist working for People and Plants International, a non-profit organization, and for Charles Darwin University. He has worked widely in tropical and subtropical Africa, Asia and Australia.

Edmond Dounias, PhD, is a French ethno-ecologist at the Institut de Recherche pour le Développement (IRD) in Montpellier, France. He is also seconded to the Center for International Forestry Research, Bogor, Indonesia.

Richard G. Dudley, PhD, is an American fisheries biologist specializing in system dynamics modelling of resource management policy issues. He has worked in Asia, Africa and Latin America, as well as the US, and is an independent consultant, sometimes working for the Center for International Forestry Research.

Pablo Eyzaguirre, PhD, a Chilean anthropologist, is Senior Scientist in Anthropology and Socioeconomics at the Diversity for Livelihoods Programme, Bioversity International, Rome, Italy.

Cynthia Fowler, PhD, is an American ecological anthropologist and Assistant Professor at Wofford College (Spartanburg), South Carolina, USA.

Alexander Fröde, MSc, is a German ecologist with a specialization in sustainable land use and adaptation to climate change in developing countries. He currently works as a Natural Resource Management Advisor for the German Development Service (DED) in Harare, Zimbabwe.

Alain Froment, MD, PhD, is a French biological anthropologist and medical doctor, Research Director at IRD (Institut de Recherche pour le Développement, France), and Curator of Anthropology at the Musée de l'Homme, Paris, France.

Robert W. Gardner, PhD, is an American demographer who does freelance writing and editing for the United Nations Population Division and for Johns Hopkins Center for Communication Programs, Baltimore, MD.

Lisa Gollin, PhD, is an American medical anthropologist, ethno-botanist and Affiliate Researcher with the Ecology and Health Group at the John A. Burns School of Medicine, University of Hawaii, Honolulu, Hawaii, USA.

Gale Goodwin Gómez, PhD, is an American Professor in the Department of Anthropology at Rhode Island College, Providence, Rhode Island, USA. She has a PhD in anthropological linguistics from Columbia University and has worked with the Yanomami in Brazil since 1984.

Jean-Paul J. Gonzalez, MD, PhD, is a French virologist and Research Director at IRD (Institut de Recherche pour le Développement, France). He is presently Visiting Professor of Microbiology at Mahidol University's Faculty of Science, Center of Excellence for Vectors and Vector-Borne Diseases, in Bangkok, Thailand.

Meriadeg Ar Gouilh is a French mammalogist with an MSc in Systematics, Taxonomy and Phylogeny from the Muséum National d'Histoire Naturelle de Paris (MNHN). Since taking his PhD at the MNHN (ecology and evolution of viruses carried by Chiroptera and wildlife in Thailand) he has become a research scientist specializing in evolutionary biology at the French Institute for Research and Development (Faculty of Science, Mahidol University, Thailand) and collaborates with the CIBU unit of Pasteur Institute Paris (Biological Emergency Response Unit).

Margaret Gyapong, PhD, is a Ghanaian medical anthropologist and Head of the Dodowa Health Research Centre on the southern coast of Ghana, in the Greater Accra region.

Timothy Johns, PhD, is a Canadian nutritionist and ethno-botanist, Professor of Human Nutrition at McGill University in Montreal, Quebec, Canada, and Honorary Fellow for Nutrition in the Diversity for Livelihoods Programme of IPGRI-Bioversity International, Rome, Italy.

Witness Kozanayi is a Zimbabwean researcher and MSc student based in the UK, specializing in people-centred research and development approaches. He works as Consultant for the Center for International Forestry Research in Harare, Zimbabwe. He is also Research Fellow at the Shanduko Trust, Zimbabwe.

Sally A. Lahm, PhD, is an American ecologist and Research Associate at the Institut de Recherche en Ecologie Tropicale, Makokou, Gabon, and the Department of Conservation and Research Center, Smithsonian Institution, Washington, DC, USA. She is Chief Environmental Scientist at Ecology and Environment, Inc., Lancaster, New York, USA.

Sarah Laird is an American doctoral candidate in anthropology at University College London, and Director of People and Plants International, New York, USA.

Eric A. Leroy, DVM, PhD, is a French immunologist and Research Director at IRD (Institut de Recherche pour le Développement, France). He is presently working at the Laboratoire des Maladies Emergentes (Laboratory of Emerging Diseases) at the Centre International de Recherche Médicale (International Medical Research Centre), Franceville, Gabon.

Godwin Limberg, MSc, is a Dutch agronomist working as Consultant at the Center for International Forestry Research (CIFOR) in East Kalimantan, Indonesia.

Pascal Lopez, PhD, is a German tropical forestry scientist and consultant with the ECO Consulting Group, Oberaula, Germany.

Heather McMillen is an American PhD candidate specializing in medical anthropology and ethno-botany at the Department of Anthropology, the University of Hawaii, Honolulu, Hawaii, USA.

Nontokozo Nemarundwe, PhD, is a Zimbabwean sociologist working as Consultant for the Center for International Forestry Research in Harare, Zimbabwe. She is also Research Associate at the Institute of Environmental Studies at the University of Zimbabwe.

Subhrendu K. Pattanayak, PhD, an Indian environmental economist, is a Fellow at RTI International and Associate Professor at North Carolina State University, both in the Research Triangle area of North Carolina, USA.

Gerard A. Persoon, PhD, is a Dutch anthropologist working at the Institute of Environmental Sciences at Leiden University, Leiden, The Netherlands.

Jean-Marc Reynes, DVM, PhD, is a French medical virologist from the Réseau International des Instituts Pasteur (Pasteur Institute International Network). He is presently Head of the Virology Unit at the Institut Pasteur de Madagascar, Antananarivo, Madagascar.

Maria Nanette Ramiscal Roble is of Philippine origin, but with long-term residence in the US and now in France. Her MSc is in Public Health with a specialization in Socio Medical Sciences, and special interests in women's health, child health and reproductive health including attention to traditional medicine.

Patricia Shanley, PhD, is an American ecologist working with the Forests and Livelihoods Program of the Center for International Forestry Research (CIFOR) based in Bogor, Indonesia.

Kirk R. Smith, PhD, is an American environmental scientist and Professor of Global Environmental Health at the School of Public Health, University of California, Berkeley, California, USA.

Anne Marie Tiani, Doctorate, is a Cameroonian ecologist who works as Consultant for the Center for International Forestry Research and other institutions, in Yaoundé, Cameroon.

Barbara Vinceti, PhD, an Italian forest ecologist, is Scientist for Forest Biodiversity in the Understanding and Managing Biodiversity Programme of IPGRI-Bioversity International, Rome, Italy.

Junko Yasuoka, DSc, MPH, is a Japanese ecologist and Assistant Professor in the Department of International Community Health, Graduate School of Medicine, University of Tokyo, Japan.

Acknowledgements

This book has had a chequered history. It was originally conceived in the late 1990s by Tony Cunningham and Tim Johns. It got a shot in the arm in October 2003, when the Center for International Forestry Research convened a mini-workshop on health and forests, inviting Erin Sills, Lisa Gollin and Tony Cunningham to Bogor, Indonesia as resource persons to discuss the issues with interested staff. Then in a June 2004 workshop, organized by Misa Kishi in Canterbury, England, an outline for the book was actually produced. Various obstacles were hurdled, and finally in August 2005 Carol Colfer agreed to take on the editing task, which proved to be considerably more complicated than originally anticipated. Almost none of the authors who had originally intended to contribute were able to do so. So we have a near-totally new group of authors from a wide range of institutions, disciplines, and countries, who bring exciting perspectives to bear on the subject.

We want to thank our respective institutions for their support during the writing and editing process. CIFOR has borne the cost of Carol Colfer's time as she coordinated this process between 2005 and 2007. She is appreciative of the support she has received from Doris Capistrano, Director of the Governance Programme; Bruce Campbell, Director of the Forests and Livelihoods Programme; and Markku Kanninen, Director, Forests and Environmental Services; as well as our Publications Committee, led by Michael Hailu. Thanks also go to CIFOR geographic information system specialist, Atie Puntodewo, for her map- and figure-making skills, and to Rahayu Koesnadi for her administrative backup. Sally Atwater, an independent technical editor, turned our sometimes ponderous prose into much more readable form, and we thank her sincerely. We would also like to thank SwedBio, which (through funds from Sida) has supported CIFOR in the conduct of stakeholder meetings for disseminating these and other research findings.

We would like to thank Tony Cunningham for his support in including this book in the People and Plants Series. Thanks also to Rob West for his continuing interest, even as the process took longer than originally anticipated, and to Rob's colleagues at Earthscan: Gudrun Freese, Hamish Ironside and Alison Kuznets. We also thank Peter Mayer of IUFRO, and Hannu Raitio, coordinator of the Task Force on Forests and Human Health, and Eeva Karjalainen, who also works on this Task Force, for their support of this effort. This book contributes to the work of this task force.

List of Acronyms and Abbreviations

$\mu g/m^3$	micrograms per cubic metre
ACM	adaptive collaborative management
ADB	Asian Development Bank
AIDS	acquired immune deficiency syndrome
ALRI	acute lower respiratory infections
ASEAN	Association of South-East Asian Nations
AYUSH	Department of Ayurveda, Yoga, Unani, Siddha and Homeopathy (Indian government)
BMI	body mass index
CCPY	Pro-Yanomami Commission (Brazil)
CIFOR	Center for International Forestry Research (Bogor, Indonesia)
CITES	Convention on International Trade in Endangered Species of Wild Fauna and Flora
COPD	chronic obstructive pulmonary disease
CORI	Community Outreach Initiatives (Indonesian NGO)
DALY	disability-adjusted life year
DDT	dichlorodiphenyltrichloroethane
DED	German Development Service
DEET	N,N-diethyl-3-methylbenzamide
EPA	Environmental Protection Agency (USA)
FAO	Food and Agriculture Organization of the United Nations
FUNAI	National Indian Foundation (Brazil)
FUNASA	Brazilian National Health Foundation
GDP	gross domestic product
GIS	geographic information system
GTZ	Gesellschaft für Technische Zusammenarbeit, a German development agency
HIV	human immunodeficiency virus
HTLV	human T-lymphocyte virus
ICDP	integrated conservation and development project
IRD	Institut de Recherche pour le Développement, France
ISSC-MAP	International Standard for Sustainable Wild Collection of Medicinal and Aromatic Plants
KCHP	Kelay Conservation Health Program (Indonesia)
LPG	liquefied petroleum gas
NGO	non-governmental organization

OAPI	African Intellectual Property Organization for French-Speaking Africa
PKDL	post-*kala-azar* dermal leishmaniasis
ppm	parts per million
R&D	research and development
SARS	severe acute respiratory syndrome
TAWG	Tanga AIDS Working Group (Tanzania)
TB	tuberculosis
THETA	Traditional and Modern Health Practitioners Together Against AIDS (Uganda)
TNC	The Nature Conservancy
TRAMIL	Traditional Medicine in the [Caribbean] Islands
UNAIDS	Joint United Nations Programme on HIV/AIDS
UNESCO	United Nations Educational, Social and Cultural Organization
UNICEF	United Nations Children's Fund
URIHI	Yanomami Health (Brazilian NGO)
USAID	United States Agency for International Development
WHO	World Health Organization
WWF	World Wide Fund for Nature

1

Introduction[1]

Carol J. Pierce Colfer

Every author in this book has a story to tell about the beginning of his or her interest in human health and forests, and about the experiences that have led them to make the analyses they provide here. I begin with a short account of what gave birth to my own interest in this topic.

In September 1979, I made my first trip up a wide, winding, muddy river in East Kalimantan. I was enchanted by the lush greenery along the banks, the monkeys jumping from branch to branch above, and the pure adventure of it. I was headed for the village of Long Segar. Some two days and a night by slow riverboat from the provincial capital, Long Segar was inhabited by about 1000 Uma' Jalan Kenyah Dayaks, who had moved there voluntarily, beginning in 1963, from the very remote Apo Kayan in Central Borneo. They had come seeking better access to consumer goods, education and medical care; they had left their homeland because of religious and land-tenure conflicts (see Colfer and Dudley, 1993; Colfer et al, 1997; Colfer, in press). In 1975, the Javanese-dominated Indonesian government had classified the new village as a 'resettlement' village. The many resettlement programmes were designed to bring 'primitive' (*terasing*) peoples out of the hinterlands, to give them the benefits of 'development' and to 'civilize' them.

The resettlement project changed things for the Dayaks. The government provided materials for housing, contingent on the people changing from their traditional longhouses (rather like modern condominiums or apartment blocks) to individual homes. This was a significant change with implications for community integrity, social capital, communications and even gender equity. The government also gave planting materials, with specific instructions and maps showing where each plant should go. This resulted in dramatic changes in the settlers' home gardens, with implications for diet as well. Governmental efforts to increase people's incomes involved the distribution of 'seed' cattle – beasts with which the people were completely unfamiliar – with the idea that one calf could be kept and the next distributed, until everyone had a cow. The people went some distance to their rice fields, however, leaving their cattle in the village for days at a time. Free-ranging cattle wreaked havoc in the gardens and failed to prosper. The government supplied an extension agent, whose mandate (as in all Indonesian forested areas) was to convert the people from swidden to permanent agriculture, specifically paddy rice (the agricultural form most valued on Java). However, the topography did not allow for natural irrigation, there were no pumps and the soil was acid and infertile.

The surrounding area was granted as a concession to a foreign timber company, which also tried, with little success, to minimize swidden agriculture – even though it was the only viable system for growing subsistence foods in that context – and to control the people's use of forest products. Besides removing valuable timber and destroying other forest products, the activities of the company increased the sediment load in the river, with adverse effects on water quality and, probably, fish catches. Toxic chemicals used to protect the felled trees from insects and disease leached into the rivers.

The local people's ill-health was obvious even to me, an anthropologist. During my first month there, nine children died of measles; infant deaths were commonplace. Although not competent to make definitive diagnoses and having no laboratory diagnostic tools, I noted the light brown hair, skinny arms and pot-bellies apparent on many of the children, symptoms that my public health training had taught me often indicate nutritional deficiencies and worms. Women complained of aching legs and fatigue, which I suspected indicated calcium deficiencies and anaemia, among other things. Dramatic infections that spread across people's skin were common. The people were surprised at my tears when their children died; death was too common for people to grieve seriously for the fate of other people's children.

Then, after some eight months in Long Segar, I went to Long Ampung, the village from which Long Segar residents had gradually moved, starting in 1962. It was far inland, near the Malaysian border with Sarawak in the Apo Kayan. Long Ampung had a small population: about 500 people (in 1980) had lived there for decades, using the lands around their community in a 10–15-year rotation, which maintained soil fertility and reasonable harvests of rice and other crops. Their longhouses were surrounded by plants with edible fruits, and the forests were full of animals, particularly the valued Borneo pig. The river that flowed through the village was crystal clear. The people appeared much healthier. The children had well-rounded bodies, and people said that few died in childhood (none died during any of my visits there). Health problems existed, but apparently to a lesser degree than in Long Segar.

This experience made me ponder the interactions among people, health and forests. The links between people and forest foods was obvious, as was the role of forest regrowth in returning fertility and organic matter to the soil, enabling it to produce sufficient rice. Some diseases were more common in forests. Local people had forest medicines, the efficacy of which I was unsure. The intricacy with which the forest was interwoven with their cultural system was clear. The meaning in their lives – and thus their mental health – was dependent on the forest in very profound ways.

Some of the authors represented in this book had similar formative experiences. Others have approached human health and forests from ecological perspectives, starting with the plants and their habitats and then tracing the links to human beings. Still others began with an interest in specific diseases but gradually realized how ecology and human characteristics and behaviour interact with diseases. In many of the chapters, co-authors have used their disciplinary differences to bring to light interactions often neglected by single disciplinary analyses. One of the strengths of this book is the diversity of perspectives that it brings to bear on the broad topic of human health and forests.

THE GENESIS OF THIS BOOK

In this book, we examine people, health and forests, and provide some sense of the universe of experience about the connections, both positive and negative. The interaction between forests and human health seems intuitively to be a reasonable topic to investigate. Yet we have found that the many papers and books written on the subject are widely scattered in a whole host of disciplines, making it difficult for researchers and students alike to get a handle on the issues. This book is designed to highlight the central issues, as well as expose readers to the wealth of literature on the subject.

We initially approached this topic from the forestry angle, where health issues have only recently emerged as worthy of attention. In the mid-1990s, a series of interdisciplinary and international field teams in Austria, Brazil, Cameroon, Côte d'Ivoire, Germany, India, Indonesia, the USA, and later in Australia and South Africa identified human health as a central issue in the sustainable management of forests (summarized in, for example, CIFOR, 1999).

A separate stream of results on the health of people living in forests has come out of the anthropological literature. Anthropologists conducting ethnographic studies around the world often found forest peoples' health conditions to be abysmal (e.g., Harper, 2002); yet they also often found many positive aspects of indigenous forest-based health-care systems, particularly in terms of nutrition, medicinal plants and approaches to caregiving (for example, see Leaman, 1996). Within the discipline of medicine, investigations of people's health in forests tend to focus on a single disease rather than the health of the population at large, and formal medical practitioners are likely to neglect traditional medicine. A further complication, of course, is the common lack of political will in developing countries to address the health needs of ethnic minorities (who often live in forested areas); donors are also guilty of neglecting these people.[2]

A small team at the Center for International Forestry Research (CIFOR) therefore convened a series of expert meetings and conducted a literature survey on the topic of forests and health (Colfer et al, 2006). The report identified four main topics of relevance: food and nutrition, disease, medicinal plants and forest-based cultures. However, the survey in many ways raised more questions than it answered, and more research is needed to understand the complex links between forests and human health. I argue that the *kind* of research needed is action research that links researchers and study populations in shared efforts to solve forest health problems.

A further impetus for this book was the Millennium Development Goals, which are symbolic statements of global values. Four of the goals focus specifically on health-related topics: Goal 1 (to eradicate extreme poverty and hunger); Goal 4 (to reduce child mortality); Goal 5 (to improve maternal health); and Goal 6 (to combat HIV/AIDS, malaria and other diseases). Goal 7 (to ensure environmental sustainability) provides an additional stimulus for our work; many have argued that improvements in human health (as part of human wellbeing) are a crucial prerequisite for accomplishing this goal.

Two more Millennium Development Goals stress or imply gender equity: Goal 2 (to achieve universal primary education); and Goal 3 (to promote gender equality and empower women). These goals also have fairly direct implications for human health, given the central role women typically play in family health. In most places, it is women who provide families with nutritious meals and maintain standards of hygiene. In forested areas, women often gather non-timber forest products, collect water and engage in forest agriculture. As the primary caregivers when other family members fall ill, forest women often use medicinal forest products. Finally, women's decisions about family size affect not only their own health and that of their offspring, but also the health of forests as well.

With all this in mind as background, this book offers a collection of essays on topics that we consider central to the study of human health. The book pulls together diverse approaches to this field of study and highlights the critical need for improved interdisciplinary communication among those concerned with forests and human health.

FOUR CRITICAL ISSUES

In thinking about forests and human health, I would like to highlight four observations that I consider important to bear in mind while reading this book. The first is the comparative lack of attention paid in the health field to people living in forests. The rationale for this neglect always boils down to the facts that forests have few people, resources are limited and health planners are seeking 'the biggest bang for the buck'. These are legitimate points. However, this collection shows clearly that there are serious health problems in forests, and that forest diseases, vectors and reservoirs affect populations far beyond the forest edge.

In the aforementioned survey of health and forests, I was struck by the preponderance of health-related studies conducted in and around cities, even in the medical anthropological literature, where I expected to find a wealth of health findings related to forests. The urban orientation of highly trained personnel, such as medical doctors and nurses, anthropologists, foresters and ecologists, is another factor that cannot be disregarded. The lack of power, wealth and prestige accruing to most forest-dwelling groups ensures that they do not have the political clout to make their desires and needs felt. A related contributing factor is the fact that professional prestige rises as one is able to distance oneself from the poor and the uneducated. New recruits in forestry are sent to the field, while successful senior foresters work in urban-based headquarters; young anthropologists go to the field, but older ones are likely to teach or supervise from afar; in some countries young doctors are required to spend a few years in rural areas, but their reluctance is widely recognized, and few remain as their careers progress. The findings from this collection suggest that real thought should be given to this pattern. Senior people should at least periodically spend time in villages (including forest villages), and greater efforts should be made to institutionalize better health care in forested areas.

A second observation is the significance (and neglect) of forest peoples' cultures. Important aspects of these peoples' cultures involve forests, aspects that give meaning to their lives (see collections by Dove, 1988; Hladik et al, 1990; Croll and Parkin, 1992; Redford and Padoch, 1992). Just as 'civilized' Americans and Europeans value their Christmas trees, so the 'uncivilized' forest-dwellers attach potent symbolic meaning to the forests. The difference is that Westerners have the power and media control to make their forest symbols visible; not so most forest-dwellers. Few outsiders know the meaning of the forest to any given forest community; we routinely 'manage' tropical forests knowing virtually nothing of their significance to local people. That needs to change, if we are collectively to manage forests and improve people's lives and health more effectively. This is most dramatically relevant for mental health, but of course mental health has implications for physical health.

I am not here suggesting some form of 'cultural tailoring', but rather an iterative process that involves ongoing back-and-forth communication between forest peoples and those who would manage their lives (foresters, ecologists, medical practitioners, governments). Such a process would require attitudinal changes among the educated and powerful (e.g., more respect for local communities, openness to their input) and changes in access to information and other opportunities for forest peoples that would make the playing field more level and empower them.

Meanings vary enormously from culture to culture and time to time. The cultures of adjacent groups in Borneo or Cameroon can be nearly as different as either of them is from French or Japanese culture. We cannot in good conscience paint the 'culture of forest peoples' with one broad stroke, any more than we can so paint communities or women. The variations among cultures, and within a given community or a given gender, are enormous, and they are always changing. Policy-makers throw up their hands at the thought of dealing with such dynamism and complexity, but they need not do so. Governments (and other actors affecting human and forest health) must devise more flexible, bottom–up mechanisms that invite local people more effectively into the policy-making process. No one knows better what is important to local people than they do themselves (though there are also obvious benefits from marrying local and outsider knowledge).

The third observation builds on the second. The contributions in this book show clearly the complexity of ecological forest systems, and the matrix or network of connections between them and the complex human systems that have evolved within forest contexts. The interactions – among seasonality, habitats, eating habits and nutrition (Chapter 4); bats, their behaviour, SARS and other diseases (Chapter 8); or malaria, gold mining, forest clearing and near-genocide (Chapter 11), among many others – demonstrate the complexity of the interconnections (see also Patz et al, 2000). Whether we are looking at health and forests from the perspective of an anthropologist, a virologist, a public-health specialist or a forester, the same level of complexity emerges. What this observation implies is the strong need for expanding our repertoire of research approaches. There is no question that conventional science – in which a specific problem is identified, a hypothesis is formed, experiments are planned and conducted, and the hypothesis is accepted or rejected – has contributed enormously to human welfare, and it will continue to do so. However, for the pressing human and environmental problems in forests, we need some new tools. We need to look at those patterns and interconnections, we need to observe processes and structures, we need to include holistic approaches that take into account total systems. Based on such detailed understanding, we can then hone in on specific conventional research topics. To do this effectively, we – the researchers, project implementers, government officials and service providers – need to communicate our findings and problems more effectively across disciplinary and administrative boundaries. We need to learn each other's languages, professional traditions and norms of interaction (all of which differ markedly); we need to publish in and read each other's journals; we need to speak at each other's professional meetings.

I realize that policy is often *not* evidence-based, as implied by the course of action proposed here. Certainly, broader economic and political forces play important roles in determining what actually happens on the ground, and vested interests can thwart any efforts to insert rational or humane decision-making into policy spheres. However, that does not mean that we should not continue to pursue truth and human and environmental wellbeing through better science and other means at our disposal. We are not impotent.

Finally, this collection brings home dramatically the need to expend more effort trying to *anticipate* change (again, holistically). The authors provide many examples of well-meaning efforts to improve health or human wellbeing that have had exactly the opposite effect. This is not a new observation, but as global warming increases the pressures on the Earth (Menne and Ebi, 2006), we need to think more seriously about the implications of these changes and try to anticipate, rather than just react. Several authors have addressed the possible health implications of climate change.

We now have new techniques to help us anticipate. 'Future scenarios', used interactively, allow groups to discuss current situations and their hopes and projections for the future (e.g., Butler et al, 2003; Martens et al, 2006). System dynamics modelling, another method for incorporating qualitative variables (e.g., Dangerfield et al, 2001), allows iterative approximation of current situations as well as investigation of possible future implications, given particular changes. We need to take full advantage of any methods that can help us anticipate what the future holds, and then make health and forest plans that address these probable problems.

OVERVIEW OF THE BOOK

We begin with a series of synthetic overviews (Part I) to set the stage and expose some of the central issues and their interconnections. This section also conveys the diversity of terminology and approaches that characterize investigations into health and forests. We have tried to simplify language as much as possible, but different disciplines approach issues differently, and that is patently clear throughout this multidisciplinary collection.

In Chapter 2, Colin Butler, a physician who worked on the Millennium Ecosystem Assessment, observes that relationships between forests, health and other aspects of human wellbeing are complex, interlinked, bidirectional and multiscaled, over both space and time. He builds on my brief introduction by drawing out a correspondingly broad range of issues that pertain to health and forests. He sets the stage for the more detailed investigations of specific topics that follow.

Chapter 3 (by non-timber forest products specialist Anthony Cunningham, ecologist Patricia Shanley and economist Sarah Laird) examines the role of medicinal plants in the health care of people living in and beyond tropical forests around the world. One important contribution of this chapter is its emphasis on the ubiquity and value of medicinal forest plants outside the forest. Another is its analysis of the threats to medicinal plants and to the forests where they grow.

The focus on medicinal plants is followed in Chapter 4 by a focus on foods. Forest ecologist Barbara Vinceti, anthropologist Pablo Eyzaguirre and nutritionist Tim Johns provide an illustrative survey of the relevance of forest foods in maintaining the nutritional status of people living in and around forests, stressing the importance of biodiversity maintenance for human beings. Besides documenting the nutritional value of many forest foods, the authors emphasize the particular relevance of such foods for vulnerable groups as a recurrent seasonal safety net, and during times of crisis.

Many of the foods discussed in Chapter 4 require cooking to make them palatable, and in most forests of the world, firewood is the fuel. In Chapter 5, Kirk Smith, an environmental scientist and public health specialist, provides a worrying overview of decades of research on the adverse effects of the routine use of firewood for cooking, particularly on women and small children. He provides details of the constituent parts of smoke and the exposure levels of various groups within society. He divides his survey of the epidemiological evidence on the effects of smoke inhalation into 'well accepted', 'highly suggestive' and 'speculative', and discusses the disease implications of each.

Chapter 6, by myself (an anthropologist), Richard Dudley (a biologist and system dynamicist) and Robert Gardner (a demographer), follows up on Smith's concern with women. After a discussion of global population issues and an introduction to causal loop diagramming, we focus on the 'win–win' aspects of attention to population, both for women and for the health of forest peoples and forests. We argue that access to birth control technology can increase women's income-generating potential, access to education, participation in community action and family health, as well as reducing rates of population growth, with often positive effects on both local environments and women's status.

Our emphasis on women's health continues in Chapter 7 (by Pascale Allotey, a public health specialist, and Margaret Gyapong and myself, both anthropologists). This chapter surveys the literature on common diseases found in forested areas, highlighting the differences based on gender. The authors examine eight of the important diseases affecting peoples living in and near forests, concluding that in many cases, critical (and often ignored) differences in exposure to disease, disease presentation, cultural preferences, treatment by medical personnel, effects of the disease on health and other factors have important gender implications.

Chapter 8 (by virologists Jean Paul Gonzalez and Jean-Marc Reynes, phylogenist and taxonomist Meriadeg Ar Gouilh and immunologist Eric Leroy) uses a disease vector, bats, as an entry point. The authors introduce the lives of bats, showing how their forest contexts and behaviour contribute to the spread of a variety of diseases. They then hone in on SARS, Ebola fever and Nipah virus, all scourges that affect forest peoples (among others), providing detailed analyses of the interactions among environment, vector, disease and humans for these three diseases. The complexity of these interactions for even this one vector, bats, is indicative of the scientific challenge before us as we seek to understand the relationships among forests, people and health.

Continuing the emphasis on disease, Chapter 9 (by economist Subhrendu Pattanayak and epidemiologist Junko Yasuoka) uses human ecological insights to examine the multifarious interconnections between disease and the processes of ecosystem change. Their analyses, which focus primarily on malaria, proceed from global through regional (Brazil) to local level (Indonesia), in each case looking at the interactions among variables and emphasizing the importance of human behaviour in understanding human health and disease.

In Part II, the focus changes from global synthesis to thematic and regional issues pertaining to forests and health. The intention is to dig deeper into a few examples that represent wider problems.

Focusing again on a specific disease, Chapter 10 (by forester Pascal Lopez), looks at the links between HIV/AIDS and forests in southern Africa. Lopez describes the ravages that the disease has wrought on the natural resource management establishment – similar to reports from Madagascar by participants at a 2004 USAID (United States Agency for International Development) workshop in Bangkok on linking health, population and the environment, who considered HIV/AIDS infection rates among natural resource managers a serious constraint to sustainable management – as well as the important roles the forest plays in providing food and income to families impoverished by illness. The author makes a strong plea for greater attention to HIV/AIDS within the forest sector and for greater interaction between forestry and public health establishments in the region.

Chapter 11 (by anthropologist Gale Gómez) is a careful, longitudinal case study of how forest-cover change has affected the Yanomami, a quintessential forest-dwelling group in South America. Gómez documents the intrusion of gold miners, disease, ranchers and others, and the resulting deterioration of health and lifespan among the indigenous peoples. This case, a more dramatic version of the Borneo experience mentioned at the beginning of this book, clarifies the interconnections among health, forests and people's lives in one context.

In Chapter 12, biological anthropologist and physician Alain Froment takes us to Central Africa, where he examines the health of another group of forest-dwelling peoples, the Pygmies. His perspective is evolutionary, evaluating the fit between these forest people's way of life, level of disease and their environment. He provides useful comparisons between the health of the hunter-gatherer Pygmies and that of the more settled swidden agriculturalists to whom the Pygmies are tied economically; and he provides longitudinal data on the changes that have adversely affected the true forest-dwellers. Because of such changes, the traditional foraging way of life can probably no longer be sustained, and indeed, there are now just three families of one such group, the hunter-gatherer Penan of Sarawak, still living their traditional lifestyle in the forest (Brosius, 2006).

Ethno-botanist Edmond Dounias and I continue this thread in Chapter 13, building on Dounias's experience in Central Africa and complementing it with our respective research in Indonesia. We compare cultural aspects of the lives of Indonesian and Central African forest-dwellers, examining particularly the sociocultural dimensions of diet and health. Two boxes are presented, one methodological and one ethnographic, to provide a fuller understanding of the cultural aspects of health and forests, an issue that is difficult to convey succinctly because of the complexity of human systems.

Part III focuses on health-care service in remote, forested areas, using two specific cases from Indonesia to illustrate the difficulties. In Chapter 14, anthropologist Cynthia Fowler provides a global survey of attempts to integrate traditional medicine with national public health initiatives. She begins with the historical background of the effort to develop an integrative medicine, and proceeds to describe some national public health policies that strive to make this goal a reality. After describing integrative policies and how they have worked, she discusses the barriers to and conducive conditions for such integration, concluding with some thoughts on post-colonialism, globalization and integrative medicine.

Chapter 15, by physician Robbie Ali, provides a case study of an attempt to integrate comprehensive health care and conservation in Borneo, in Indonesia. He begins by discussing the different conceptual approaches to combining health care and conservation. After describing the context, he describes what The Nature Conservancy and its Indonesian partners (Community Outreach Initiatives and the Ministry of Health) have done in East Kalimantan to address the serious health needs of the Punan, the local hunter-gatherers. His analysis includes assessment of the impacts of their work, on both people's health and the environment.

Anthropologist Gerard Persoon, in Chapter 16, provides another case from Indonesia, this time from Siberut (a remote island just west of Padang, West Sumatra). He compares the people's health status in the past and the present, and analyses what has gone wrong in previous attempts to improve people's health there. He then describes a new effort that involves a bottom–up approach, about which he is cautiously hopeful.

The final chapter pulls together seven features and issues that recur throughout this collection: gender, cultural difference, the roles of insects and animals in disease transmission, emerging diseases, medicinal plants, forest foods, and deforestation and land-use change. This chapter concludes with some recommendations for action, focused particularly on suggestions for more action research and improved interdisciplinary collaboration.

In this book, we have tried to capitalize on a diversity of perspectives to provide a well-rounded view of human health and forests. Our contributors are 15 women and 20 men, representing 14 countries (Australia, Canada, Chile, France, Germany, Ghana, India, Italy, Japan, Netherlands, the Philippines, the UK, the USA and Zimbabwe). We collectively have expertise in anthropology (including ecological, physical, biological, cultural, linguistic and medical), botany (including ethno-botany), demography, ecology, economics, environmental sciences, epidemiology, evolutionary biology, fisheries biology, forestry, forest ecology, immunology, medicine, microbiology, non-timber forest products, nutrition, phylogeny and taxonomy, public health, system dynamics and virology.

Our primary goal is to alert researchers, as well as anyone concerned broadly about forest conservation, development or human health, to the overlapping concerns that we share. Our hope is that this book can provide insights that will lead decision-makers to useful, effective entry points for improving both human and environmental wellbeing.

NOTES

1 This chapter has benefited from the helpful critiques of Peter Kunstadter, Mike Arnold, Pablo Eyzaguirre and Subhrendu Pattanayak, though they bear no responsibility for remaining errors.
2 Thanks to Peter Kunstadter for these observations (January 2007).

References

Brosius, P. (2006) personal communication, 16 November

Butler, C.D., Chambers, R., Chopra, K., Dasgupta, P., Duraiappah, A., Kumar, P., McMichael, A.J. and Wen-Yuan, N. (2003) 'Ecosystems and human well-being', in *Ecosystems and Human Well-being: A Framework for Assessment*, Island Press, Washington, DC, pp71–84

CIFOR (1999) *C&I Toolbox*, CIFOR, Bogor, Indonesia

Colfer, C.J.P. (in press) *Longhouse of the Tarsier: Changing Landscapes, Gender and Well Being in Borneo*, Borneo Research Council, Williamsburg, VA

Colfer, C.J.P. and Dudley, R.G. (1993) *Shifting Cultivators of Indonesia: Managers or Marauders of the Forest? Rice Production and Forest Use among the Uma' Jalan of East Kalimantan*, Food and Agriculture Organization of the United Nations, Rome

Colfer, C.J.P., Peluso, N.L. and Chin, S.C. (1997) *Beyond Slash and Burn: Building on Indigenous Management of Borneo's Tropical Rain Forests*, New York Botanical Garden, New York, NY

Colfer, C.J.P., Sheil, D. and Kishi, M. (2006) 'Forests and human health: Assessing the evidence', CIFOR Occasional Paper 45, CIFOR, Bogor, Indonesia

Croll, E. and Parkin, D. (eds) (1992) *Bush Base, Forest Farm: Culture, Environment and Development*, Routledge, London

Dangerfield, B.C., Fang, Y. and Roberts, C.A. (2001) 'Model-based scenarios for the epidemiology of HIV/AIDS: The consequences of highly active antiretroviral therapy', *System Dynamics Review* 17(2): 119–150

Dove, M. (1988) *The Real and Imagined Role of Culture in Development: Case Studies from Indonesia*, University of Hawaii Press, Honolulu, HI

Harper, J. (2002) *Endangered Species: Health, Illness and Death among Madagascar's People of the Forest*, Carolina Academic Press, Durham, NC

Hladik, C.M., Bahuchet, S. and de Garine, I. (eds) (1990) *Food and Nutrition in the African Rain Forest*, UNESCO/MAB, Paris, France

Leaman, D.J. (1996) *The Medicinal Ethnobotany of the Kenyah of East Kalimantan (Indonesian Borneo)*, Ottawa-Carleton Institute of Biology, University of Ottawa, Ottawa, Canada

Martens, P., Lorenzoni, I. and Menne, B. (2006) 'Implications of the SRES scenarios for human health in Europe', in B. Menne and K.L. Ebi (eds) *Climate Change and Adaptation Strategies for Human Health*, Steinkopff Verlag Darmstadt, Heidelberg, Germany, pp395–407

Menne, B. and Ebi, K.L. (eds) (2006) *Climate Change and Adaptation Strategies for Human Health*, Steinkopff Verlag Darmstadt, Heidelberg, Germany

Patz, J.A., Graczyk, T.K., Geller, N. and Vittor, A.Y. (2000) 'Effects of environmental change on emerging parasitic diseases', *International Journal for Parasitology* 30(12–13): 1395–1405

Redford, K.H. and Padoch, C. (1992) *Conservation of Neotropical Forests: Working from Traditional Resource Use*, Columbia University Press, New York, NY

PART I – SYNTHETIC ANALYSES

2

Human Health and Forests: An Overview

Colin D. Butler

Although the causal pathways are complex and indirect, human health has in general benefited from the clearance and modification of forests for agriculture and cities. For example, the vast forests of Europe have largely been cleared, and the population supported by that land today is mostly prosperous and healthy. It is also much larger. Although the health of at least some hunter-gatherer populations appears to have surpassed that of many early and even some recent agriculturalists (Sahlins, 1972; Diamond, 2002; Holden, 2006), the larger populations facilitated by forest clearance have stimulated and permitted many of the specializations and technologies that make modern civilization possible. Some people (including some readers) may aspire to the allegedly simpler and perhaps richer life of our long-dead forest-dwelling ancestors, but forest clearance has generally been associated with improvements in human wellbeing. In China, where forests have been largely cleared, the health and prosperity of its truly vast population are improving, and a future that provides abundant ecosystem services, high technology, a long life expectancy and wellbeing for the average citizen remains possible. Nevertheless, the clearance and manipulation of forests have often harmed people, and this phenomenon continues today. The proportion of the global population directly harmed by forest clearance and modification remains small, but as an absolute number it is substantial.

The relentless increase in human population may continue to drive the conversion of forests to cropland, especially in the tropics (for example, in Brazil; see Naylor et al, 2005). Although population growth, as a percentage, crested in the late 1960s, the world's total population, now about 6.7 billion, continues to rise by at least 70 million per annum (discussed further in Chapter 6).

The harm to human health and wellbeing caused by forest clearance is often disguised by scale, time and the socioeconomic and cultural distance between the policy-makers whose decisions facilitate forest clearance and those who suffer. There is a paucity of appropriate economic and other feedbacks that percolate to policy-makers and wealthy populations. Consequently, the agents of privileged populations continue to make policy and purchasing choices that harm forests, as well as many other ecosystems. In the short run, such decisions are likely to continue without obvious harm to wealthy populations. In the long run, however, even wealthy populations may be placed at risk if present trends continue.

This chapter considers some of the immediate and local health hazards resulting from forest conversion. It discusses the relationship between forests and health, especially for people living in tropical countries, and reviews the major diseases associated with forest environments. It also looks at the benefits to health and wellbeing that flow from the three main forms of ecosystem services that forests provide for human beings (Millennium Ecosystem Assessment, 2003): broadly, provisioning services (e.g., food, fibre and pharmaceutical products); regulating services (e.g., water-flow, climate regulation and – perhaps – infectious disease mediation); and cultural services (e.g., the aesthetic, psychological and spiritual benefits that accrue to some people who dwell in forests).

FROM FORESTS TO AGRICULTURE

Originally, humans evolved from hominids, who inhabited forest and savannah in tropical regions for millennia. On leaving Africa, human expansion is thought to have mainly been coastal at first (Stringer, 2000). However, population pressure, new technologies, and human curiosity and ingenuity gradually allowed and provoked humans to explore and to colonize many inland regions, much of which was forested.

Until soon after the start of the current interglacial, the Holocene, all humans, including those dwelling in forests, were hunter-gatherers (Kealhofer, 2003; see Chapters 12, 13, 15 and 16, this volume). Many (perhaps most) of these populations appear to have had reasonably good nutrition (Eaton and Konner, 1985; Cordain et al, 2005), although many forest-dwelling hunting populations were prone to infection. At least some also suffered from chronic parasitic infections, as a consequence of eating poorly cooked meat, especially *Trichinella*-infected pork (Owen et al, 2005). They probably also had knowledge of medicinal plants.

Today, no known populations who live in or near forests survive exclusively as hunter-gatherers. Most rely on gardens and/or fishing (Webb et al, 2005) to supplement their diet. Thus many of the diseases they experience may be associated with agricultural land use, with an allied forest fringe effect or with river pollution.

Where agriculture developed, it allowed higher human population densities. Several early agricultural sites (in Mesoamerica, Amazonia and – probably – New Guinea and West Africa) were in tropical forests (Diamond, 2002). The limited evidence available suggests that the health of many early agriculturalists (including, presumably, swidden farmers) was inferior to their hunter-gather predecessors (Diamond, 1991; Boyden, 2004). Health is believed to have deteriorated as the quality of diet fell and leisure decreased. The average farmer consumed more carbohydrates but ate less iron and protein-rich meat and fish, and consumed fewer micronutrient-rich plants (and perhaps insects) than his or her hunter-gatherer predecessor. However, this nutritional disadvantage was outweighed, on a population basis, by the larger human populations that could be supported in a cultivated area than in an equivalent area of 'natural' forest (Noble and Dirzo, 1997). Bigger populations permitted greater specialization and more privileged elites, as well as bigger raiding parties and, eventually, standing armies. In many cases, neighbouring populations must have been forced to either migrate or to also adopt agriculture to compete and survive.

In some regions agriculture took the form of silviculture: the deliberate planting of food-bearing trees, such as Brazil nuts (Mann, 2002), coconuts and fruit trees. In other cases, human occupation and land management may have increased tree diversity, as in the rainforests of Thailand (Kealhofer, 2003).

HEALTH: A MULTIDIMENSIONAL CONTINUUM

Soon after its founding, the World Health Organization (WHO) defined health as a state of 'complete physical, mental and social wellbeing and not merely the absence of disease or infirmity' (World Health Organization, 1978). This highly idealized goal is rarely if ever attainable over any but the briefest period. Even states of physical, mental and social wellbeing that fall short of perfection rely on numerous contributing, interacting factors. An adequate physical foundation is essential for health, in the form of sufficient nutrition, shelter and energy. Safety and security are also vital for good health, ideally coupled with love and nurturing (especially in childhood) and freedom from excessive anxiety. Health also requires a good social fabric. Humans cannot survive as individuals but are social animals, whose wellbeing is supported by their social position, which often entails both duties and entitlements. Our duties often are to provide security and care for other people, and our socially determined entitlements often guarantee us a minimum of care and insurance in times of hardship. We depend on the services and duties of a large social network. We may sometimes grumble about our onerous social duties, but without at least some duties, our health is likely to suffer.

In addition to those foundations, health increasingly depends on knowledge, technology and science, which in turn depend on an adequate social capacity for their delivery. Knowledge (such as knowing the benefits of drinking clean water) and health services (such as the administration of a vaccine, medication or surgical intervention) require a long chain of individuals for their production, dissemination and preservation. Interestingly, some ancient populations learned of potential health harm from certain exposures and managed to embed protective knowledge in their culture, and in some cases, their religion. Two examples are the risk of parasitism from eating pork, and the possibility of acquiring leprosy from close contact. Other societies have preserved cultural knowledge concerning 'famine foods' (Diamond, 2001), including their safe preparation (such as cyanogen-containing cassava) (Tylleskär et al, 1992).

Analogously, a rapidly expanding literature describes traditional means of protecting ecosystem services and resources, including forests, through myriad means, including customs, songs, rituals, laws, and spiritual and religious practices (Berkes et al, 1998; Folke, 2004; Xu et al, 2006).

Many developing countries with populations that inhabit and partly depend upon tropical forests have limited governmental and societal capacity to deliver (and, in some cases, even to receive) new knowledge and health services (see Part III, this volume). Even worse, some traditional societies have become so demoralized and fragmented that their capacity to maintain traditional knowledge and health-care systems has been damaged (see Chapters 11 and 13, this volume). Many indigenous Australian populations, for example, suffer from high levels of violence (including towards children) and substance abuse, especially alcohol use and petrol sniffing. Traditional laws, customs and behaviours in these societies have weakened, and the laws and society of the white newcomers have often been poor substitutes.

Such a fate is not confined to marginalized people in developed countries (Ohenjo et al, 2006). Xu et al (2006) describe the plight of the Kuchong people, until recently nomadic hunter-gatherers who lived in the tropical forests on the border of Yunnan (China) and Vietnam. The forests within the mosaic landscapes that had been preserved by local people and their cultural beliefs for generations have been largely replaced by plantations of rubber, tea and sugar cane. The worship of holy hills and watersheds has frequently been regarded as superstitious and suppressed by the state. Other attempts by the modern Chinese state to 'civilize' the Kuchong people include the building of permanent homes and the introduction of fried foods. Xu et al (2006) report that the changes have had a significant and harmful impact on the Kuchong, and that social problems, including alcoholism, are now prevalent in the community.

Similar self-reinforcing processes of social disintegration may be underway in many parts of sub-Saharan Africa, particularly because of HIV/AIDS (see Chapter 10, this volume). This epidemic has had a devastating and disproportionate impact on the health and vitality of young adults. In some countries, more nurses and teachers are dying of HIV/AIDS than are graduating (Piot, 2000; Lewis, 2006). Countless African children are being raised in orphanages or by grandparents (Save the Children, 2006). Many children are bringing themselves up with little adult guidance. Perhaps, after another generation or so, this tragic situation will improve. In the meantime, despite extraordinary human resilience, the collective social capacity of many of these countries is in decline, further eroding their ability to deal effectively with their crisis.

INFECTIOUS DISEASES IN THE TROPICS

Many people in tropical countries carry the double burden of poverty and infectious disease, as well as a high incidence of chronic non-infectious diseases (cf. Chapters 7 through 10, this volume). Some of the infectious diseases are unique to the tropics; others occur there with greater frequency and virulence. Though more common in the tropics, measles, acute respiratory illness and tuberculosis can legitimately be considered diseases of poverty rather than of rainfall and latitude. But this is not the case with malaria, which was ranked in 1990 as the 11th leading global cause of disability-adjusted life years (DALYs) in a study of the global burden of disease (Murray and Lopez, 1997; see also Chapter 9, this volume). *Falciparum* malaria, which causes 200 million to 300 million infections and 1 million to 3 million deaths annually (Breman, 2001), is probably the most important disease confined to tropical regions. Although some forms of malaria once occurred in more temperate regions, such as the UK and the northern US states, it existed at these latitudes only in one of its milder forms, *Plasmodium vivax*. Unlike *P. falciparum*, *P. vivax* is debilitating and (unless properly treated) chronic, but rarely fatal.

Although some epidemics of *P. falciparum* have occurred in extra-tropical regions (such as the Indian Punjab; Zurbrigg, 1994), large-scale transmission in these areas was 'unstable', rarely if ever lasting for more than a few months a year. Sometimes droughts limited transmission for more than a year, making control, and perhaps even eradication, far more feasible (Coluzzi, 1999).

In contrast, many parts of the tropics, especially in Africa, have long experience of 'holo-endemic' malaria – the relentless transmission of malaria on such a scale that virtually every child, in the absence of treatment, becomes chronically infected. Combined with poverty and inadequate funding, holo-endemicity continues to thwart serious attempts to eradicate malaria in most of tropical Africa (Kager, 2002).

The debility of such chronic diseases is not merely a consequence of poverty but also a cause of poverty: it saps stamina, reduces learning capacity and leads to economic underperformance. This effect of malaria has been recognized for at least a century (Bynum, 2002) but was recently quantified and publicized by the WHO Commission on Macroeconomics and Health (Sachs and Malaney, 2002).

Tropical forests and vector-borne infectious diseases

Many infectious diseases that are unique to the tropics have a special relationship with forests. Their life cycles can involve three species (the pathogen, a vector and humans) and even a fourth (a 'reservoir' species). Most vectors are insects, including mosquitoes, *reduviid* (kissing) bugs, sand flies, tsetse flies and ticks. Water-inhabiting snails, a form of mollusc, are essential for the transmission of schistosomiasis, a disease found near water in Africa, southern China, southeastern Asia and parts of Brazil. Some forest-associated diseases are transmitted directly by mammals, such as rabies (not a purely tropical disease; see Chapter 8, this volume, for a broad discussion of animals as vectors).

Nipah virus, which causes encephalitis, has a more complex ecological causation. The first known outbreak, in Malaysia in 1998, killed more than 100 people and temporarily crippled the Malaysian pig industry (Chua et al, 2000; Epstein et al, 2003). Infected humans were closely associated with pigs, which in turn appear to have been infected by close contact with bats, which were eating fruit from trees that shaded the pig farms (Dobson, 2006). The bats appear to have been displaced from their usual habitat by the haze and smoke of the particularly severe deforestation and fires associated with the severe El Niño event of 1998. (Such fires, which can be very extensive and last weeks, are also clearly harmful to human health: see Chapter 5, this volume.)

The more recent outbreaks of Nipah encephalitis in humans in Bangladesh have not involved pigs; here, the disease has mainly affected young boys, who may have been directly exposed to bat droppings (Enserink, 2004). However, the ecological drivers of the Bangladeshi outbreaks have been less well characterized (Hsu et al, 2004). Nipah virus has also been identified in Cambodia and Thailand, but not yet in humans in these countries (Reynes et al, 1995; Wacharapluesadee et al, 2005).

Bats have also been implicated in the ecological epidemiology of Ebola virus (Leroy et al, 2005) and sudden acute respiratory syndrome (SARS) (Dobson, 2006). Ebola is a significant cause of death among gorillas, chimpanzees and duikers, as well as humans (Leroy et al, 2004; 2005). Most, if not all, primary Ebola infections have occurred after the handling of carcasses of infected bushmeat (described in Box 12.1). Secondary (person-to-person) Ebola infections have occurred mainly within health-care settings and are caused by inadequate infection control measures (Bennett and Brown, 1995). This is also the case for Marburg, a related haemorrhagic fever (Enserink, 2005). Contact with infected primates in Africa is also believed to be the mechanism by which HIV and several less virulent infections have entered the human population (Wolfe et al, 2004).

Rodent urine is the source of the tropical arenaviruses, including Lassa fever, a haemorrhagic fever first described in eastern Nigeria and Cameroon (Richmond and Baglole, 2003). Many other haemorrhagic fevers have also been described. Some, such as Junin, Guanarito and Machupo virus, appear to have a particular relationship with forest-inhabiting rodent species in Latin America.

Because these diseases all involve other species, they can be viewed as having an ecological dimension. (At a microscopic scale, of course, all infectious diseases are ecological, in the sense that they involve at least two species: humans and pathogens.) Many of these tropical illnesses exist near to, and are sometimes dependent upon, tropical forests. Some, such as HIV, have long ago left their forest origins behind and occur mostly in urbanized populations. Ebola is one that remains restricted to the forest environment and to patients, staff and caregivers in close contact with infected patients.

Tick-borne diseases

Several tick-borne diseases, including Lyme disease, Rocky Mountain spotted fever and tick-borne encephalitis, occur in temperate forests. Lyme disease responds to antibiotics, but chronic sequelae, including debilitating musculoskeletal, cardiac and neurological ailments, can occur if treatment is not prompt. The incidence and prevalence of Lyme disease have increased greatly in recent years, especially in the USA, partly because of the increased use of forests for low-density housing and recreation by humans seeking the cultural benefits of forest exposure.

Ecological changes also play an important role in the observed increase in Lyme disease. Tick density has increased alongside the populations of the preferred hosts for its larval and adult forms: the white-footed mouse and the white-tailed deer, respectively. Populations of those hosts, in turn, may have risen with the extermination of the passenger pigeon (once the most common bird in the USA, numbering an estimated 5 billion), which once competed with the mouse and the deer for acorns (Blockstein, 1998). Another cyclic feedback involves increased populations of gypsy moths, which defoliate oak trees. Mice eat moth pupae. Therefore, high or low mouse population density, at low gypsy moth population density, can respectively suppress or release moth populations through altered pupal predation (Jones et al, 1998). Jones et al (1998) caution against attempts to reduce Lyme disease by reducing acorn masting because it might lead to reduced oak populations, perhaps with other harmful effects. Ostfeld and Keesing (2000) also point out that a more diverse range of vertebrates can, at least theoretically, reduce Lyme disease, because other hosts may be less competent incubators of the causal agent for Lyme disease than are mice. They call this a dilution effect and speculate that it could be widespread, implying that biodiversity may help reduce disease severity.

Tick-borne encephalitis is a potentially severe viral disease that occurs in central and northern Europe. Although it has no cure, it can be prevented by a vaccine. This disease has been closely studied as an indicator of climate change, with growing consensus that its range is moving northwards (Lindgren and Gustafson, 2001). The total area affected may be little changed, however, perhaps because of shrinkage along its southern border. Like Lyme disease, the incidence of tick-borne encephalitis is affected by changing human behaviour and, probably, altered ecological conditions. However, its preventable nature complicates its eco-climatic-epidemiological analysis (Randolph, 2001).

Two viral 'forest' diseases

Two viral diseases are so associated with forests that the word appears in their common names. Kyasanur forest disease, a tick-borne haemorrhagic fever, was first reported in Karnataka, India (Banerjee, 1996) and has since been identified in the Andaman and Nicobar islands (Padbidri et al, 2002). Barmah forest disease is a mosquito-borne alphavirus, similar to Ross River fever. It has a wide distribution within Australia, far wider than the forest from which its name is derived (Lindsay et al, 1995).

PROVISIONING AND REGULATING SERVICES

Forests provide many goods essential for human health and wellbeing, such as timber, fuelwood, game meat, medicinal plants and fodder. But through silviculture, conversion to agriculture and the establishment of plantations to provide oil, fibre or rubber, forests' provision of ecosystem services has been increased by human intervention (Daily, 1997; Millennium Ecosystem Assessment, 2003).

The influence that forests can have on local climate and salinity can be considered examples of 'regulating' ecosystem services. The mitigation of catastrophic landslides and the amelioration of drought by standing forests may be another important regulating service, although the extent of this has recently been questioned (FAO and CIFOR, 2005). Supporting the idea that forests affect flooding is the recent finding that plantations lower the water table in many areas. This effect, documented in some cases to last at least two decades (Jackson et al, 2005), can reduce stream flow and impair irrigation-dependent agriculture in nearby areas. In China, the catastrophic floods along the Yangtze River valley in 1998 contributed to a major rethinking of Chinese forest policy, a slowing of deforestation in the river's catchment and active afforestation (Zhang et al, 2000). In the former case, forest plantations can, at least where the water supply is marginal, harm wellbeing. In the latter case, excessive forest clearing can contribute to flooding.

Trade in forest products, such as paper, lumber, palm oil (for food and increasingly fuel), rubber, nuts and oils for cosmetics, is now a global business. This has many implications for health, some of which could be truly dramatic and transcontinental. For example, the use of Asian gangs, accompanied by Asian sex workers, to log the rich stands of forests that remain in Liberia may introduce yellow fever to Africa (Nisbett and Monath, 2001) if the logging company fails to immunize workers against this viral haemorrhagic fever. Similarly, African strains of HIV could be easily transmitted to Asia. Centuries ago, the transatlantic slave trade is believed to have introduced both malaria and yellow fever into the immunologically naïve populations of the Americas (McNeill, 1976).

The provisioning services that flow from forest conversion generally facilitate better human health because the change in land cover increases not only the supply of food, fibre and fuel, but also employment and trade. But such improvements need not be inevitable, and they are unlikely to be uniform. Whether ecosystem services bring widely shared improvements in health and wellbeing depends critically on human factors (including governance) and the rules, mores and institutions that influence their distribution (Butler and Oluoch-Kosura, 2006). Large-scale forest conversion has often been imposed on forest-dwelling peoples without their informed consent or adequate compensation (Dauvergne, 1997; White et al, 2006), and when this happens, their health and wellbeing are very likely to suffer.

Tropical forests and infectious disease 'mediation'

The ecological aspects of some infectious diseases were recognized centuries ago, well before the understanding of their parasitological or microbiological nature, or their vectors or animal sources. An early name for malaria was 'paludism', from the Latin word *palus*, 'swamp'. According to the Chinese environmental historian Mark Elvin, the Bai people of Yunnan province and the Han Chinese who lived near and among them were aware of the link between malaria and the anopheles mosquito from at least the 14th century (Elvin, 2004). Ronald Ross, awarded the Nobel Prize for showing (500 years later) that mosquitoes transmitted malaria, devoted much of his subsequent career to attempts to change environmental and ecological conditions to reduce mosquito numbers.

The idea that entering, disturbing or modifying forest ecosystems can increase the transmission of many infectious diseases, including malaria, is probably old. A more recent corollary is the concept that undisturbed ecosystems can retard infectious disease transmission. This has been called infectious disease 'regulation' (Patz et al, 2005) or infectious disease 'mediation' (Foley et al, 2005).

Patz et al (2005) list many fascinating and suggestive examples in support of the general concept. Although the theory cannot be extrapolated to all ecosystems, infectious diseases or populations, it warrants publicity and further research. Modification of ecosystems, including forests, *beyond a threshold* in many cases may facilitate the widespread transmission of new diseases. If such thresholds could be identified, and avoided, then huge costs – both human and financial – might be saved, because our environment could be harnessed in ways that would enable us to reduce or even block the spread of disease.

Four caveats regarding infectious disease mediation

The theory of infection disease mediation is, however, subject to four caveats.

Confounding factors

Many factors are involved in the epidemiology of tropical diseases, including poverty, human population density, human migration, climatic changes and the presence or scarcity of human capability (Sen, 1999). Modification of forest ecosystems may be *necessary*, but is rarely *sufficient* to increase disease transmission or to generate an epidemic.

Time and scale

The increased disease potential from ecosystem modification has dimensions of both time (chronotones) (Bradley, 2004) and scale. In many cases, the increased risk of disease transmission may be temporary. As development proceeds, the likelihood of many infectious diseases declines, as a function of both increased human capacity and – at least in some cases – of additional ecological changes that prove unfavourable to disease transmission.

In the Amazon, de Castro et al (2006) describe a rise and fall of malaria risk as forest clearance and agricultural settlement proceed. They describe how weak institutions, low community cohesion, politically marginalized settlers and high rates of in- and out-migration combine to thwart malaria control programmes in newly deforested regions, with the revealing exception of corporate-sponsored forest clearance. In these cases, the transformation of the landscape was faster, and the personnel involved had more knowledge of the risk of malaria and were better able to take protective measures against mosquito exposure. As a result, malaria was minimal.

The Punjab of pre-Independence India, which straddled present-day India and Pakistan, experienced a vast expansion of irrigation, population inflow and deforestation from 1860 until the Second World War. The death toll from a succession of malaria epidemics increased until 1908 but then greatly declined, despite the ongoing expansion of irrigation and the lack of effective pesticides. This decline has been attributed to an early warning system, the development of organized civil society, and the judicious use of limited supplies of quinine, targeted especially at immunologically naïve children (Swaroop, 1949; Butler, 1997). Better nutrition may have also played a role in reducing mortality (Zurbrigg, 1994).

The term 'paddies paradox' has been used to describe situations in Africa where irrigation has increased vector populations without any increase in malaria (Ijumba and Lindsay, 2001). For instance, villages surrounded by irrigated rice fields in Kenya showed a 30- to 300-fold increase in the number of the local malaria vector, *Anopheles arabiensis,* compared with those without rice irrigation, and yet were found to have had a significantly lower malaria prevalence (0–9 per cent versus 17–54 per cent) (Mutero et al, 2004). This has been attributed to the zoophilic biting preference of *A. arabiensis,* rather than any deliberate intervention from better-educated or more affluent humans (Patz et al, 2005). If so, the introduction of cattle could be seen as an ecological mediator, in this case operating within a highly modified ecosystem. However, even if this malaria 'zooprophylaxis' is valid in Kenya, it was not found to be in Pakistan (Bouma and Rowland, 1995).

Opposite effects

The third caveat is that ecosystem modification is sometimes used to *reduce* disease transmission. Two clear illustrations come from Africa. The three forms of African trypanosomiasis (sleeping sickness) are transmitted by savannah- and scrub-dwelling tsetse flies. Game animals and cattle constitute the reservoir for two of these forms (*Trypanosoma brucei rhodesiense* and *T. brucei brucei*). A large region of savannah in Africa, otherwise suitable for the rearing of cattle, consequently has a very low human population density (Robinson, 1985).

Onchocerciasis (river blindness) is caused by a filarial worm transmitted by species of the black fly (*Simulium* spp). This problem is particularly severe in savannah regions of West Africa, where the black fly breeds in the white-water rapids of fast-flowing rivers. In recent years, intensive and generally promising efforts to eradicate this disease have been made by attempting to exterminate the black fly. Serious setbacks in this programme have been attributed to tropical deforestation, facilitating the expansion of the savannah species into the newly deforested areas, and perhaps expanding the zone of onchocerciasis transmission (Patz et al, 2005).

Besides HIV and Ebola, other diseases appear to have crossed into the human population as a consequence of intimate contact with comparatively undisturbed forest ecosystems. In the Brazilian Amazon, 32 distinct arboviruses associated with human disease have so far been described. Almost all of these are maintained within complex cycles in the forest, but epidemics are mostly reported in newly cleared areas adjacent to forests, or in the vicinity of dams (Patz et al, 2005).

A large literature, especially from South and Southeast Asia, describes high rates of malaria transmission among tribal peoples living in heavily forested regions. This literature consistently identifies forest exposure as a risk factor for malaria (e.g., Erhart et al, 2004) but makes few comments about the presence of gardens, the degree of nearby agricultural expansion or the increased light at the forest edge. Because no peoples are today purely hunter-gatherers, it is possible that much of the risk of malaria in these populations may more accurately be attributed to forest garden exposure rather than purely to forest exposure.

However, mosquitoes do occur in forests, sometimes in high concentrations. Both *Aedes* and *Haemagogus* species breed in forest canopies, and *Aedes* mosquitoes breeding in tree holes in moist savannah can sometimes reach very high densities (Monath, 2001). In support of the concept of infectious disease mediation, genetic evidence suggests that the modern form of *Plasmodium falciparum* is comparatively recent and may have co-evolved with humans and vectors following the introduction of slash-and-burn agriculture in Africa 5000 or 6000 years ago (Volkman et al, 2001).

In the highlands of Uganda, Lindblade et al (2001) studied two populations living near swamps. Slightly higher (though not statistically significant) rates of malaria were found in villagers living near swamps that had been drained and cultivated, compared with villagers near unmodified papyrus swamps. Rather than attributing this to the 'mediating' function of the intact swamp, the researchers thought that malaria might have increased near the drained swamps because the ecosystem modification had inadvertently caused a slightly higher temperature more favourable to the breeding of mosquitoes.

Romantic notions

The idea of infectious disease 'regulation' by 'pristine' forests could be conflated with the conceit of an intrinsically benign 'nature'. A corollary is that 'wild' nature, rich in biodiversity, should be preserved to reduce outbreaks of human disease. It seems more plausible that numerous species, including humans, vectors and pathogens, have long been driven by co-evolutionary forces involving competition, synergism and cooperation. Few environments are likely to be intrinsically benign to humans, though human ingenuity and intelligence have been able to modify almost all terrestrial environments to accommodate human habitation. Wherever there has been animal life in forests where people dwell, there have been hunters, and human hunters are always likely to have accepted the reward of nutritious (and often status-enhancing) meat in exchange for health risks, whether from injury, parasites or strange zoonotic infections.

Indeed, to support a larger human population, the limited food-provisioning services in 'undisturbed' forests provide an incentive for their modification that appears far stronger than any drive to preserve forests intact to reduce hypothetical epidemics. Viewed this way, the continuing transformation of forest ecosystems to grow food or feed is unlikely to slow as long as the human population continues to expand and becomes affluent enough to demand more meat (Naylor et al, 2005; McMichael et al, 2007).

It is highly likely that human actions have at times modified ecosystems in ways that inadvertently favoured vectors and pathogens and thus promoted infectious disease transmission. But at other times, humans have modified not only ecosystems but also their own behaviour to reduce the risk of infectious disease, including lowering the populations of vectors and pathogens. Many natural forests may host comparatively little transmission of many vector-borne diseases. But these forests are also places with a low human population density, and in the absence of non-human reservoir species, they therefore have low concentrations of pathogens. Large-scale forest modification may increase opportunities for vectors to breed, particularly by creating puddles and irrigation canals. Similarly, the in-migration of workers, who are often disproportionately poor and lack both health care and knowledge of vector-borne diseases, is likely to create conditions that favour pathogens (such as malarial parasites). Combined, these effects set the stage for the outbreak of many epidemics (MacDonald, 1973).

The suggestion that an intact forest regulates or even mediates most or even many vector-borne diseases, in my view, over-interprets the evidence. No one would suggest that the absence of an epidemic on an ocean liner is evidence that the boat, or indeed the ocean, regulates or mediates the epidemic, when a simpler explanation is simply that insufficient conditions exist for such an outbreak. At the same time, newly cleared alpine forest will not lead to a malarial epidemic even if it is irrigated and tilled by the most deprived population. The temperature will simply be too low for the mosquito vectors to become sufficiently numerous. Some (perhaps many) intact, sparsely inhabited forests are unsuitable environments for large-scale vector-borne epidemics, but I am less convinced that it therefore follows that these forests regulate such outbreaks.

In many such cases, the forests may be a potent reservoir of latent infection, ready to erupt if disturbed. When considered this way, tropical forests appear to be sources of danger if disturbed to improve their provisioning services. One can see how the idea of infectious disease regulation has developed, but I hope the reader will also see its limitations.

West Nile virus, yellow fever and 'ecosystem immunity'

Although there are limits to the mediation or regulation of infectious disease by intact ecosystems, there are nonetheless several intriguing examples in which the spread of infectious disease appears to be reduced by large-scale ecological factors. One example may be the failure (to date) of West Nile virus to become established in Australia. Antibodies to a similar flavivirus, called Kunjin virus, have been suggested as cross-protective to West Nile. It has been suggested that these antibodies are so widely distributed among the potential host species in Australia that West Nile is unlikely to take hold there (Mackenzie et al, 2003).

Conversely, yellow fever has not been introduced to Asia despite the presence of large numbers of its vector, the *Aedes* species. This risk is now well recognized and guarded against, mainly by scrupulous enforcement of yellow fever vaccination for travellers. Nevertheless, the long history of human contact and trade between East Africa (where yellow fever has long occurred) and India, going back millennia, suggests that an as-yet unidentified protective factor in Asia prevents the introduction of yellow fever. Indeed, as with West Nile virus, Monath (2001) has suggested that cross-protective antibodies from dengue fever, which is widely distributed, may be protective against yellow fever.

A second point is that the ecosystems in both East Africa and Asia have been extensively changed. If the disease-mediating hypothesis of an intact ecosystem is valid in this case, then the threshold of ecosystem modification required to permit the establishment of yellow fever in Asia may be very high – though perhaps vaccination is the critical protective factor. It certainly seems prudent to continue this programme, particularly since the vaccine is comparatively cheap, safe and effective, and the disease has a high fatality rate and no treatment.

The Amazon jungle provides several well-documented cases of introduced diseases that have failed to have a severe impact (at least to date). These include schistosomiasis, *kala-azar* (*Leishmania donovani*) and cholera (Patz et al, 2005). *Kala-azar*, a disease common in India and in arid areas such as parts of Sudan (Thomson et al, 1999), has become established only in two geographically restricted areas of the Brazilian Amazon: a savannah area in the northern part and a peri-urban setting in the central part (Confalonieri, 2000). Schistosomiasis has been introduced to parts of Brazilian Amazonia but has not been established, apparently because a widespread lack of mineral salts necessary for shell formation has limited populations of the snail species *Biomphalaria* (Sioli, 1953, cited in Patz et al, 2005).

On the other hand, both yellow fever and malaria are thought to have been successfully introduced to Brazil – including, presumably, its forests – hundreds of years ago as a result of the slave trade (McNeill, 1976). Here the comparatively undisturbed ecosystem proved an inadequate defence. Dengue and yellow fever now co-exist in Brazil, unlike in many parts of Asia. This suggests that the explanation proposed by Monath for the absence of yellow fever in Asia – that it is inhibited by dengue – is unlikely to be complete.

CULTURAL SERVICES

From an ecosystemic perspective, the conversion of many unique forest ecosystems (e.g., to fields of rice or soy) is an enormous loss. Many populations, species and – in some cases – entire ecosystems have been destroyed or irrevocably changed. Elvin (2004) remarks how, in China, the cultural admiration of whole forests is now expressed as veneration for an individual tree, an apparent symbol of an entire ecosystem.

Many indigenous peoples have long engaged in practices that have had the effect, if not the conscious intent, of ecosystem protection and preservation (Berkes et al, 1998; Folke, 2004; Xu et al, 2006). For such people, the loss must often be demoralizing. Self-esteem and health are also intimately connected to culturally determined livelihoods and responsibilities. Among the Kenyah Dayaks in East Kalimantan, for instance, women's roles are intimately connected with swidden rice cultivation. Although men also work in the fields, the women derive status and satisfaction from this work (Colfer, 1991). Similarly, in parts of Amazonia many men gain community respect and self-esteem from participating in a successful hunt (Siskind, 1973).

One of the many problems with conventional measures of economic growth is that they accord no value to self-esteem or social cohesion. Sometimes, mechanization, globalization or other external factors can deprive both women and men of core responsibilities in ways that hurt self-worth and satisfaction and can lead to alcoholism, depression and despair.

The loss of biodiversity from wide-scale forest conversion (and increasingly from climate change) is also keenly felt by many scientists and conservationists. E. O. Wilson (2002) has suggested that humans are biophilic, in having a deep sense of connection with other living beings. The spontaneous emergence of the 'green' movement in the 20th century may illustrate an expression of population-scale biophilia. The loss and degradation of species and entire ecosystems are undoubtedly painful for many people; they may also impair human health.

The psychological and spiritual relationships between many humans and ecosystems and their species (and in some cases wild individuals) illustrate the third main class of ecosystem services: cultural services (Butler et al, 2003). There is growing evidence that these cultural services provide substantial health benefits and, in some cases, spiritual connection and fulfilment (Frumkin, 2002). Some indigenous peoples describe cultural relationships with 'sacred groves' (Ramakrishnan et al, 1998). Others describe links with totemic species, sometimes in mosaics that appear to encourage the conservation of biodiversity over large areas. Although this subject is poorly researched, it is likely that the psychological – and hence physical – health of many indigenous people who lose contact with sacred groves, or whose totemic species is endangered or extinct, is impaired. There is also growing evidence that the health of many urbanized people is improved by contact with nature, including forests (Frumkin, 2002; Maller et al, 2006). Some of these issues are explored in greater detail in Chapters 11, 13 and 16.

FORESTS, CLIMATE CHANGE AND THRESHOLDS

The rate of forest clearance, whether tropical (Achard et al, 2002; Hansen and DeFries, 2004), temperate or boreal, is debatable. Globally, the forested area may be stabilizing because of an expansion of plantations and temperate forests. However, there is consensus that both the quality and the quantity of tropical forests are continuing to decline. The ongoing clearance of tropical forests is particularly important for both biodiversity (including as-yet undiscovered pharmaceuticals; see Chapter 3, this volume) and global climate change.

Modelling shows feedback cycles in which climate change may lead to drought and fires within tropical forests, including the Amazon, and this forest loss could then exacerbate climate change, causing further forest loss (Cox et al, 2000).

From a purely human perspective, the conversion of forests to generate increased provisioning services, even if at the expense of some regulating services (including perhaps mediation of some infectious diseases) has clearly enhanced human wellbeing and health. This is true in the narrow sense that the total number of humans – supported by intact forests, modified forests and lands that were formerly forested – is larger than ever. Many of these people also have, on average, a comparatively long lifespan. Again, there are caveats.

First, the conversion of natural capital to support an ever-expanding population of fairly long-lived humans has arisen as a consequence of history and geography, rather than because of any general consensus that this trade-off is desirable, even from an anthropocentric view. Second, the aggregation of populations and effects to calculate averages masks the large number of comparative losers, including many indigenous and tribal populations whose traditional property rights and management strategies are discounted, if not entirely ignored. Last, the comparative success of the human species does not guarantee that the process can continue indefinitely. Evidence is mounting that the transformation of forests and other ecosystems is on a scale sufficient to cause many unexpected effects that appear harmful to human wellbeing and, in the long run, human health.

Policy implications

Not all of the land that once supported forest and now supports agriculture is as fertile as is most of Europe. In addition, there is more to human welfare than the total number of people who can be supported at a given lifespan. Critics of deforestation argue that through indirect effects, particularly upon global climate and biodiversity, the current rate and scale of tropical forest clearance has the potential to harm the health and wellbeing of many more people than those who are currently being displaced.

It is difficult to conclude that large-scale forest transformation will – at least over the next few decades – cause net harm to human health, but it will clearly damage some desirable forest products, many non-human species and cultural systems (some unique), many ecosystems, and biodiversity as a whole. More importantly, the conversion of forests into agricultural land cannot continue indefinitely without great harm, not only to the forests themselves, but also to human health and wellbeing.

Any driver aware of a sharp bend in the road ahead knows to slow down to safely negotiate the curve. Analogously, the undeniable consequences of infinite forest clearance should modify the views and intentions of far-sighted policy-makers. Unfortunately, misleading economic signals obstruct their vision and domestic, international and intergenerational inequality constrain action.

In some ways, forests are comparable to other poorly regulated environmental public goods (McMichael et al, 2003), such as the global climate and ocean fisheries. The 'tragedy of the commons' (Gordon, 1954; Hardin, 1968) has been extensively lamented at a small scale. Yet in numerous cases, functioning, intact societies have successfully recognized, practised and enforced institutions to protect common resources, including forests, grazing land and near-shore and freshwater fisheries (Buck, 1985; Adams et al, 2003).

On the global scale, however, institutions to protect resources from plunder remain embryonic. Despite increasing recognition among academics that sanctions are required to protect global public goods (Berkes et al, 2006; Gürerk et al, 2006), effective and widely accepted international laws are lacking. Instead, comparatively untrammelled market forces continue to fuel a race to the environmental bottom.

The Millennium Development Goals, if they can be achieved, will provide a modest but important stimulus to protect common resources, including forests. This is likely because realization of the goals will help level the playing field between those who disproportionately profit from exploiting these resources and those who lose, thus increasing the leverage that the poor – those who are most immediately vulnerable to catastrophic loss of environmental public goods – have on global policy. However, progress toward the goals is very slow. There is a real risk that global society will continue to evolve towards an unsustainable and highly dangerous 'fortress world', in which growing inequalities and the resulting violence prompt wealthy populations to increase their own security, creating a negative feedback loop that further exacerbates the problems and eventually plunges the world into chaos (Butler and Oluoch-Kosura, 2006).

MISLEADING INDICATORS OF PROGRESS

Conventional measures of economic growth, such as gross domestic product and personal income, account for the market goods that flow from the provisioning services of forests and formerly forested land. They ignore the change in forms of non-financial capital, such as natural capital (e.g., mineral stocks or standing forests) and social capital (or social cohesiveness) (Butler, 1994; Dasgupta, 1996; Arrow et al, 2004). They also fail to capture non-market provisioning services, such as those used by subsistence populations. They ignore changes in health and the distribution of wealth and income, and they fail to measure externalities that follow from the change of ecosystem regulation and cultural services, each of which has implications for human wellbeing. Policy-makers need to consider these complex dimensions if their goal is to maximize sustainable human health and wellbeing.

Despite the problems with conventional economic measures, the dominant economic theorists make the implicit assumption that human welfare is directly proportional to them. Because the harm from forest clearance rarely affects policy-makers or consumers, they have little incentive to support policies that might limit this example of the tragedy of the commons.

In some cases, policies have created a dual harm, not only destroying the qualities of the pre-existing forest (with consequent harm to the peoples dependent on them), but also failing to produce viable agricultural land in exchange. The best example of this is probably the failure to convert substantial amounts of forest in Kalimantan, Indonesia to rice. The peaty underlying soil was unsuitable for rice, and problems were exacerbated by translocation and cultural differences, especially between the Madurese newcomers and the indigenous Dayak population (Carey, 2001; Aldhous, 2004).

The harm to human health and wellbeing is often disguised by scale, time and the socioeconomic and cultural distance between the policy-makers whose decisions facilitate forest clearance and those who suffer. There is a paucity of appropriate economic and other feedbacks that percolate to policy-makers and wealthy populations. Consequently, these privileged populations continue to make policy and purchasing choices that harm forests and other ecosystems. In the short run such decisions are likely to continue, without obvious harm to wealthy populations. In the long run, however, even wealthy populations are likely to be placed at risk.

CONCLUSION

In the past century, the process of converting forests to meet the needs and wants of a substantial part of the global human population has reached a scale that was once unimaginable. One question is whether the scale of global forest conversion could exceed a threshold beyond which the quality of modern civilization significantly deteriorates.

Such a possibility may seem far-fetched to some, but there are several well-accepted cases in which pre-modern humans altered their forest environment on such a large scale that their civilization – and, by implication, their health – was undermined. Three examples are Easter Island, the Indus Valley and Mesopotamia.

In the first, a frenzy of statue-building led to the cutting of almost all of the island's forest, including timber needed for boat-building. Thus, indirectly, deforestation led to significant food shortages, especially of fish (Hunt and Lipo, 2006). In the Indus Valley, the vast quantities of wood used to bake bricks required the clearing of forest on such a large scale that the local climate is thought to have changed. In Mesopotamia, extensive deforestation – probably mainly to permit the growing of wheat – is thought to have contributed to salinization. As the salt built up, wheat was replaced by comparatively salt-tolerant barley (Jacobsen and Adams, 1958), but eventually this strategy also faltered.

With greater attention to equity, education and basic health, many of the health problems associated with forest clearance and agricultural settlement can be lessened. The incidence of strictly forest-associated diseases, such as Ebola, can also be reduced by better education and nutrition so that hunters do not handle and cook infected bushmeat carcasses. The burden of large-scale vector-borne diseases, including malaria, can be reduced by a combination of improved treatment and prevention, including the judicious use of insecticides, such as dichlorodiphenyltrichloroethane (DDT). The recent decline in the incidence of HIV, observed in southern India and in many African countries, shows that education, assistance and governance can improve health. Although governance need not be perfect to effect these improvements, leadership at all scales is needed if the health of populations dependent upon forests is to be improved. Ultimately, it is in the interest of us all for this to occur.

REFERENCES

Achard, F., Eva, H.D., Stibig, H.-J., Mayaux, P., Gallego, J., Richards, T. and Malingreau, J.-P. (2002) 'Determination of deforestation rates of the world's humid tropical forests', *Science* 297: 999–1002

Adams, W.M., Brockington, D., Dyson, J., and Vira, B. (2003) 'Managing tragedies: Understanding conflict over common pool resources', *Science* 302: 1915–1916

Aldhous, P. (2004) 'Borneo is burning', *Nature* 432: 144–146

Arrow, K., Dasgupta, P., Goulder, L., Daily, G., Ehrlich, P., Heal, G., Levin, S., Mäler, K.-G., Schneider, S., Starrett, D. and Walker, B. (2004) 'Are we consuming too much?', *Journal of Economic Perspectives* 18(3): 147–172

Banerjee, K. (1996) 'Emerging viral infections with special reference to India', *Indian Journal of Medical Research* 103: 177–200

Bennett, D. and Brown, D. (1995) 'Ebola virus: Poor countries may lack the resources to prevent or minimise transmission', *British Medical Journal* 310: 1344–1345

Berkes, F., Kislalioglu, M., Folke, C. and Gadgil, M. (1998) 'Exploring the basic ecological unit: Ecosystem-like concepts in traditional societies', *Ecosystems* 1: 409–415

Berkes, F., Hughes, T.P., Steneck, R.S., Wilson, J.A., Bellwood, D.R., Crona, B., Folke, C., Gunderson, L.H., Leslie, H.M., Norberg, J., Nyström, M., Olsson, P., Österblom, H., Scheffer, M. and Worm, B. (2006) 'Globalization, roving bandits, and marine resources', *Science* 311: 1557–1558

Blockstein, D.E. (1998) 'Lyme disease and the passenger pigeon?' [letter], *Science* 279: 1831

Bouma, M.J. and Rowland, M. (1995) 'Failure of zooprophylaxis: Cattle ownership in Pakistan is associated with higher prevalence of malaria', *Transcripts Royal Society Tropical Medicine & Hygiene* 89: 351–353

Boyden, S. (2004) *The Biology of Civilisation*, UNSW Press, Sydney

Bradley, D. (2004) 'An exploration of chronotones: A concept for understanding the health processes of changing ecosystems', *EcoHealth* 1: 165–171

Breman, J. (2001) 'The ears of the hippopotamus: Manifestations, determinants, and estimates of the malaria burden', *American Journal of Tropical Medicine and Hygiene* 64: 1–11

Buck, S.J. (1985) 'No tragedy on the commons', *Environmental Ethics* 7: 49–61

Butler, C.D. (1994) 'Overpopulation, overconsumption and economics', *The Lancet* 343: 582–584

Butler, C.D. (1997) 'Malaria in the British Punjab: 1867–1943', unpublished MSc (epidemiology), London School of Hygiene and Tropical Medicine, University of London

Butler, C.D., Chambers, R., Chopra, K., Dasgupta, P., Duraiappah, A., Kumar, P., McMichael, A.J. and Wen-Yuan, N. (2003) 'Ecosystems and human well-being', in *Ecosystems and Human Well-Being: Current State and Trends*, Island Press, Washington, DC, pp391–415

Butler, C.D. and Oluoch-Kosura, W. (2006) 'Linking future ecosystem services and future human well-being', *Ecology and Society* 11(1): 30, www.ecologyandsociety.org/vol11/iss1/art30/

Bynum, W.F. (2002) 'Mosquitoes bite more than once', *Science* 295: 47–48

Carey, P. (2001) 'Indonesia's heart of darkness', *Wall St Journal Asia*

Chua, K.B., Bellini, W.J., Rota, P.A., Harcourt, B.H., Tamin, A., Lam, S.K., Ksiazek, T.G., Rollin, P.E., Zaki, S.R., Shieh, W.-J., Goldsmith, C.S., Gubler, D.J., Roehrig, J.T., Eaton, B., Gould,

A.R., Olson, J., Field, H., Daniels, P., Ling, A.E., Peters, C.J., Anderson, L.J. and Mahy, B.W.J. (2000) 'Nipah virus: A recently emergent deadly paramyxovirus', *Science* 288: 1432–1435

Colfer, C.J.P. (1991) 'Indigenous rice production and the subtleties of culture change', *Agriculture and Human Values* VIII: 67–84

Coluzzi, M. (1999) 'The clay feet of the malaria giant and its African roots: Hypotheses and inferences about origin, spread and control of *Plasmodium falciparum*', *Parassitologia* 41: 277–283

Confalonieri, U. (2000) 'Environmental change and human health in the Brazilian Amazon', *Global Change and Human Health* 1: 174–183

Cordain, L., Eaton, S.B., Sebastian, A., Mann, N., Lindeberg, S., Watkins, B.A., O'Keefe, J.H. and Brand-Miller, J. (2005) 'Origins and evolution of the Western diet: Health implications for the 21st century', *American Journal of Clinical Nutrition* 81: 50–54

Cox, P.M., Betts, R., Jones, C.D., Spall, S.A. and Totterdell, I.J. (2000) 'Acceleration of global warming due to carbon-cycle feedbacks in a coupled climate model', *Nature* 408: 184–187

Daily, G.C. (ed) (1997) *Nature's Services: Societal Dependence on Natural Ecosystems*, Island Press, Washington, DC, pp1–392

Dasgupta, P. (1996) 'The economics of the environment', *Proceedings of the British Academy* 90: 165–221

Dauvergne, P. (1997) *Shadows in the Forest: Japan and the Politics of Timber in Southeast Asia*, MIT Press, Cambridge, MA

de Castro, M., Monte-Mor, R., Sawyer, D. and Singer, B. (2006) 'Malaria risk on the Amazon frontier', *Proceedings of the National Academy of Science of the USA* 103: 2452–2457

Diamond, J. (1991) *The Rise and Fall of the Third Chimpanzee*, Radius, London

Diamond, J. (2001) 'Unwritten knowledge', *Nature* 410: 52

Diamond, J. (2002) 'Evolution, consequences and future of plant and animal domestication', *Nature* 418: 700–707

Dobson, A.P. (2006) 'Linking bats to emerging diseases' [response], *Science* 311: 1084

Eaton, S. and Konner, M. (1985) 'Paleolithic nutrition', *New England Journal of Medicine* 312: 283–288

Elvin, M. (2004) *The Retreat of the Elephants: An Environmental History of China*, Yale University Press, New Haven, CT

Enserink, M. (2004) 'Nipah virus (or a cousin) strikes again', *Science* 303: 1121

Enserink, M. (2005) 'Crisis of confidence hampers Marburg control in Angola', *Science* 308: 489

Epstein, P.R., Chivian, E. and Frith, K. (2003) 'Emerging diseases threaten conservation' [editorial], *Environmental Health Perspectives* 111: A506–507

Erhart, A., Thang, N.D., Hung, N.Q., Toi, L.V., Hung, L.X., Tuy, T.Q., Cong, L.D., Speybroeck, N., Coosemans, M. and D'alessandro, U. (2004) 'Forest malaria in Vietnam: A challenge for control', *The American Journal of Tropical Medicine and Hygiene* 70: 110–118

FAO and CIFOR (2005) *Forests and Floods: Drowning in Fiction or Thriving on Facts?*, Food and Agriculture Organization of the United Nations, Rome/Center for International Forestry Research, Bogor

Foley, J.A., DeFries, R., Asner, G.P., Barford, C., Bonan, G., Carpenter, S.R., Chapin, F.S., Coe, M.T., Daily, G.C., Gibbs, H.K., Helkowski, J.H., Holloway, T., Howard, E.A., Kucharik, C.J., Monfreda, C., Patz, J.A., Prentice, I.C., Ramankutty, N. and Snyder, P.K. (2005) 'Global consequences of land use', *Science* 309: 570–574

Folke, C. (2004) 'Traditional knowledge in social-ecological systems' [editorial], *Ecology and Society* 9: 7, www.ecologyandsociety.org/vol9/iss3/art7/

Frumkin, H. (2002) 'Beyond toxicity: Human health and the natural environment', *American Journal of Preventive Medicine* 20: 234–240

Gordon, H.S. (1954) 'The economic theory of a common property resource: The fishery', *Journal of Political Economy* 62: 124–142

Gürerk, Ö., Irlenbusch, B. and Rockenbach, B. (2006) 'The competitive advantage of sanctioning institutions', *Science* 312: 108–111

Hansen, M.C. and DeFries, R.S. (2004) 'Detecting long-term global forest change using continuous fields of tree-cover maps from 8-km advanced very high resolution radiometer (AVHRR) data for the years 1982–99', *Ecosystems* 7: 695–716

Hardin, G. (1968) 'The tragedy of the commons', *Science* 162: 1243–1248

Holden, C. (2006) 'Long-ago peoples may have been long in the tooth', *Science* 312: 1867

Hsu, V.P., Hossain, M.J., Parashar, U.D., Ali, M.M., Ksiazek, T.G., Kuzmin, I., Niezgoda, M., Rupprecht, C., Bresee, J. and Breiman, R.F. (2004) 'Nipah virus encephalitis reemergence, Bangladesh', *Emerging Infectious Diseases* 10: 2082–2087

Hunt, T.L. and Lipo, C.P. (2006) 'Late colonization of Easter Island', *Science* 311: 1603–1606

Ijumba, J.N. and Lindsay, S. (2001) 'Impact of irrigation on malaria in Africa: Paddies paradox', *Medical and Veterinary Entomology* 15: 1–11

Jackson, R.B., Jobba'gy, E.G., Avissar, R., Roy, S.B., Barrett, D.J., Cook, C.W., Farley, K.A., Maitre, D.C.L., McCarl, B.A. and Murray, B.C. (2005) 'Trading water for carbon with biological carbon sequestration', *Science* 310: 1944–1947

Jacobsen, T. and Adams, R.M. (1958) 'Salt and silt in ancient Mesopotamian agriculture', *Science* 128: 1251–1258

Jones, C.G., Ostfeld, R.S., Richard, M.P., Schauber, E.M. and Wolff, J.O. (1998) 'Chain reactions linking acorns to gypsy moth outbreaks and Lyme disease risk', *Science* 279: 1023–1026

Kager, P.A. (2002) 'Malaria control: Constraints and opportunities', *Tropical Medicine and International Health* 7: 1042–1046

Kealhofer, L. (2003) 'Looking into the gap: Land use and the tropical forests of southern Thailand', *Asian Perspectives* 42: 72–95

Leroy, E.M., Rouquet, P., Formenty, P., Souquière, S., Kilbourne, A., Froment, J.-M., Bermejo, M., Smit, S., Karesh, W., Swanepoel, R., Zaki, S.R. and Rollin, P.E. (2004) 'Multiple Ebola virus transmission events and rapid decline of Central African wildlife', *Science* 303: 387–390

Leroy, E.M., Kumulungui, B., Pourrut, X., Rouquet, P., Hassanin, A., Yaba, P., Délicat, A., Paweska, J.T., Gonzalez, J.-P. and Swanepoel, R. (2005) 'Fruit bats as reservoirs of Ebola virus', *Nature* 438: 575–576

Lewis, S. (2006) *Race against Time*, Text Publishing, Melbourne

Lindblade, K.A., Walker, E.D., Onapa, A.W., Katungu, J. and Wilson, M.L. (2001) 'Land use change alters malaria transmission parameters by modifying temperature in a highland area of Uganda', *Tropical Medicine and International Health* 5: 263–274

Lindgren, E. and Gustafson, R. (2001) 'Tick-borne encephalitis in Sweden and climate change', *The Lancet* 358: 16–18

Lindsay, M.D.A., Johansen, C.A., Broom, A.K., Smith, D.W. and Mackenzie, J.S. (1995) 'Emergence of Barmah forest virus in Western Australia', *Emerging Infectious Diseases* 1: 22–26

MacDonald, G. (1973) 'The analysis of malaria epidemics', in L. Bruce-Chwatt and V. Glanville (eds) *Dynamics of Tropical Disease*, Oxford University Press, London, pp146–160

Mackenzie, J.S., Smith, D.W. and Hall, R.A. (2003) 'West Nile virus: Is there a message for Australia?' [editorial], *Medical Journal of Australia* 178: 5–6

McMichael, A.J., Butler, C.D. and Ahern, M.J. (2003) 'Global environment', in R. Smith, R. Beaglehole, D. Woodward and N. Drager (eds) *Global Public Goods for Health*, Oxford University Press, Oxford, pp94–116

McMichael, A.J., Powles, J., Butler, C.D. and Uauy, R. (2007) 'Food, agriculture, energy, climate change and health', *The Lancet* 370: 1253–1263

McNeill, W.H. (1976) *Plagues and Peoples*, Anchor Press, Garden City, NY

Maller, C., Townsend, M., Pryor, A., Brown, P. and Leger, L.S. (2006) 'Healthy nature healthy people: "Contact with nature" as an upstream health promotion intervention for populations', *Health Promotion International* 21: 45–54

Mann, C.C. (2002) '1491', *Atlantic Monthly* 298: 41–53

Millennium Ecosystem Assessment (2003) *Ecosystems and Human Well-being*, Island Press, Washington, DC

Monath, T. (2001) 'Yellow fever: An update', *Lancet Infectious Diseases* 1: 11–20

Murray, C.J.L. and Lopez, A.D. (1997) 'Alternative projections of mortality and disability by cause 1990–2020: Global Burden of Disease Study', *The Lancet* 349: 1498–1504

Mutero, C.M., Kabutha, C., Kimani, V., Kabuage, L., Gitau, G., Ssennyonga, J., Githure, J., Muthami, L., Kaida, A., Musyoka, L., Kiarie, E. and Oganda, M. (2004) 'A transdisciplinary perspective on the links between malaria and agroecosystems in Kenya', *Acta Tropica* 89: 171–186

Naylor, R., Steinfeld, H., Falcon, W., Galloway, J., Smil, V., Bradford, E., Alder, J. and Mooney, H. (2005) 'Losing the links between livestock and land', *Science* 310: 1621–1622

Nisbett, R.A. and Monath, T.P. (2001) 'Viral traffic, transnational companies and logging in Liberia, West Africa', *Global Change and Human Health* 2: 18–19

Noble, I.R. and Dirzo, R. (1997) 'Forests as human-dominated ecosystems', *Science* 277: 522–525

Ohenjo, N., Willis, R., Jackson, D., Nettleton, C., Good, K. and Mugarura, B. (2006) 'Health of indigenous people in Africa', *The Lancet* 367: 1937–1946

Ostfeld, R.S. and Keesing, F. (2000) 'The function of biodiversity in the ecology of vector-borne zoonotic diseases', *Canadian Journal of Zoology* 78: 2061–2078

Owen, I.L., Morales, M.A.G., Pezzotti, P. and Pozio, E. (2005) '*Trichinella* infection in a hunting population of Papua New Guinea suggests an ancient relationship between *Trichinella* and human beings', *Transactions of the Royal Society of Tropical Medicine and Hygiene* 99: 618–624

Padbidri, V., Wairagkar, N., Joshi, G., Umarani, U., Risbud, A., Gaikwad, D., Bedekar, S., Divekar, A. and Rodrigues, F. (2002) 'A serological survey of arboviral diseases among the human population of the Andaman and Nicobar Islands, India', *Southeast Asian Journal of Tropical Medicine and Public Health* 33: 794–800

Patz, J.A., Confalonieri, U.E.C., Amerasinghe, F.P., Chua, K.B., Daszak, P., Hyatt, A.D., Molyneux, D., Thomson, M., Yameogo, L., Malecela-Lazaro, M., Vasconcelos, P., Rubio-Palis, Y., Campbell-Lendrum, D., Jaenisch, T., Mahamat, H., Mutero, C., Waltner-Toews, D., Whiteman, C., Epstein, P., Githeko, A., Rabinovich, J. and Weinstein, P. (2005) 'Human health: Ecosystem regulation of infectious diseases', in R. Hassan, R. Scholes and N. Ash (eds) *Conditions and Trends*, Island Press, Washington, DC, pp391–415

Piot, P. (2000) 'Global AIDS epidemic: Time to turn the tide', *Science* 288: 2176–2178

Ramakrishnan, P., Saxena, K. and Chandrashekara, U.M. (1998) *Conserving the Sacred: For Biodiversity Management*, UNESCO, Oxford/IBH, New Delhi

Randolph, S. (2001) 'Tick-borne encephalitis in Europe' [letter], *The Lancet* 358: 1731

Reynes, J.-M., Counor, D., Ong, S., Faure, C., Seng, V., Molia, S., Walston, J., Georges-Courbot, M.C., Deubel, V. and Sarthou, J.-L. (1995) 'Nipah virus in Lyle's flying foxes, Cambodia', *Emerging Infectious Diseases* 11: 1042–1047

Richmond, J.K. and Baglole, D.J. (2003) 'Lassa fever: Epidemiology, clinical features, and social consequences', *British Medical Journal* 327: 1271–1275

Robinson, D. (1985) *Epidemiology and the Community Control of Disease in Warm Climate Countries*, Churchill Livingstone, Edinburgh

Sachs, J. and Malaney, P. (2002) 'The economic and social burden of malaria', *Nature* 415: 680–685

Sahlins, M.D. (1972) *Stone Age Economics*, University of Chicago Press/Walter De Gruyter, Chicago, IL

Save the Children (2006) *Missing Mothers*, Save the Children Fund, London

Sen, A.K. (1999) *Development as Freedom*, Oxford University Press, Oxford and New Delhi

Sioli, H. (1953) 'Schistosomiasis and limnology in the Amazon region', *American Journal of Tropical Medicine and Hygiene* 2: 700–707

Siskind, J. (1973) *To Hunt in the Morning*, Oxford University Press, New York

Stringer, C. (2000) 'Coasting out of Africa', *Nature* 405: 24–27

Swaroop, S. (1949) 'Forecasting of epidemic malaria in the Punjab, India', *American Journal of Tropical Medicine* 29: 1–17

Thomson, M., Elnaiem, D.A., Ashford, R.W. and Connor, S.J. (1999) 'Towards a kala azar risk map for Sudan: Mapping the potential distribution of *Phlebotomus orientalis* using digital data of environmental variables', *Tropical Medicine and International Health* 4: 105–113

Tylleskär, T., Banea, M., Bikangi, N., Fresco, L., Persson, L. and Rosling, H. (1992) 'Cassava cyanogens and konzo, an upper motor neuron disease found in Africa', *The Lancet* 339: 208–211

Volkman, S.K., Barry, A.E., Lyons, E.J., Nielsen, K.M., Thomas, S.M., Choi, M., Thakore, S.S., Day, K.P., Wirth, D.F. and Hartl, D.L. (2001) 'Recent origin of *Plasmodium falciparum* from a single progenitor', *Science* 293: 482–484

Wacharapluesadee, S., Lumlertdacha, B., Boongird, K., Wanghongsa, S., Chanhome, L., Rollin, P., Stockton, P., Rupprecht, C.E., Ksiazek, T.G. and Hemachudha, T. (2005) 'Bat Nipah virus, Thailand', *Emerging Infectious Diseases* 11: 1949–1951

Webb, J., Mainville, N., Mergler, D., Lucotte, M., Betancourt, O., Davidson, R., Cueva, E. and Quizhpe, E. (2005) 'Mercury in fish-eating communities of the Andean Amazon, Napo River Valley, Ecuador', *EcoHealth* 1 (supplement 2) 59–71.

White, A., Sun, X., Canby, K., Xu, J., Barr, C., Katsigris, E., Bull, G., Cossalter, C. and Nilsson, S. (2006) *China and the Global Market for Forest Products: Forest Trends*, CIFOR, Bogor, Indonesia

Wilson, E.O. (2002) 'Nature matters', *American Journal of Preventive Medicine* 20: 241–242

Wolfe, N.D., Switzer, W.M., Carr, J.K., Bhullar, V.B., Shanmugam, V., Tamoufe, U., Prosser, A.T., Torimiro, J.N., Wright, A., Mpoudi-Ngole, E., McCutchan, F.E., Birx, D.L., Folks, T.M., Burke, D.S. and Heneine, W. (2004) 'Naturally acquired simian retrovirus infections in central African hunters', *The Lancet* 363: 932–937

World Health Organization (1978) *Report of the International Conference on Primary Health Care, Alma-Ata, USSR, 6–12 September 1978*, World Health Organization and United Nations Children's Fund, Geneva

Xu, J., Ma, E.T., Tashi, D., Fu, Y., Lu, Z. and Melick, D. (2006) 'Integrating sacred knowledge for conservation: Cultures and landscapes in Southwest China', *Ecology and Society* 10: 7

Zhang, P., Shao, G., Zhao, G., Master, D.C.L., Parker, G.R., Dunning, J.B., Jr. and Li, Q. (2000) 'China's forest policy for the 21st century', *Science* 288: 2135–2136

Zurbrigg, S. (1994) 'Re-thinking the "human factor" in malaria mortality: The case of Punjab, 1868–1940', *Parassitologia* 36: 121–135

3
Health, Habitats and Medicinal Plant Use

Anthony B. Cunningham, Patricia Shanley and Sarah Laird

Medicinal plants are the roots of medical practice. Of the 12,807 species used in traditional Chinese medicine, for example, 11,146 are plant species (Zhao, 2004). Medicinal plant uses range from anti-microbial 'chewing sticks' for dental care and the treatment of internal parasites to symbolic uses. In fact, of the global total of 422,000 flowering plant species, more than 50,000 are used for medicinal purposes, with an estimated 2500 species of medicinal and aromatic plants traded worldwide, most still collected from wild sources (Schippmann et al, 2003).

Across all cultures, for most of human history, all doctors effectively were botanists, using medicinal plants as the primary source of medicines to treat disease. In Europe and North America, however, commercial pharmaceutical production since the 1950s, a medical 'effectiveness revolution' (Stevens and Milne, 1997) and randomized clinical trials have altered attitudes towards the botanical roots of medicine. Nevertheless, public health-care programmes involving traditional medicines and traditional healers have been implemented in many parts of the world (see Chapter 14, this volume). In China and Vietnam, for example, a long, well-documented history of medicine linked with significant policy support has resulted in public health programmes that use herbal treatments for many common illnesses (Wahlberg, 2006). These programmes often include collaborations with local universities and research institutes, which study and sometimes standardize traditional medicines (Balick et al, 1996). In Nigeria, collaborations between local universities, NGOs, traditional healers' associations and government have helped create the scientific and legal foundation for broader dissemination of traditional medicine as part of primary health care (Iwu and Laird, 1998). In the Caribbean, a programme called TRAMIL (Traditional Medicine in the Islands) has focused on medicinal plants used in households, promoting the safest and most significant species for primary health care (Lagos-Witte et al, 1997). In Brazil, primary health-care programmes are being developed using traditional medicines to meet the needs of communities that lack basic pharmaceutical medicines (Silva et al, 2005), and in the Peruvian Amazon, a regional indigenous federation known as FENAMAD is addressing local health-care needs (Alexiades and Lacaze, 1996). The legal and ethical implications of researching and commercializing traditional knowledge associated with medicinal plants have received extensive attention in recent years from indigenous peoples' groups, researchers, NGOs, governments and others, including as part of the policy process growing from the Convention on Biological Diversity (e.g., Posey, 1999; ten Kate and Laird, 1999; Laird, 2002). But it is beyond the scope of this chapter to review these issues here.

This chapter focuses on links between health care and medicinal plants, dealing with both demand and supply for herbal medicines. In terms of demand, we ask the following questions: What roles do herbal medicines play in public health-care systems? Do plants and their habitats have a role in psychosocial health? What about toxicity and adverse reactions to herbal medicines? Why do urban populations around the world continue to seek out herbal medicines, even when they have access to pharmaceuticals? We then deal with supply factors: What types of medicinal plants are most widely used, and which habitats are favoured sources of medicinal plants? Why should we worry about continued access to forests for slow-growing and more threatened medicinal species if weeds are such an important source of medicinally active ingredients? What role is played by 'conservation through cultivation'? And finally, what policy changes are needed?

DEMAND FOR MEDICINAL PLANTS

Herbal medicine in public health-care systems

Worldwide, the skewed distribution of medical doctors is a weakness in public health care. Typically, high numbers of medical doctors practise in the large cities of developed countries; few are found in the rural areas of developing countries (Wibulpolprasert and Pengpaibon, 2003). The situation in many developing countries offers little ground for optimism. First, the ratio of medical doctors, psychologists, dentists and midwives to total population is extremely low. Second, many skilled, locally trained medical personnel emigrate to Europe, North America or Australia (Hagopian et al, 2004). In sub-Saharan Africa, emigration has reduced the number of formally trained health professionals below the threshold needed to achieve the health-related Millennium Development Goals (United Nations, 2006). In the 1980s, the ratio of physicians to total population was 1:10,000, compared with a ratio of 1:110 for traditional healers to population (Green, 1985). In 1996, based on data in Padarath et al (2004), the physician-to-population ratio in Swaziland had increased to 1:6600. Access to medical doctors is often much worse. In Ghana, for example, the physician–population ratio was 1:16,100, and just 1:24,390 in Tanzania. Access to dentists is even more limited, but midwife numbers are better. In the mid-1990s, dentist–population ratios in Tanzania and Ghana, respectively, were 1:143,000 and 1:500,000 for dentists and 1:2230 and 1:1900 for midwives (Padarath et al, 2004). Equally scarce are psychologists with cross-cultural skills. As a result, traditional medicines continue to be the main form of health care for an estimated 80 per cent of people in developing countries (WHO, 2002).

In many developing countries, traditional birth attendants and traditional healers form important links in the chain of health personnel providing primary health care. Despite the establishment of hospitals and health centres, it is to these traditional healers and birth attendants that the majority of people turn in times of sickness and childbirth. It is therefore important that due regard be paid to the activities of these traditional practitioners.

Across the world, diverse local health-care systems to treat parasitic diseases and diarrhoea and improve oral hygiene have developed over hundreds or thousands of years. For example, in the Riau area of Sumatra, Indonesia, of the 114 medicinal plant species used (from 51 plant families), 50 per cent were used to treat fevers, 33 per cent diarrhoea, and 31 per cent other gastrointestinal problems (Grosvenor et al, 1995). These are common ailments across the tropics. Use of medicinal plants is also widespread in developed countries. In Australia, for example, 48 per cent of the population use 'complementary and alternative medicine', and in the USA, 42 per cent of the population use it (Eisenberg et al, 1998), with use levels increasing significantly over the past decade (Pagán and Pauly, 2005).

Traditional healers also fill an important gap in psychological health-care systems where there are few psychologists with cross-cultural skills. Determining the perceived root causes of misfortune or conflict within communities and guarding against them is a role for the diviner rather than the herbalist, with diviners using plant and animal species for their symbolic value. In Tanzania (Rwiza et al, 1993) and Ethiopia (Alem et al, 1999), mental ill-health is widely considered to have supernatural causes that are best treated by traditional healers. In their study of epilepsy among rural Tanzanian residents, for example, Rwiza et al (1993) found that 36.8 per cent of respondents believed epilepsy could not be cured, and 17.1 per cent believed it could not even be controlled. On the other hand, 45.3 per cent believed epilepsy could be treated by traditional healers, and only 50.8 per cent believed hospital drugs were of any use.

Traditional treatment for parasitic diseases

In Tanzania, Gessler et al (1995) showed that most traditional healers were very familiar with the signs and symptoms relating to malaria, as defined by Western medicine. Many healers were aware of different manifestations of malaria and called them by different names; the types corresponded to the scientific terms describing the different types of *Plasmodium falciparum* malaria, such as cerebral malaria, clinical malaria or febrile type, and the gastrointestinal type.

The use of herbal medicines to treat parasitic diseases illustrates their importance in public health both locally and internationally. In tropical developing countries, parasitic diseases affect hundreds of millions of people (Tagboto and Townson, 2001), yet access to synthetic drugs is often difficult because of cost or remoteness. Best known of all treatments for parasitic disease is quinine, from *Cinchona* bark (Honigsbaum, 2002; Lee, 2002), but many other species are used locally. In Bolivia, for example, a bark extract of *Ampelocera edentula* (Ulmaceae) is used by the Chimanes Indians to treat cutaneous leishmaniasis (Fournet et al, 1994). Plants with insect-repellent properties are used preventively. Although inferior to the common synthetic repellent DEET (N,N-diethyl-3-methylbenzamide), andiroba oil, from the seeds of *Carapa guianensis* (Meliaceae), is an effective, low-cost and locally available mosquito repellent (Miot et al, 2004), widely used in South America. In West Africa, smoke from *Daniellia olivieri* (Caesalpinaceae) bark was shown to be more effective in reducing the incidence of malaria than the use of mosquito nets (Palsson and Jaenson, 1999).

Parasites also affect people's lives indirectly, by infesting their livestock. In some cases, the same herbal medicine species are used to treat both livestock and people, such as bark from the aptly named *Albizia anthelmintica*, which is effective against nematode parasites, such as *Haemonchus* and *Heligmosomoides*. Some parasitic diseases can also be controlled through their hosts. In parts of the tropics where bilharzia parasites are common, there is considerable interest in herbal molluscides for controlling the snails that are the hosts for this parasitic disease: for example, preparations from *Phytolacca dodecandra, Warburgia salutaris* (Appleton et al, 1992), *Lawsonia inermis* (Singh and Singh, 2001) and *Azadirachta indica* (Osuala and Okwuosa, 1993).

Dentists are scarce in many parts of Africa, particularly in rural areas. Although diet plays a major role in the incidence of dental caries, the practice of dental hygiene is also important. Toothpaste use is low and chewing sticks are still in common use in many parts of Africa, particularly West Africa. In Nigeria, for example, most suburban children (72.5 per cent) used toothbrushes and toothpaste to clean their teeth, whereas in rural areas, 49.8 per cent used chewing sticks (Otuyemi et al, 1994), mostly from forest trees, with *Garcinia* species particularly favoured (Figure 3.1). In other areas, even if people prefer toothbrushes, high cost or remoteness prevents access to toothpaste and modern brushes (Figure 3.1).

Herbal medicines and health-seeking behaviour

Reasons for the use of traditional and alternative medicine include the perceived efficacy of traditional systems, the high cost of allopathic medical care, and cultural beliefs, as well as the lack of available medical doctors. The holistic, philosophical character of many medical systems strongly influences people's health-seeking behaviour, even when Western medicines are available. Good examples include the continued use of Chinese, South Asian (Ayurvedic and Unani), Thai, Tibetan and Vietnamese medical systems as well as diverse healing practices in Africa and Latin America. The continued importance of traditional medicine to the estimated 1.24 billion people in China is a well-known example, with herbal preparations accounting for 30–50 per cent of total medicinal consumption in China (WHO, 1996). Traditional medicine takes a holistic approach where disease or misfortune results from an imbalance between the individual and the social environment, whereas modern biomedicine takes a technical and analytical approach. This is also a reason behind the often-uneasy relationship between traditional and modern medicine. Both rural and urban patients commonly consult traditional healers before, after or simultaneously with consultation with medical doctors, often switching between different medical systems according to the ailment. In Guinea, West Africa, for example, 33 per cent of 397 diabetes patients surveyed used herbal medicine, motivated by factors that we suggest are also common in other countries: belief in the efficacy of herbal medicines (74 per cent), ease of access to medicinal plants (70 per cent), lower cost (48 per cent) and search for a complete cure (37 per cent). Hearing that others have had positive experiences using herbal medicines was a major factor, persuading 78 per cent of the respondents to use medicinal plants; 85 per cent of the users were satisfied with the results (Balde et al, 2006).

Medicinal plants and psychosocial health

Although this chapter concentrates on medicinal plant species used to treat topical diseases, two additional links to health are important: first, the wider cultural and psychosocial

Figure 3.1 *Medicinal plants in trade*

a) Trunks of the forest tree *Garcinia* being sawn for processing into chewing sticks at a market occupied by 400 chewing-stick traders in Accra, Ghana. b) Traditional medicines, many of them from forest species, for sale at Bouake market, Côte d'Ivoire. c) *Plectranthus grallatus* being harvested by a Zulu diviner (*isangoma*), for whom forests play a crucial role as a source of 'wild power'. d) Yohimbe bark (*Pausinylstalia johimbe*) being processed at a factory in Cameroon for export to Europe as an aphrodisiac. e) *Stangeria eriopus*, a 'living forest fossil' and the only living representative of its family (Stangeriaceae), highly threatened by the traditional medicines trade. f) Seeds from *Carapa* trees, the source of medicinal oil in both South America and tropical Africa. g) Seeds of 'Sichuan pepper', from *Zathoxylum yunnanensis*, a medicinal spice harvested from agroforestry systems and from wild populations. h) Seeds from *Hagenia abyssinica*, a montane forest tree, locally called *kosso* and used for their anti-parasitic activity; this is the top-selling medicinal in Ethiopian markets.

meaning of habitats and landscapes, and second, the symbolic and religious role of plants in psychosocial health.

Forests have complex cultural meanings that are directly linked to the wellbeing, culture and belief systems of forest peoples (Reichel-Dolmatoff, 1996: Posey, 1999). Forests can also be viewed as therapeutic landscapes, as in Japan, where *shinrin-yoku* (walking or staying in forests to promote health) is a major form of relaxation (Shirakawa, 2006). Similarly, in the 19th century, many urban parks and protected forests were established in industrializing countries because of the belief that trees and nature promoted a sense of tranquillity (Melnyk, 2001).

Plants are also used symbolically and for religious and cultural purposes, all of which are connected to psychosocial health. Worldwide, aromatic plants, many of them weedy species, are used to ward off misfortune. In Zulu and Xhosa homes in southern Africa, the plaited stems, leaves and flowers of *Helichrysum odoratissimum* (Asteraceae), known as *imphepho* (a word also used for incense burned in Christian churches), are widely burned as incense to honour ancestor spirits. Plants with red, black or white attributes – such as red or white sap or black fruits – are used symbolically because of the widespread importance of the colour triad red, white and black (see Jacobson-Widding, 1979).

Ritual use of psychoactive plants has been recently reviewed by Shepard (2005) and is summarized in Table 3.1. Here, it is sufficient to say that several species play pivotal cultural roles and can truly be considered 'cultural keystone species' (Garibaldi and Turner, 2004). They also have contemporary significance. In the western Pacific, kava *(Piper methysticum)* is a crucial component of community meetings. In South and Southeast Asia, betel *(Areca catechu)* 'nuts' are not only widely chewed, but, like kava, are also offered to ancestor spirits.

The most potent psychotropic plants are from forests in tropical Africa *(Tabernanthe iboga),* Mesoamerica *(Salvia divinorum)* and tropical South America (*Banisteriopsis caapi, Psychotria viridis, Brugmansia aurea* and *Virola*) (Table 3.1). Most widespread is the use of psychoactive infusions from a mix of the forest vine *Banisteriopsis caapi* (Malphigiaceae) with leaves from *Psychotria viridis* (Rubiaceae), commonly known as *ayahuasca.* The leaves are used as a religious sacrament in many Amerindian societies (Schultes and Raffauf, 1992) as well as by two syncretic religious groups, Santo Daime and União do Vegetal (which blend *ayahuasca* use with elements of Christianity) and the Afro-Brazilian church, Barquinha. Neither *ayahuasca* nor *Tabernanthe iboga* is a recreational drug. Instead, these are potent, purgative, psychoactive plants that cause profound insights, introspection and value change. Finding a spiritual turning point in life can play an important role in psychosocial health (Fiori et al, 2004), including people whose lives have been affected by addiction and substance abuse. In ceremonial context, potent psychoactive plants catalyse spiritual experience and behavioural change. Grob et al (1996) found that regular ceremonial *ayahuasca* use had helped União do Vegetal adherents overcome alcoholism and drug addiction. Similarly, ibogaine from *Tabernanthe iboga* root bark is effective in the treatment of heroin, cocaine and amphetamine addiction, a use that may be increasingly important in the future (Mash et al, 1998). In a global context, in line with the widespread Amerindian belief that *ayahuasca* communicates with people who have taken this sacrament, McKenna (2005) suggests that *ayahuasca* has an urgent message for all of us in terms of health and biodiversity links: the need to avoid global ecological catastrophe.

Table 3.1 *Leading psychoactive plant species in Africa, Latin America and Asia*

Scientific name, family	*Common name*	*Life form, habitat and country*	*Part used traditionally*	*Source*	*Uses and trade*
Areca catechu, Arecaceae	Betel nut	Palm, moist forest, South and SE Asia	Fruit, with admixture of lime and *Piper betel* leaves	Cultivated in agroforestry systems	Ritual and recreational use, widely traded
Banisteriopsis caapi, Malphigiaceae	Ayahuasca, daime, hoasca, yage	Vine, tropical moist forest, South America	Stem, with admixture from *Psychotria viridis* or *Diplopterys cabrerana*.	Cultivated in home gardens and managed forests	Ritual, small-scale trade
Boophane disticha, Amaryllidaceae	Incotho	Geophyte, coastal renosterveld and montane grasslands, southern Africa	Corm	Wild harvest and cultivated by healers in home gardens	Divination and interpretation of dreams
Cannabis sativa, Cannabaceae	Cannabis	Shrub, disturbed areas, Himalaya	Seeds, resin	Widely cultivated	Recreational drug, fibre and pickled seeds as vegetable
Catha edulis, Celastraceae	Khat, qat, miraa	Tree, forest margins, Africa and Arabian peninsula	Leaves	Cultivated in agroforestry systems	Prevents fatigue, widely traded
Duboisia hopwoodii, *D. leichardtii* and *Nicotiana species*, Solanaceae	Pituri	Shrub, desert, Australia	Resin, with *Acacia* ash admixture	Wild, but hybrid cultivated for commercial scopolamine production	Ritual (e.g. male puberty), widely traded
Ephedra species (E. sinica, E. gerrardiana), Ephedraceae	Ephedra	Shrub, semi-arid steppe and arid Himalaya	Stems	Wild harvest for international commercial trade	Harvested commercially (Pakistan, China). Colds, fatigue, high-altitude sickness
Erythroxylum coca, Erythroxylaceae	Coca	Shrub, Andean highlands, South America	Powdered or fresh leaves, with *Cecropia* leaf ash admixture	Cultivated, including in the Amazon	Prevents fatigue and high-altitude sickness. Illegal trade in processed form (cocaine)
Mitragyna speciosa, Rubiaceae	Kratom	Tree, moist tropical forest, SE Asia	Leaves	Wild harvest and home gardens	Prevents fatigue, treatment of morphine addiction
Piper methysticum, Piperaceae	Kava	Shrub, moist forest and agroforestry systems, western Pacific	Root	Cultivated, many varieties	Ceremonial and recreational use, widely traded
Salvia divinorum, Labiatae	Diviner's sage	Perennial herb, cloud forest, Mexico	Leaves	Cultivated, home gardens	Ritual use by Mazatec Indians, Mexico. Small-scale trade.
Sceletium tortuosum, Mesembryan-themaceae	Kanna, kougoed	Succulent, semi-desert, South Africa	Leaves	Wild harvest	Small-scale trade
Synaptolepis kirkii, Thymeleaceae	Uvuma-omhlophe	Scandent climber, dry coastal forest, southern Africa	Lignotuber	Wild harvest	Commercial trade in South Africa
Virola species, especially *V. theiodora*, Myristicaceae		Tree, moist tropical forest, South America	Resin from bark, with *Justicia pectoralis* leaves	Managed forests	Local use and barter

Note: Many of these plants are highly toxic and their mention here does not mean we advocate their use.

Sources: Kalix (1991), Schultes and Raffauf (1992), Smith et al (1996), unpublished data and Callaway et al (1999), de Rios and Stachalek (1999), Prisinzano (2005) and de Feo (2005)

Risks associated with herbal medicines

Herbal medicines have natural origins, but natural does not mean non-toxic. Like pharmaceutical drugs, herbal medicines need to be harvested, stored, prepared and prescribed with attention to safety, quality and efficacy. In rural areas where healthy stocks of favoured species remain, healers are able to harvest quality products locally. Safety, quality and efficacy are more difficult to achieve in urban areas, or where rural healers have to buy favoured species from traders. In Nigeria, for example, all herbal medicines tested contained heavy metals (Obi et al, 2006). A similar problem was found with Asian herbal medicines, which contained arsenic, lead and mercury at levels ranging from toxic (49 per cent) to levels higher than public health guidelines (74 per cent) (Garvey et al, 2001). High price and scarcity due to rarity or over-harvest pose additional problems, since substitution with similar-looking species with different properties becomes common.

In a recent literature review, Yang et al (2006) identified 32 pharmaceutical drugs that interacted with herbal medicines, primarily anti-coagulants, sedatives and anti-depressants, oral contraceptives, anti-HIV agents, cardiovascular drugs, immunosuppressants and anti-cancer drugs. Preventing these adverse reactions is important. Until South Africa's recent policy changes on supplying anti-retrovirals to HIV-positive people, use of herbal medicines such as *Hypoxis* and *Sutherlandia* were being promoted. The recent study by Mills et al (2005) showed that the use of these African herbal medicines might put patients at risk for drug toxicity, treatment failure or viral resistance.

Adverse reactions are not restricted to herbal medicines; they also occur with pharmaceuticals. In 1995, for example, the US Food and Drug Administration's surveillance system MedWatch recorded 6894 fatalities due to adverse drug reactions (Chyka, 2000). In France, a study of more than 300 'adverse drug events' by Queneau et al (2007) showed that 410 types of synthetic drugs were involved, most commonly psychotropic agents (20.5 per cent), diuretics (11.7 per cent), analgesics (13.9 per cent) and anti-coagulants (9.3 per cent) or other cardiovascular drugs (15.4 per cent). Strategies to avoid adverse reactions to herbal medicines can be developed, just as they can for pharmaceuticals. This includes understanding not only which pharmaceuticals interact with which herbal medicines, but also what drives demand for traditional medicines.

Urbanization, epidemiology and cultural preferences

In the past, sustainable harvest of medicinal plants was facilitated by several inadvertent or indirect controls and some intentional management practices. Taboos, seasonal and social restrictions on gathering medicinal plants, and the technology of harvesting equipment all served to limit medicinal plant harvesting. Before metal machetes and axes were widely available, plants were collected with a pointed wooden digging-stick or small axe, which limited the quantity of bark or roots gathered. Pressure on medicinal plant resources continues to remain low in remote areas without road access. In rural areas, medicinal plant use involves self-medication or traditional healers. In urban areas, however, the herbal medicines trade involves self-employed commercial harvesters and formal-sector traders who supply the large demand. As a result, herbal medicine trade is a booming business.

Rapid urbanization characterized the 20th century. In 1900, just 13 per cent (220 million people) of the world's population was urban, growing to 29 per cent in 1950, 49 per cent in 2005 and projected to reach 60 per cent (or 4.9 billion people) in 2030 (United Nations, 2006). Urbanization will be particularly rapid in developing countries and, combined with cultural preferences in health care, will continue to drive commercial trade in medicinal plants. It has not only changed patterns of where and how medicinal plants are sourced, but has also increased the incidence of lifestyle diseases, such as diabetes and hypertension, which in developing countries are often treated with herbal medicines. In South Asia, for example, the incidence of Type 2 diabetes, commonly treated with *Momordica charantia* (Curcurbitaceae) (Saxena and Vikram, 2004) was higher in urban (11.6 per cent) than rural (2.4 per cent) populations (Ramachandran et al, 1999). In South Africa, a study of baTswana urban migrants similarly showed how hypertension increased because of factors associated with urbanization (van Rooyen et al, 2000). The epidemiology of insect-borne disease is also creating demand for herbal medicines. Belém, Brazil, where *Carapa guineensis* (andiroba) oil is used to prevent mosquito bites (Shanley and Luz, 2003), has seen an increase in malaria since the 1970s and the reappearance of *Anopheles darlingi* in the mid-1990s, thought to have been eradicated in 1968, probably because the city expanded into surrounding forest (Póvoa et al, 2003).

Urban markets create easy access to medicinal plants for city-dwellers, with cultural association coupled to commercial marketing fuelling demand (Figure 3.2). In addition, some native medicinal plants are considered powerful and effective in treating common and chronic diseases (Table 3.2) and have been widely commoditized. In India,

> *... market capitalism has shaped, constrained and transformed Indian traditional medicine over the last 25 years. ... nowadays approximately 90 per cent of the Ayurvedic and Unani formulas are over-the-counter brands that are marketed to urban middle-class consumers. The rise in the last decade of the 20th century of a relatively affluent urban consumer class of about 100 million people explains the proliferation of relatively expensive Ayurvedic and Unani brands. Because of their propagation in the public media, commoditized medicines increasingly determine image and substance of Ayurveda and Unani Tibb, India's largest medical traditions. For many Indians both forms of Indian medicine are no longer the tailor-made formulas made by humoral experts, the cheap alternatives of the poor and the medical betel nuts sold on city pavements, but modern looking medicines attractively packed and sold as remedies against common ailments, degenerative and chronic diseases and as important assets in fighting the stress of modern city life.* (Bode, 2006)

A similar situation has developed in China, with many of the same high-altitude genera or species harvested from the Himalayas for trade to India, which like China is a major exporter and consumer of medicinal plants. Nepal exports between 7000 and 27,000 tonnes of medicinal plants a year, most of them to India, worth US$7 million to 30 million per year (Olsen, 2005). At a global scale, China is the largest exporter of medicinal and aromatic plants, mainly to Hong Kong (140,500 tonnes), and is also a significant importer

Table 3.2 *The habitats, uses and main sources of the leading indigenous medicinal plant species in Latin America (eastern Brazil (moist forest), northeastern Brazil (caatinga)), North America (USA), Asia (China, Nepal) and Africa (South Africa, Ethiopia)*

Scientific name, family	*Common name*	*Life form, habitat and country*	*Part used*	*Source*	*Uses*
Alepidea amatymbica, Apiaceae	Ikhathazo	Perennial herb, montane grassland, South Africa	Rhizome	Wild	Coughs and colds
Boweia volubilis, Liliaceae	Igibisila	Perennial geophyte, montane grassland, South Africa	Bulb	Wild	Protective charm
Carapa guianensis, Meliaceae	Andiroba	Tree, moist forest, Brazil	Seed oil	Wild	Insect repellent, sprains, arthritis
Copaifera reticulata, Leguminosae	Copaíba	Tree, moist forest, Brazil	Oleoresin	Wild	Wounds, sore throat
Echinacea angustifolia, Compositae	Echinacea	Perennial herb, grasslands, USA	Root	Cultivated	Immuno-stimulant, wound healing
Echinacea purpurea, Compositae	Echinacea	Perennial herb, grasslands, USA	Root	Cultivated	Immuno-stimulant, wound healing
Embelia schimperi, Myrsinaceae		Scandent climber, forest margins, Ethiopia	Seeds	Wild	Anthelmintic
Eucomis autumnalis, Liliaceae	Umathunga	Perennial geophyte, montane grasslands, South Africa	Bulb	Wild	Enema preparation to cleanse colon
Eucommia ulmoides, Eucommiaceae	Duzhong	Tree, montane forests, western China	Bark	Cultivation, probably extinct in the wild	Dizziness due to hypertension
Gastrodia elata, Orchidaceae	Tianma	Perennial geophyte, southwest China	Tuber	Mainly from cultivation	Vertigo, dizziness, hypertension
Ginkgo biloba, Ginkgoaceae	Baiguo, ginkgo	Tree, temperate forest, China	Leaves, seeds	Culivation	Cough, coronary heart disease, slows cognitive deterioration
Hagenia abyssinica, Rosaceae	Kosso	Tree, montane forest, Ethiopia	Seeds	Wild	Anthelmintic
Himatanthus sucuuba, Apocynaceae	Sucuúba	Tree, moist forest, Brazil	Bark/ exudate	Wild	Worms, herpes, uterine inflammation
Hydrastis canadensis, Ranunculaceae	Goldenseal	Perennial herb, deciduous temperate forest, Canada and USA	Root	Wild	Masks illicit drugs in urine, anti-diarrhoeal

Table 3.2 *(cont'd)*

Scientific name, family	*Common name*	*Life form, habitat and country*	*Part used*	*Source*	*Uses*
Nardostachys jatamansi, Valeriancaeae	Jatamansi	Perennial herb, montane pastures, Himalaya	Roots	Wild	Stomach ache, anorexia, toothache, essential oils
Panax ginseng, Araliaceae	Ginseng	Perennial herb, temperate forest, China	Root	Wild and cultivated	Cognitive function, endurance enhancement, anorexia
Panax quinquefolius, Araliaceae	American ginseng	Perennial herb, temperate forest, USA and Canada	Root	Wild	Cognitive function, endurance enhancement
Parachornia fascicuIta, Leguminosae	Ipê roxo	Tree, moist forest, Brazil	Exudate	Wild	Respiratory illness
Picrorhiza scrophulariiflora, Scrophulariaceae	Huhuanglian, kutki	Montane Himalaya	Rhizomes	Wild	Fever, dysentery, conjunctivitis
Rheum palmatum, Polygonaceae	Wild rhubarb, dahuang	Montane pastures, Himalaya	Rhizome and roots	Wild	Dysentery, constipation, intestinal bleeding
Scilla natalensis, Liliaceae	Inguduza	Perennial bulb, montane grasslands, South Africa	Bulb	Wild	Enema preparation to cleanse colon
Serenoa repens, Areaceae	Saw palmetto	Palm savanna, Florida, USA	Fruits	Wild	Benign prostatic hypertrophy
Siphonochilus aethiopicus, Zingiberaceae	Indungulo, wild ginger	Montane grasslands, South Africa	Rhizome	Wild	Coughs and colds, supernatural protection
Stryphnodendren barbatima, Leguminosae	Barbatimão	Tree, moist forest, Brazil	Bark	Wild	Haemorrhage, uterine and vaginal infections
Tabebuia impetiginosa, Bignoniaceae	Pau d'arco	Tree, moist forest, Brazil	Bark	Wild	Inflammations, ulcers, skin ailments
Terminalia chebula, Combretaceae	Chebula	Tree, *terai* woodlands, lowland South Asia	Fruits	Wild	Diarrhoea, uterine bleeding
Warburgia salutaris, Canellaceae	Isibaha, pepper bark	Tree, forest/savanna ecotones, South Africa	Bark	Wild	Coughs, colds and opportunistic candida infections due to HIV

Figure 3.2 *China: The world's biggest medicinal trade*

a) Massive demand for herbal medicines in China is apparent in the bustling marketplaces like Hehuachi market, in Chengdu, China, which has an annual trade turnover of 1.2 billion yuan (around US$148 million) in more than 200,000 tonnes of traditional medicine per year. b) Market values have encouraged cultivation of 25 to 30 per cent of Chinese medicinal species, including *Gingko biloba* (for seeds and leaves). c) Over-harvesting of wild populations of forest species, such as *Eucommia ulmoides*, a monotypic species, has forced a shift to plantation production. d) For women in ethnic minorities, such as Yao people, providing herbal-based health care is an important cross-cultural niche for their skills. e) Some species cannot be cultivated, however, such as this parasitic species, *Cynomorium songaricum* (Cynomoriaceae), from Xinjiang, China. f) To enhance production of *Ammomum tsåo-ko*, forest trees are felled to keep the canopy open, affecting montane forests of northern Vietnam and southwestern China. g) Medicinal fungi, often from forests, are also highly valued, such as *Ganoderma* (chizi), which are traded internationally.

(80,550 tonnes) (Lange, 1998). According to a report published by Hong Kong Trade Development Council, the global sales for Chinese medicines have grown 8 per cent a year since 1994. In 2002, the total global sales of traditional Chinese medicines were US$23.2 billion (Phillip Securities Research, 2003). Sales may increase even faster with the formal, industrialized production and export of traditional preparations. In Asia, particularly China, India, Pakistan and Vietnam, government support for the development and modernization of traditional medical systems is likely to increase harvest levels from wild stocks. In India, where the Ayurvedic industry is worth an estimated US$1 billion per year, 7500 factories produce thousands of Ayurvedic and Unani formulas (Bode, 2006). In China, clinical trials for traditional preparations are now frequent (e.g., Cao et al, 2005; Qiong et al, 2005; Taixiang et al, 2005), and there are plans to establish standards for these products and a competitive, modern industry in traditional Chinese medicine. In Africa and South America, production is less formalized and branding is less sophisticated, yet the scale of the trade is nevertheless large. In South Africa, for example, 1.5 million informal traders sell about 50,000 tonnes of medicinal plants annually in a region with an estimated 450,000 traditional healers (Mander, 2004).

What is urgently needed is closer attention to the resource base of herbal health-care systems, because wild populations of medicinal plants remain the main sources of supply. The growing demand from urban areas, especially for the most favoured medicinal species, has catalysed the trade, drawing in remote resources to towns and cities (see Figure 3.2 a–d).

SUPPLY OF MEDICINAL PLANTS

Threats to continued availability

Poor people tend to rely on environmental services and harvested products that provide 'green social security'. As a result, they are particularly vulnerable to environmental degradation (Cavendish, 1999). The resource base of the herbal medicines trade is being affected by multiple factors simultaneously, at different spatial and time scales. The most serious of these are habitat loss and fragmentation, global climate change, species-specific over-exploitation and invasive species. Global projections of plant diversity loss show that the largest losses of habitat and diversity will occur in tropical ecosystems (forest, woodland, savannah), accounting for a projected loss of 25,000 to 40,000 plant species by 2050 (van Vuuren et al, 2006). In the African and Indo-Malayan tropics, regions where herbal medicines are extensively used, the worst losses in biodiversity are projected to occur primarily through loss of habitat. In contrast, it is climate change that will drive species loss in tundra, boreal forests and cool conifer forest (van Vuuren et al, 2006). What also needs to be taken into account in these global predictions is species-specific over-harvesting in what appears to be intact habitat.

Worldwide, it is estimated that of 422,000 flowering plants, 12.5 per cent (52,000) are used medicinally, and 8 per cent (4160 species) of these are threatened (Schippmann et al, 2003). Major biological factors that influence vulnerability or resilience include life form, age to reproductive maturity, productivity, density, resprouting potential and plant part harvested. In general, species that are most susceptible to over-harvesting are habitat-specific, slow-growing and destructively harvested for their roots, bark or whole plant (Peters, 1994; Cunningham, 2001). Many wild species supplying medicinal plant markets show declining availability (Cunningham, 1991; 1993; Shanley and Luz, 2003; Botha et al, 2004). In South and Central America as well as tropical Africa, numerous valuable trees are 'conflict-of-use' species: trees with both a timber and a non-timber use. Eight of the most valuable fruit and medicinal species used and traded in eastern Amazonia, for example, are extracted as timber (Shanley and Luz, 2003). In Amazonia, two long-lived, low-density medicinal tree species, *Hymenaea courbaril* and *Tabebuia impetiginosa*, are particularly affected. The medicinal bark of *Tabebeuia impetiginosa*, locally and internationally used for internal inflammations, skin diseases and tumors, is now collected as a by-product from sawmills. In Central America, *Metroxylon balsamum*, which is a favoured medicinal bark, is also heavily logged, as is *Prunus africana*, logged from Afromontane forests of West and East Africa. Ecological studies of both species indicate poor capacity for regeneration after logging, with some scientists recommending that like mahogany, these species merit listing by the Convention on International Trade in Endangered Species of Wild Fauna and Flora (CITES). Such a policy change has already occurred for *Prunus africana*, with mixed results (Cunningham, 2005).

Of most concern in terms of conservation, phytochemistry and health care are species that are not only slow-growing and vulnerable to over-harvesting but also phylogenetically distinctive – that is, just one species is the only living representative of an entire plant family or an entire genus (Table 3.3). The conservation of such monotypic species is important, compared with large medicinal plant genera, such as *Astragalus* (1750 species) and *Euphorbia* (2000 species), where many species share similar active ingredients.

At the local level, communities in the eastern Amazon that have experienced exploitative logging no longer have access to valuable barks or the medicinal oils from *Carapa guianensis* and *Copaifera* species, which are cut for timber. Certification of the timber of threatened medicinal bark species, such as *Tabebuia impetiginosa* and *Hymenaea courbaril*, and lively sales to 'green' consumers in the USA and Europe signify a substantial lack of understanding and communication regarding the livelihood benefits of species (Shanley et al, 2006).

In general, most botanicals companies are largely unaware of the sources of their raw material, including the geographic origins, production systems, and the social and environmental impacts. Many gatherers are poorly paid for their labour. King et al (1999) report that harvesters receive between US$0.30 and $0.65 per kilogram for raw cat's claw (*Uncaria* spp) in Peru, while the price of bulk, unprocessed cat's claw in the USA is US$11 per kilogram. In part this is due to the common practice of sourcing raw material as a bulk commodity, and the physical and cognitive distance between most companies and their sources (Pierce and Laird, 2003; Laird et al, 2005).

At the same time, medicinal plants feed large international markets for botanical medicine, nutraceuticals, and personal care and cosmetic products, which are part of a larger 'supplement' market valued at more than US$50 billion annually (*Nutrition Business Journal*, 2003). Medicinal plants and other natural products, like micro-organisms, insects and marine organisms, are also the basis of many of our pharmaceutical drugs (Farnsworth et al, 1985; Grifo et al, 1997; Laird and ten Kate, 2002; Newman et al, 2003; Cragg et al, 2005).

Table 3.3 *Examples of commercial medicinal plants with high conservation values*

Phylogenetic distinctiveness, family	*Scientific name*	*Common name, life form, part used, main use*	*Geographic region and comments on conservation status*
MONOTYPIC FAMILIES			
Eucommiaceae	*Eucommia ulmoides*	*Duzhong*, tree, bark, hypertension	**China**, endemic family to China. Very rare (or possibly extinct) in the wild (montane forest, western China). Rare/endangered in Sichuan (Zhang and He, 2002) and Rare (Walter and Gillet, 1998).
Ginkgoaceae	*Ginkgo biloba*	*Yinxingye* (leaves), *baiguo* (seeds), tree	**China**, endemic family to China. Very rare in the wild. Dioecious. Rare/endangered in Sichuan (Zhang and He, 2002) and Rare (Walter and Gillet, 1998). Widely cultivated for ornamental and medicinal purposes.
Stangeriaceae	*Stangeria eriopus*	*Imfingo*, protective charm in Zulu traditional medicine	**South Africa**. 1 species, endemic to South Africa. Dioecious. Wild harvest from forest or forest margin grasslands.
MONOTYPIC GENERA			
Ranunculaceae	*Circaeaster agrestis*	Perennial herb, roots, Chinese traditional medicine	**NW Himalayan region**, endemic to W and NW China. Rare/endangered in Sichuan (Zhang and He, 2002).
Ranunculaceae	*Hydrastis canadensis*	Perennial herb, herbal antibiotic, inflammation	**Eastern North America**, endemic. Temperate forests.
Ranunculaceae	*Kingdonia uniflora*	Perennial herb, roots, Chinese traditional medicine	**China**. Rare/endangered in Sichuan (Zhang and He, 2002).
Asteraceae	*Lamprachaenium microcephalum*	Perennial herb, Chinese traditional medicine	**India**, endemic. Considered vulnerable.
Valerianaceae	*Nardostachys grandiflora*	*Jatamansi*, perennial herb, roots, stomach ache, vomiting and massage oils	**Himalayan** endemic, monotypic genus. High-altitude yak pastures and mountain slopes.
Scrophulariaceae	*Neopicrorhiza scrophulariiflora*	*Kutki*, perennial herb, roots, fever, dysentery, diarrhoea	**W. Himalayan** endemic. 1 species. High-altitude yak pastures and mountain slopes. Wild harvest. Rare/endangered in Sichuan (Zhang and He, 2002).

Table 3.3 *(cont'd)*

Phylogenetic distinctiveness, family	*Scientific name*	*Common name, life form, part used, main use*	*Geographic region and comments on conservation status*
MONOTYPIC GENERA			
Monimiaceae	*Peumus boldus*	*Boldo*, small tree, leaves, antioxidant, rheumatism	**Chile**, endemic. 1500 tonnes of leaves exported, 80% within Latin America (Argentina, Brazil), 18% Europe (France, Germany).
Dioscoreaceae	*Trichopus zeylanicus*	*Arogya pacha,* perennial herbs, fruits, anti-fatigue, anti-microbial.	**Indo-Malayan region**, considered critically endangered in South India.
Asclepiadaceae	*Utleria salicifolia*	Perennial herb, rhizomes, anti-ulcer	**South India**, endemic, considered critically endangered in South India.

Note: These species are the only representatives of their families, or should be rated highly for conservation action because of their phylogenetic (and in many cases, phytochemical) distinctiveness.

These drugs now contribute significantly to pharmaceutical company revenues and make up a significant portion of top-selling drugs, particularly in categories like infectious disease and cancer (Newman and Laird, 1999; Newman et al, 2003). Industry interest in natural products is a response to technological, scientific, legal and market developments, and as such is cyclical and constantly changing. Industry continues to return to natural products as a source of novelty and diversification, as today's surging interest in micro-organisms and species from extreme environments attests (Laird and Wynberg, 2005).

Favoured medicinal plant habitats

The origins and management of species used in health care can inform our understanding of the relationship between biodiversity, medicinal plants and health. Old-growth tropical forests are widely considered the most important source of medicinal plants because of their diversity and the occurrence of plant families rich in active ingredients (such as the Apocynaceae and Menispermaceae). In other areas, secondary or disturbed habitats are preferred. In Ghana, for example, Falconer (1990; 1994) found that primary forest was not the main source of plant medicines for households (in terms of the percentage of species used), and that fallow areas and the village periphery are the most important sources. Of those interviewed, 82 per cent gathered their own medicines from these areas, with 12 per cent from the forest, 65 per cent from farm fallow and 46 per cent from village areas. However, healers make greater use of forest species, even if many are sourced from disturbed habitats or planted in compounds. Laird et al (2007) similarly found that indigenous Bakweri healers around Mount Cameroon use significantly more medicinal forest species than most members of communities, and that many forest species are transplanted from the forest into healers' home gardens. As a result, the location of species collections is not necessarily an indication of the conservation value of the species or of the value of forest to local communities. In addition, forest species are often considered the most powerful and used to treat the most severe problems. Although disturbed sites (e.g., village, farm, fallow and secondary forest areas) offer a wide range of species, forest species are critical to the specialist medicine practised by healers. Healers also collect almost six times the number of native and wild medicinal species as the village average, followed by hunters.

Introduced weeds are widely adopted into local pharmacopoeia. In a thought-provoking paper, Stepp (2004) pointed out that many indigenous peoples harvest medicinal plants from non-forested, disturbed habitats, and that 36 of the 101 plant species from which 119 contemporary pharmaceuticals are derived come from plants that could be classed as weeds. This analysis was not focused solely on short-lived, annual weeds of disturbed areas, but also included forest trees, such as *Taxus brevifolia* (the source of taxol), and long-lived clonal species, such as *Convallaria majalis* (whose individuals can live longer than 670 years). In the *caatinga* dry forests of northeastern Brazil, for example, only 56 per cent of the medicinal species used are native (de Albuquerque, 2006).

There are also significant differences between communities that have resided in a forest area for many generations and continue to rely on the forest, and those that recently arrived in an area or live more independently of forest resources. In Cameroon, Laird et al (2007) found that of the more than 400 useful plant species documented in Bakweri villages around Mount Cameroon, a quarter had been introduced. The indigenous Bakweri communities in this area have long histories of contact with outside groups and have incorporated many introduced species into their livelihoods. However, total livelihood contributions from native and wild species to indigenous households are more than five times the contribution of native and wild species to migrant households; this includes their use of medicinal species.

In some cultural situations, habitat selection is also influenced by the gender of the traditional healers. Among many African farming societies, it is more acceptable for men (as hunters) to spent time on their own in old-growth forests, and less so for women. In interview surveys with 714 healers using 295 plant species in southwestern Uganda, Kyashobire (1998) found that men (herbalists and non-specialists who collected plants for home remedies) used the forest more extensively than women (both traditional birth attendants and female herbalists) (Table 3.4). Men also used more bark and roots; for women, leaves were the common plant part most commonly used.

Table 3.4 *Habitats of plant species collected in or around Bwindi forest, Uganda*

Herbalist category	*Garden*	*Early fallow*	*Mature fallow*	*Bushy thicket*	*Forest*	*Healers interviewed (n)*
Traditional birth attendant	9.5%	22.7%	24.8%	21.3%	21.7%	184
Female herbalist	8.6%	23.2%	24.1%	24.1%	20.0%	179
Male herbalist	9.4%	18.0%	21.1%	22.8%	28.7%	258
Male, non-specialist	3.9%	18.1%	15.2%	25.0%	37.8%	93

Note: Some 295 plant species are used in this area.
Source: Kyashobire, 1998

Cultivation of medicinal plants

'Conservation through cultivation' has long been promoted as a means of ensuring future supplies of medicinal plants (e.g., Cunningham, 1993), yet most medicinal plants are still harvested from wild populations. Based on figures from Europe, China and India, the number of medicinal and aromatic species cultivated worldwide does not exceed a few hundred (Schippmann et al, 2003). From an economic perspective, cultivation can be desirable, guaranteeing a regular supply and adherence to quality standards (Pierce et al, 2002). What is needed is an objective assessment of where and why conservation through cultivation has been successful (or not).

For the rural poor without land or livestock, harvesting of wild plant resources, including medicinal plants, is a common option, particularly in ecosystems with low arable potential. In theory, medicinal plant cultivation also offers an economic opportunity for rural smallholder farmers and an answer to declining wild stocks. Cultivation of medicinal species for illegal but lucrative processing for the drug trade certainly takes place: examples include *Cannabis sativa* (Australia, Africa, Asia, North America), *Catha edulis* (Ethiopia, Kenya, Yemen), *Erythroxlon coca* for cocaine (South America) and opium from *Papaver somniferum* (Afghanistan and Southeast Asia).

In practice, economically viable, legal production of medicines is difficult to achieve on a large enough scale to meet commercial demand. Small-scale producers commonly face high risks and transaction costs, and lack guaranteed markets and trust among different actors along the value chain (van de Kop et al, 2006). In Africa, *Prunus africana* is a good example of the failure of cultivated production to date. Bark production can be an economic proposition (Cunningham et al, 2002), and plantation trials were implemented in Kenya as early as 1920. In northwestern Cameroon, an estimated 3200 farmers grow *Prunus africana* in small-scale agroforestry systems, yet export continues to be based on wild harvest, with many populations over-exploited (Cunningham, 2005). There have been successes elsewhere, such as in China, where a much higher proportion of medicinal plants (10 to 25 per cent, or 100 to 250 of the estimated 1000 commonly traded species) are cultivated, compared with less than 1 per cent of medicinal plants sourced from cultivation globally. We suggest that cultivation in China has been a response to several factors, including price increases linked to major declines in wild stocks and the attraction of conservation through cultivation from a policy perspective. Progress towards policy goals, for example, is evaluated in China according to targets, which are usually set by the next-highest level of the Communist Party (for party leaders) or government (for government officials). A target of a certain area of a medicinal plant species under cultivation is much easier to measure than sustainable harvest of dispersed, wild populations.

In Peru, the realized or expected increases in demand for such plants as camu-camu *(Myrciaria dubia)* and cat's claw *(Uncaria guianensis)* have led state agencies and NGOs to promote the cultivation or intensification of the production system through direct intervention, technical assistance or subsidies (Panduro and de Jong, 2004; Nalvarte and de Jong, 2004). In Brazil, demand has catalysed innovation that favours sustainable practices, especially in cases where land tenure is clear. Particularly in peri-urban areas close to markets, demand has encouraged farmer-led innovations not only in palms (Anderson and Jardim, 1989) but also in long-lived exudate-producing species. Local management practices leading to improved production in peri-urban areas include germplasm selection, selective weeding of fallows and enrichment planting. Results illustrate that species useful for nutritional fruits, medicinal exudates and roots are more likely to be managed than long-lived species used for medicinal bark. For example, in areas close to markets in the eastern Amazon of Brazil, demand has encouraged farmer-led innovations in the management of *Parahancornia fasciculata* (amapa). The latex is used for respiratory diseases and to fortify the body after malaria. Farmers are currently experimenting with extraction techniques and incipient attempts at enriching forests with seedlings. Where land tenure is clear, innovations of sustainable practices are incipient, whereas on communal or invaded private properties, exploitative practices are widespread.

In other cases, cultivation can worsen the situation by contributing to habitat destruction. In southwestern China, for example, cultivation of commercially traded medicinal plants, such as *Ammomum villosum* (sharen), prevents lowland rainforest regeneration (Liu et al, 2006). The area of *Ammomum villosum* cultivation in Xishuangbanna, Yunnan, is now 58km^2, much of it within conservation areas (Liu et al, 2006). Similarly, cultivation of *Ammomum tsao-ko* (caoguo) (Figure 3.2f) is affecting montane forest conservation areas of high conservation priority, such as forests in Fengshuiling Nature Reserve, Yunnan, and the cross-border area in northern Vietnam. After the understorey is cleared, the *Ammomum* plants that are cultivated underneath canopy trees create intense shade that suppresses forest regeneration.

Despite an economic rationale, consumer preferences for wild-harvested material are widespread (Cunningham, 1993; Pierce et al, 2002). Cultivation is expensive, not always feasible and can take a long time (Schippmann et al, 2003). For families without land and with few livelihood options, wild-harvesting is a safety net. Large and medium-scale plantations generally favour elites, excluding small farmers and those without access to farmland (Scheffer, 2004). Public–private partnerships similar to those implemented in India for *Neopicrorhiza kurroa* cultivation are likely to support small-scale producers in the future (van de Kop et al, 2006).

POLICY RECOMMENDATIONS

As Bodeker and Kronenberg (2002) point out, the Traditional Medicines Strategy (WHO, 2002) focuses on four areas that need implementation to get the best pubic health benefits from the use of traditional, and complementary and alternative medicine:

1. policy changes leading to better recognition of non-Western medicine and its integration into health-care systems;
2. attention to safety, efficacy and quality of medicinal plant use;
3. improved access to traditional medicine, including sustainable harvesting of materials, which in turn affects access issues such as cost, since resource depletion of popular species leads to price increases and scarcity; and
4. improving the quality of information on the rational use of traditional medicine.

These areas of focus raise questions:

- How can empirical knowledge of medicinal plant uses, often held by an older generation of healers in remote areas, be accumulated, stored and transmitted to next generations without compromising their intellectual property rights?
- How can the providers of medicine, the innovators of the trade and the transmitters of knowledge be adequately compensated?
- How can consumers be protected from false information or the use of products with negative side-effects?
- How can sustainable wild sourcing be implemented – or the medicinal plants be 'domesticated' – to secure supplies of quality products before the over-harvesting of wild stocks depletes the resource?

For this last question, policy changes are needed on several fronts: first, the implementation of recent recommendations for sustainable wild harvest made by Schippmann et al (2003), and second, development of public–private partnerships for smallholder cultivation that follow initial successes in India (van de Kop et al, 2006). Sustaining supplies of slow-growing medicinal plants is difficult to achieve but essential for the achievement of the other goals of the Traditional Medicines Strategy, since declining wild stocks are accompanied by adulteration and species substitutions, which in turn reduce efficacy, quality and safety. In India, for example, scarcity and high prices result in *Podophyllum hexandrum* (Araliaceae) roots being substituted or adulterated by *Ainsliaea latifolia* (Asteraceae) roots (Shah, 2006).

Practical steps include the following:

- Develop risk management strategies to prevent adverse reactions to herbal medicines by identifying safe species (as TRAMIL has done) and adverse reactions with pharmaceuticals (Yang et al, 2006).
- Sustain wild harvests. Several approaches can be taken:
  - Develop management plans that link traditional ecological knowledge and sustainable harvest methods, such as those already developed for non-timber forest products (Peters, 1994; Cunningham, 2001).
  - Strengthen capacity through recognizing the skills of knowledgeable local people in conducting inventories, and monitoring and assessing the effects of medicinal plants, as has been done in northwestern Nepal for management of *Nardostachys grandiflora* and *Neopicrorhiza scrophulariiflora* (Ghimire et al, 2004).
  - Provide international and national support for regional training courses and curriculum development in applied ethno-botany and resource management.
- Develop appropriate certification systems. For medicinal plants to be certified in developing countries, producers must weigh the costs and benefits of different certification schemes and decide whether a particular set of standards is a good fit for their product, consumer market, budget and organizational capacity. Several schemes are available:
  - wild harvest certification: the International Standard for Sustainable Wild Collection of Medicinal and Aromatic Plants (ISSC-MAP) is the most recent initiative (Patzold et al, 2006);

  - good agricultural and collection practices: established guidelines that set standards for the proper handling and sanitation of products during harvest, storage and shipping;
  - good manufacturing practices: guidelines for infrastructure, staff and processing procedures, including for food and herbal products;
  - organic certification; and
  - certification of geographic origin, which has potential for application to some medicinal plant species under an approach termed Appellation Origine Protégée, adopted by the European Union in 1992, which includes protection not only of the product and the know-how related to it, but also the conservation of its habitat.
- Develop marketing networks so that certified producer associations or companies can capitalize on certification, including 'matchmaking' between producer associations and industry partners committed to buying sustainable harvested or high-quality cultivated products at fair prices to producers.
- Lower the barriers to smallholder cultivation on a larger scale, following recommendations made by van de Kop et al (2006).
- Expand medicinal plant domestication programmes based on policy agreements between 'range states', the countries in which a species commonly grows; such programmes build up provenance collections for regional benefit that take advantage of the genetic and chemical diversity within a species over its distribution range.
- Develop secure *ex situ* field gene banks for the conservation of high-priority species.

CONCLUSIONS

Tropical ecosystems (forests and woodlands) and high mountain areas are major harvesting areas for the leading medicinal plants in Ayurvedic, Chinese, Tibetan, Unani and diverse African and Latin American health systems. Many of these favoured medicinal plants are long-lived, often habitat-specific perennial species. These species are threatened by habitat loss, climate change, and species-specific, multipurpose over-harvesting and logging. Thus the resource base for traditional health-care systems used by at least 2 billion people is also threatened. Stepp (2004) correctly concluded that many weeds in degraded habitats contain medically active ingredients, but this should not distract attention from chemically and phylogenetically distinctive medicinal species in tropical forests. The world's major psychotropic plants (e.g., *Banisteriopsis caapi, Psychotria viridis, Tabernanthe iboga*), which come from forests, are central to ancient cultural rituals but also have contemporary significance. It is also in forest and montane ecosystems that medicinal species occur in monotypic families (Eucommiaceae, Ginkgoaceae, Stangeriaceae), monotypic genera (*Peumus, Neopicrorhiza*) and very small genera of two to five species (*Ekmanianthe, Holostemma, Notopterygium, Podophyllum, Warburgia*). These habitats need to be the focus of improved policy and practice for medicinal plant conservation and resource management.

REFERENCES

Alem, A., Jacobsson, L., Araya, M., Kebede, D. and Kullgren, G. (1999) 'How are mental disorders seen and where is help sought in a rural Ethiopian community? A key informant study in Butajira, Ethiopia', *Acta Psychiatrica Scandinavica (Supplement)*, 397: 40–47 [abstract only]

Alexiades, M. and Lacaze, D. (1996) 'FENAMAD's program in traditional medicine: An integrated approach to health care in the Peruvian Amazon', in M.J. Balick, E. Elizabetsky and S.A. Laird (eds) *Medicinal Resources of the Tropical Rainforest*, Columbia University Press, New York, NY, pp341–365

Anderson, A.B., and Jardim, M.A.G. (1989) 'Costs and benefits of floodplain forest management by rural inhabitants in the Amazon estuary: A case study of açaí palm production', in J.O. Browder (ed) *Fragile Lands in Latin America: The Search for Sustainable Uses*, Westview Press, Boulder, CO

Appleton, C.C, Drewes, S.E. and Cunningham, A.B. (1992) 'Observations on the molluscicidal properties of (-) warburganal South African *Bulinus africanus*', *Journal of Applied and Medical Malacology* 4: 37–40

Balde, N.M., Youla, A., Balde, M.D., Kake, A., Diallo, M.M., Balde, M.A. and Maugendre, D. (2006) 'Herbal medicine and treatment of diabetes in Africa: An example from Guinea', *Diabetes and Metabolism* 32(2): 171–175

Balick, M.J., Elisabetsky, E. and Laird, S. (1996) *Medicinal Resources of the Tropical Forest: Biodiversity and its Importance to Human Health*, Columbia University Press, NY

Bode, M. (2006) 'Taking traditional knowledge to the market: The commoditization of Indian medicine', *Anthropology & Medicine* 13: 225–236

Bodeker, G. and Kronenberg, F. (2002) 'A public health agenda for traditional, complementary, and alternative medicine', *American Journal of Public Health* 92: 1582–1591

Botha, J., Witkowski, E.T.F. and Shackleton, C.M. (2004) 'The impact of commercial harvesting on *Warburgia salutaris* ("pepper-bark tree") in Mpumalanga, South Africa', *Biodiversity and Conservation* 13: 1675–1698

Cao, Y., Yinyu Shi, Yuxing Zheng, Meiyu Shi and Sing Kai Lo (2005) 'Blood-nourishing and hard-softening capsule costs less in the management of osteoarthritic knee pain: A randomized controlled trial', *eCAM* 2(3): 363–368

Callaway, J.C., McKenna, D.J., Grob, C.S., Brito, G.S., Raymon, L.P., Poland, R.E., Andrade, E.N., Andrade, E.O. and Mash, D.C. (1999) 'Pharmacokinetics of *Hoasca* alkaloids in healthy humans', *Journal of Ethnopharmacology* 65: 243–256

Cavendish, W. (1999) 'Empirical regularities in the poverty–environment relationship of African rural households', Working Paper No. WPSS 99-21, Centre for the Study of African Economies, Oxford University, Oxford

Chyka, P.A. (2000) 'How many deaths occur annually from adverse drug reactions in the United States?', *American Journal of Medicine* 109(2): 122–130

Cragg, G.M., Kingston, D.G.I. and Newman, D.J. (eds) (2005) *Anticancer Agents from Natural Products*, CRC Press, Boca Raton, FL

Cunningham, A.B. (1991) 'Development of a conservation policy on commercially exploited medicinal plants: A case study from southern Africa', in V. Heywood, H. Synge and O. Akerele (eds) *Conservation of Medicinal Plants*, Cambridge University Press, Cambridge, pp337–358

Cunningham, A.B. (1993) 'African medicinal plants: Setting priorities at the interface between conservation and primary health care', People and Plants Working Paper No. 1: 1–50, UNESCO, Paris

Cunningham, A.B. (1997) *The 'Top 50' Listings and the Medicinal Plants Action Plan*, IUCN/SSC Medicinal Plant Specialist Group, Bonn, Germany

Cunningham, A.B. (2001) *Applied Ethnobotany: People, Wild Plant Use and Conservation*, Earthscan, London

Cunningham, A.B. (2005) 'CITES significant trade review of *Prunus africana*', PC16 Doc. 10.2, CITES Management Authority, Geneva, Switzerland

Cunningham, A.B., Ayuk, E., Franzel, S., Duguma, B. and Asanga, C. (2002) 'An economic evaluation of medicinal tree cultivation', People and Plants Working Paper No. 10, UNESCO, Paris

de Albuquerque, U.P. (2006) 'Re-examining hypotheses concerning the use and knowledge of medicinal plants: A study in the Caatinga vegetation of NE Brazil', *Journal of Ethnobiology and Ethnomedicine* 2(30): 1–10

de Feo, V. (2005) 'The ritual use of *Brugmansia* species in traditional Andean medicine in northern Peru', *Economic Botany* 58: 221–229

de Rios, M.D. and Stachalek, R. (1999) 'The *Duboisia* genus, Australian aborigines and suggestibility', *Journal of Psychoactive Drugs* 31(2): 155–161

Eisenberg, D.M., Davis, R.B., Ettner, S.L., Appel, S., Wilkey, S., van Rompay M. and Kessler R.C. (1998) 'Trends in alternative medicine use in the United States 1990–1997: Results of a follow-up national survey', *JAMA* 280: 1569–1575

Falconer, J. (1990) *The Major Significance of 'Minor' Forest Products: The Local Use and Value of Forests in the West African Humid Forest Zone*, FAO, Rome

Falconer, J. (1994) *Non-timber Forest Products in Southern Ghana*, Overseas Development Administration, UK/Forestry Department, Republic of Ghana/Natural Resources Institute, UK

Farnsworth, N.R., Akerele, O., Bingel, A.S., Soejarto, D.D. and Guo, Z. (1985) 'Medicinal plants in therapy', *Bulletin of the World Health Organization* 63: 965–981

Fiori, K.L., Hays, J.C. and Meador, K.G. (2004) 'Spiritual turning points and perceived control over the life course', *International Journal of Ageing and Human Development* 59(4): 391–420

Fournet A., Barrios, A.A., Munoz, V., Hocquemiller, R., Roblot, F. and Cave, A. (1994) 'Antileishmanial activity of a tetralone isolated from *Ampelocera edentula*, a Bolivian plant used as a treatment for cutaneous leishmaniasis', *Planta Medica* 60(1): 8–12

Garibaldi, A. and Turner, N. (2004) 'Cultural keystone species: Implications for ecological conservation and restoration', *Ecology and Society* 9(3): 1, www.ecologyandsociety.org/vol9/iss3/art1

Garvey, G.J., Hahn, G., Lee, R.V. and Harbison, R.D. (2001) 'Heavy metal hazards of Asian traditional remedies', *International Journal of Environmental Health Research* 11: 63–71

Gessler, M.C., Msuya, D.E., Nkunya, M.H., Schar, A., Heinrich, M. and Tanner, M. (1995) 'Traditional healers in Tanzania: The perception of malaria and its causes', *Journal of Ethnopharmacology* 3(48): 119–130

Ghimire, S., McKey, D. and Aumeeruddy-Thomas, Y. (2004) 'Heterogeneity in ethnoecological knowledge and management of medicinal plants in the Himalayas of Nepal: Implications for conservation', *Ecology and Society* 9(3): 6, www.ecologyandsociety.org/vol9/iss3/art6/

Green, E.C. (1985) 'Traditional healers, mothers and childhood diarrhoeal disease in Swaziland: The interface of anthropology and health education', *Social Science and Medicine* 20: 277–285

Grifo, F., Newman, D., Fairfield, A.S., Bhattacharya, B. and Grupenhoff, J.T. (1997) 'The origins of prescription drugs', in F. Grifo and J. Rosenthal (eds) *Biodiversity and Human Health*, Island Press, Washington, DC

Grob, C.S., McKenna, D.J., Callaway, J.C., Brito, G.S., Neves, E.S., Oberlander, G., Saide, O.L., Labigalini, E., Tacla, C., Miranda, C.T., Strassman, R.J. and Boone, K.B. (1996) 'Human psychopharmacology of Hoasca, a plant hallucinogen used in ritual context in Brazil', *Journal of Nervous and Mental Disorders* 184(2): 86–94

Grosvenor, P.W., Gothard, P.K., McWilliam, N.C., Supriono, A. and Gray, D.O. (1995) 'Medicinal plants from Riau province, Sumatra, Indonesia. Part 1: Uses', *Journal of Ethnopharmacology* 45: 75–95

Hagopian, A., Thompson, M.J., Fordyce, M., Johnson, K.E. and Hart, L.G. (2004) 'The migration of physicians from sub-Saharan Africa to the United States of America: Measures of the African brain drain', *Human Resources for Health* 2(17): 1–10

Honigsbaum, M. (2002) *The Fever Trail: In Search of the Cure for Malaria*, Farrar Straus & Giroux, New York, NY

Iwu, M. and Laird, S.A. (1998) 'The International Cooperative Biodiversity Group: Drug development and biodiversity conservation in Africa', case study of a benefit-sharing plan, UNEP, Nairobi

Jacobson-Widding, A. (1979) *Red-White-Black as a Mode of Thought: A Study of Triadic Classification by Colours in the Ritual Symbolism and Cognitive Thought of the Peoples of the Lower Congo*, Almquist & Wiksell, Uppsala, Sweden

Kalix, P. (1991) 'The pharmacology of psychoactive alkaloids from ephedra and catha', *Journal of Ethnopharmacology* 32(1–3): 201–208

King, S.R., Meza, E.N., Carlson, T.J.S., Chinook, J.A., Moran, K. and Borges, J.R. (1999) 'Issues in the commercialisation of medicinal plants', *Herbalgram* 47: 46–51

Kleinman, A., Eisenberg, L. and Good, B. (1978) 'Culture, illness, and care: Clinical lessons from anthropologic and cross-cultural research', *Annals of Internal Medicine* 88: 251–258

Kyashobire, M. (1998) 'Medicinal plants and herbalist preferences around Bwindi Impenetrable Forest, Uganda', MSc thesis, Department of Botany, Makerere University, Kampala

Lagos-Witte, S., Germosén-Robineau, L. and Weniger, B. (1997) 'Rol actual de la fitomedicina en el caribe. presentacion de la farmacopea vegetal Caribeña (TRAMIL)', *ISHS Acta Horticulturae* 503: II WOCMAP Congress Medicinal and Aromatic Plants, Part 4: Industrial Processing, Standards & Regulations, Quality, Marketing, Economics

Laird, S.A. (2002) *Biodiversity and Traditional Knowledge: Equitable Partnerships in Practice*, Earthscan, London

Laird, S.A. and ten Kate, K. (2002) 'Linking biodiversity prospecting and forest conservation', in S. Pagiola, J. Bishop and N. Landell-Mills (eds) *Selling Forest Environmental Services: Market-Based Mechanisms for Conservation and Development*, Earthscan, London, pp151–172

Laird, S.A. and Wynberg, R. (2005) *The Commercial Use of Biodiversity: An Update on Current Trends in Demand for Access to Genetic Resources and Benefit-Sharing, and Industry Perspectives on ABS Policy and Implementation*, UNEP, Nairobi

Laird, S.A., Pierce, A.R. and Schmitt, S.F. (2005) 'Sustainable raw materials in the botanicals industry: Constraints and opportunities', *Acta Horticulture* 676: 111–117

Laird, S.A., Awung, G.L. and Lysinge, R.J. (2007) 'Cocoa farms in the Mount Cameroon region: Biological and cultural diversity in local livelihoods', *Biodiversity and Conservation* 16: 2401–2427

Lange, D. (1998) *Europe's Medicinal and Aromatic Plants: Their Use, Trade and Conservation*, TRAFFIC International, Cambridge

Lee, M.R. (2002) 'Plants against malaria, Part 1: Cinchona or the Peruvian bark', *The Journal of the Royal College of Physicians of Edinburgh* 32: 189–196

Liu, H., Lei, G., Zheng, Z. and Feng, Z. (2006) 'The impact of *Amomum villosum* cultivation on seasonal rainforest in Xishuangbanna, southwest China', *Biodiversity and Conservation* 15: 2971–2985

McKenna, D.J. (2005) 'Ayahuasca and human destiny', *Journal of Psychoactive Drugs* 37: 231–234

Mander, M. (2004) 'Phytomedicines industry in southern Africa', in N. Diederichs (ed) *Commercialising Medicinal Plants: A Southern African Guide*, African Sun Media, Pretoria, South Africa

Mash, D.C., Kovera, C.A., Buck, B.E., Norenberg, M.D., Shapshak, P., Hearn, W.L. and Sanchez-Ramos, J. (1998) 'Medication development of ibogaine as a pharmacotherapy for drug dependence', *Annals of the New York Academy of Sciences* 44: 274–292

Melnyk, M. (2001) 'Community health and conservation: A review of projects', unpublished report, Biodiversity Support Program, Washington, DC

Mills, E., Foster, B.C., van Heeswijk, R., Phillips, E., Wilson, K., Leonard, B., Kosuge, K. and Kanfer, I. (2005) 'Impact of African herbal medicines on antiretroviral metabolism', *AIDS* 19(1): 95–97.

Miot, H.A., Batistella, R.F., Batista Kde, A., Volpato, D.E., Augusto, L.S., Madeira, N.G., Haddad, V. and Miot, L.D. (2004) 'Comparative study of the topical effectiveness of the Andiroba oil (*Carapa guianensis*) and DEET 50 per cent as repellent for *Aedes* sp', *Revista do Instituto de Medicina Tropical de São Paulo* 46(5): 253–256

Nalvarte, W and de Jong, W. (2004) 'Uña de gato (*Uncaria tomentosa* (Willd ex Roem & Schult) DC y *Uncaria guianensis* (Aubl.) Gmel.): Potencial y esperanzas de un bejuco Amazónico del Perú', in M.N. Alexiades and P. Shanley (eds) *Productos forestales medios de subsistencia y conservación: Estudios de caso sobre sistemas de manejo de productos forestales no maderables Volumen América Latina*, Centro para la Investigación Forestal Internacional (CIFOR), Bogor, Indonesia

Nantel, P., Gagnon, D. and Nault, A. (1996) 'Population viability analysis of American ginseng and wild leek harvested in stochastic environments', *Conservation Biology* 10: 608–621

Newman, D.J., Cragg, G.M. and Snader, K.M. (2003) 'Natural products as sources of new drugs over the period 1981–2002', *Journal of Natural Products*, 66: 1022–1037

Newman, D.J. and Laird, S.A. (1999) 'The influence of natural products on 1997 pharmaceutical sales figures', in K. ten Kate and S.A. Laird (eds) *The Commercial Use of Biodiversity: Access to Genetic Resources and Benefit Sharing*, Earthscan, London

Nutrition Business Journal (2003) *NBJ's Supplement Business Report 2003*, Penton Media Inc, San Diego, CA

Obi, E., Akunyili, D.N., Ekpom, B. and Orisakwe, O.E. (2006) 'Heavy metal hazards of Nigerian herbal remedies', *The Science of the Total Environment* 69(1–3): 35–41

Olsen, C.S. (2005) 'Valuation of commercial central Himalayan medicinal plants', *Ambio* 34: 607–610

Osuala, F.O. and Okwuosa, V.N. (1993) 'Toxicity of *Azadirachta indica* to freshwater snails and fish, with reference to the physicochemical factor effect on potency', *Applied Parasitology* 34(1): 63–68

Otuyemi, O.D., Abidoye, R.O. and Dada, D. (1994) 'Oral health knowledge, attitude and behaviour of 12-year-old suburban and rural school children in Nigeria', *African Dentistry Journal* 8: 20–25

Padarath, A., Chamberlain, C., McCoy, D., Ntuli, A., Rowson, M. and Loewenson, R. (2004) 'Health personnel in Southern Africa: Confronting maldistribution and brain drain', Equinet Discussion Paper No. 3, www.queensu.ca/samp/migrationresources/braindrain/documents/equinet.pdf (accessed 25 October 2007)

Pagán, J.A. and Pauly, M.V. (2005) 'Access to conventional medical care and the use of complementary and alternative medicine', *Health Affairs* 24: 255–262

Palsson, T and Jaenson, W.G. (1999) 'Comparison of plant products and pyrethroid-treated bed nets for protection against mosquitoes (Diptera: Culicidae) in Guinea Bissau, West Africa', *Journal of Medical Entomology* 36(2):144–8.

Panduro, M.P. and de Jong, W. (2004) 'Camu-camu (*Myrciaria dubia* McVaugh H.B.K.), Arbusto Amazónico de Áreas Inundables con Alto Contenido de Vitamina C', in M. Alexiades and P. Shanley (eds) *Productos forestales, medios de subsistencia y conservación, Volumen 3: America Latina*, CIFOR, Bogor, Indonesia

Patzold, B., Leaman, D. and Honnef, S. (2006) 'Sustainable wild collection of medicinal plants: The need for an international standard', *TRAFFIC Bulletin* 21: 41–45

Peters, C.M. (1994) *Sustainable Harvest of Non-timber Plant Resources in Tropical Moist Forest: An Ecological Primer*, WWF, Washington, DC

Phillip Securities Research (2003) *Stock Update*, Eu Yan Sang International, 17 January, pp1–3

Pierce, A.R. and Laird, S.A. (2003) 'In search of comprehensive standards for non-timber forest products in the botanicals trade', *International Forestry Review* 5(2): 138–147

Pierce, A.R., Laird, S.A. and Malleson, R. (2002) *Annotated Collection of Guidelines, Standards, and Regulations for Trade in Non-timber Forest Products and Botanicals*, Rainforest Alliance, New York

Posey, D.A. (1999) *Cultural and Spiritual Values of Biodiversity: A Complementary Contribution to the Global Biodiversity Assessment*, UNEP, Nairobi

Póvoa, M.M., Conn, J.E., Schlichting, C.D., Amaral, J.C.O.F. and Segura, M.N.O. (2003) 'Malaria vectors, epidemiology, and the re-emergence of *Anopheles darlingi* in Belém, Pará, Brazil', *Journal of Medical Entomology* 40: 379–386

Prisinzano, T.E. (2005) 'Psychopharmacology of the hallucinogenic sage *Salvia divinorum*', *Life Sciences* 78(5): 527–531

Qiong, W., Yiping, W., Jinlin, Y., Tao, G., Zhen, G. and Pengcheng, Z. (2005) 'Chinese medicinal herbs for acute pancreatitis', *The Cochrane Database of Systematic Reviews*, Issue 1, Art. No. CD003631 1–14

Queneau, P., Bannwarth, B., Carpentier, F., Guiliana, J.M., Bouget, J., Trombert, B., Leverve, X., Lapostelle, F., Borron, S.W. and Adnet, F. (2007) 'Emergency department visits caused by adverse drug events: Results of a French survey', *Drug Safety* 30(1): 81–88

Ramachandran, S.C., Latha, E. and Manoharan, M. (1999) 'Impacts of urbanisation on the lifestyle and on the prevalence of diabetes in native Asian Indian population', *Diabetes Research and Clinical Practice* 44(3): 207–213

Reichel-Dolmatoff, G. (1996) *The Forest Within: The World-view of the Turkano Amazonian Indians*, Themis Books, Dartington, Devon

Rwiza, H.T., Matuja, W.B., Kilonzo, G.P., Haule, J., Mbena, P., Mwang'ombola, R. and Jilek-Aall, L. (1993) 'Knowledge, attitude, and practice toward epilepsy among rural Tanzanian residents', *Epilepsia* 34(6): 1017–1023

Saxena, A. and Vikram, N.K. (2004) 'Role of selected Indian plants in management of type 2 diabetes: A review', *Journal of Complementary and Alternative Medicine* 10: 369–378

Scheffer (2004) 'Produção de espinheira-santa (*Maytenus ilicifolia* Mart. ex Reiss) na região metropolitana de Curitiba, Paranam Brasil', in M. Alexiades and P. Shanley (eds) *Productos Forestales, Medios de Subsistencia y Conservación. Vol. 3 – America Latina*, CIFOR, Indonesia

Schippmann, U., Leaman, D.J. and Cunningham, A.B. (2003) 'Impact of cultivation and gathering of medicinal plants on biodiversity: Global trends and issues', in *Biodiversity and the Ecosystem Approach in Agriculture, Forestry and Fisheries*, FAO, Rome

Schultes, R.E. and Raffauf, R.F. (1992) *Vine of the Soul*, Synergetic Press, Oracle, AZ

Shah, N.C. (2006) '*Podophyllum hexandrum* and its conservation status in India', *Medicinal Plants Conservation* 12: 42–46

Shanley, P. and Luz, L. (2003) 'The impacts of forest degradation on medicinal plant use and implications for health care in Eastern Amazonia', *BioScience* 53: 573–584

Shanley, P., Pierce, A. and Laird, S. (2006) *Alem de Madeira: A Certificação de Produtos Florestais Não-Madeireiros*, CIFOR/Forest Trends, Belem, Brazil

Shepard, G. (2005) 'Psychoactive botanicals in ritual, religion and shamanism', in E. Elisabetsky and N. Etkin (eds) *Ethnopharmacology*, UNESCO/Eolss Publishers, Oxford, UK, Chapter 18

Shirakawa, T. (2006) 'Psychological effects of forest environments on healthy adults: Shinrin-yoku (forest-air bathing, walking) as a possible method of stress reduction', *Public Health*, 19 October [abstract only]

Silva, M.I., Sousa, F.C. and Gondim, A.P. (2005) 'Herbal therapy in primary health care in Maracanau, Ceara, Brazil', *Annals of Pharmacotherapy* 39: 1336–1341 [abstract only]

Singh, A. and Singh, D.K. (2001) 'Molluscicidal activity of *Lawsonia inermis* and its binary and tertiary combinations with other plant derived molluscicides', *Indian Journal of Experimental Biology* 39(3): 263–268

Smith, M.T., Crouch, N.R., Gericke, N. and Hirst, M. (1996) 'Psychoactive constituents of the genus *Sceletium* N.E.Br. and other Mesembryanthemaceae: A review', *Journal of Ethnopharmacology* 50: 119–130

Stepp, R. (2004) 'The role of weeds as sources of pharmaceuticals', *Journal of Ethnopharmacology* 92: 163–166

Stevens, A. and Milne, R. (1997) 'The effectiveness revolution and public health', in G. Scalley (ed) *Progress in Public Health*, Royal Society of Medicine Press, London, pp197–225

Tagboto, S. and Townson, S. (2001) 'Antiparasitic properties of medicinal plants and other naturally occurring products', *Advances in Parasitology* 50: 199–295

Taixiang, W., Munro, A.J. and Guanjian, L. (2005) 'Chinese medical herbs for chemotherapy side effects in colorectal cancer patients', *The Cochrane Database of Systematic Reviews*, issue 1, art. no. 4540, 1–21

ten Kate, K. and Laird, S.A. (1999) *The Commercial Use of Biodiversity: Access to Genetic Resources and Benefit Sharing*, Earthscan, London

United Nations (2006) *International Migration and the Achievement of MDGs in Africa*, United Nations Secretariat, Turin

van de Kop, P., Ghayur, A. and de Steenhuijsen Piters, B. (2006) 'Developing a sustainable medicinal-plant chain in India: Linking people, markets and values', in R. Ruben, M. Slingerland and H. Nijhoff (eds) *Agro-food Chains and Networks for Development*, Springer, The Hague, pp191–202

van Rooyen, M., Kruger, H.S., Huisman, H.W., Wissing, M.P., Margetts, B.M. and Venter, C.S. (2000) 'An epidemiological study of hypertension and its determinants in a population in transition: The THUSA study', *Journal of Human Hypertension* 14: 779–787

van Vuuren, D.P., Sala, O.E. and Pereira, H.M. (2006) 'The future of vascular plant diversity under four global scenarios', *Ecology and Society* 11(2): 25

Wahlberg, A. (2006) 'Bio-politics and the promotion of traditional herbal medicine in Vietnam', *Health* 10: 123–147

Walter, K.S. and Gillet, H.J. (1998) *The 1997 IUCN Red List of Threatened Plants*, compiled by the World Conservation Monitoring Centre, IUCN, Gland, Switzwerland

Wibulpolprasert, S. and Pengpaibon, P. (2003) 'Integrated strategies to tackle the inequitable distribution of doctors in Thailand: Four decades of experience', *Human Resources for Health* 1: 12 [abstract only]

WHO (1996) World Health Organization Fact Sheet No. 134, www.who.int/intfs/en/fact134.html
WHO (2002) 'Traditional medicine strategy 2002–2005', www.who.int/medicines/library/trm/trm_strat_eng.pdf
Yang, X.X., Hu, Z.P., Duan, W., Zju, Y.Z. and Zhou, S.F. (2006) 'Drug-herb interactions: Eliminating toxicity with hard drug design', *Current Pharmaceutical Design* 12(35): 4649–4664
Zhang, Q-Y. and He, X-J. (2002) 'Conservation of rare and endangered plants in Sichuan', *Journal of Wuhan Botanical Research* 20: 387–394 [in Chinese]
Zhao, Z. (2004) *An Illustrated Chinese Materia Medica in Hong Kong*, Chun Hwa Book Company, Hong Kong

4

The Nutritional Role of Forest Plant Foods for Rural Communities

Barbara Vinceti, Pablo Eyzaguirre and Timothy Johns

Can we do without forests? Everywhere, forests prevent soil erosion, filter and regulate the flow of fresh water, control pests and buffer disease, provide shelter, harbour pollinators and mitigate global warming by sequestering carbon. They also provide a fundamental contribution to household food security, thereby combating hunger and malnutrition (Falconer and Arnold, 1988; FAO, 1989, 1990; Falconer, 1990; Falconer and Koppell, 1990; Hoskins, 1990; Townson, 1995; Arnold and Townson, 1998; Reddy and Chakravarty, 1999). Although forest foods do not usually provide a complete diet, they do make a critical contribution to food supply and are especially important during emergencies, such as drought, famine and war, particularly as a complement to agricultural crops that are only seasonally available (Richards, 1986; Falconer and Koppell 1990; Arnold and Townson, 1998; see very comprehensive bibliographic review on the contribution of wild foods to food security by Scoones et al, 1992). The increased use of forest foods during periods of food insecurity is well described (Fleuret, 1986; Asibey and Beeko, 1998), but forest foods constitute a crucial complement to the diet during less difficult times as well (Falconer and Arnold, 1988; Falconer and Koppell, 1990) by supplying particular micronutrients that are not available in staple foods.

FOOD SECURITY

People at risk of food insecurity, or hunger and malnutrition generally have the highest degree of reliance on forest products for income and food. These vulnerable groups supplement their agricultural production by gathering forest products on common-property forestlands (lands that are owned and managed collectively) or open-access forest-lands (lands that have no effective collective or private ownership status) (Jodha, 1990; Reddy and Chakravarty, 1999). Even in regions where food distribution and supply systems have been considerably improved and provide partial solutions to problems of food security, the benefits of market and welfare mechanisms can remain out of reach for more vulnerable social groups. Thus, access to forest foods and other wild resources remains fundamental to people in many areas (e.g., in India, as documented by Pralhad, 1989; FAO, 1992; Styger et al, 1999). Cases in which farmers seek to retain a significant amount of forest on their parcels as reserves or as woodlots are well described, even where the pressure posed by population growth and deforestation is considerable (Shriar, 2002).

Forest foods help people respond to seasonal stresses by buffering shortfalls in crop yields and limited availability of other food sources. Good examples include the products harvested from the oil palm (*Elaeis guianeensis*), which form a foundation for diets in West African countries (Hartley, 1988), and those derived from the sago palm (*Metroxylon* spp) in Southeast Asia and the Southwest Pacific and consumed regularly as part of the diet by 1 million people (Ulijaszek, 1983). Fruits from *Zyzyphus* spp provide food to farmers in the dryland areas of northern India when other sources are scarce (Tiwari and Banafar, 1995). The Basarwa of the Kalahari use nuts of *Ricinodendron rautanenii* as a staple food during the dry season (Lee, 1973), and in West Africa *Parkia bicolour* is used during the dry season to prepare a fermented food product (Campbell-Platt, 1980). The variation in the seasonal availability of forest foods depends on what parts of trees are harvested and what products are collected from forests and woodlands; it relates also to seasonal patterns of fruiting, leaf production and population cycles. However, some forest foods are available throughout the year. According to archaeological studies (Puleston, 1968; 1982), the 40-metre-tall rainforest emergent tree known as *ramón* or the Maya breadnut (*Brosimum alicastrum*, Moraceae), once a dietary staple of the Maya people, enabled the sophisticated Maya civilization to flourish in the lush jungles of Central America. The seeds of *B. alicastrum* were an important alternative food when yields of the staple grain and legume crops were low (Peters and Pardo-Tejeda, 1982). A common food usage consisted of boiled seeds mashed and eaten as a substitute for root crops (Thompson, 1930). The seeds contain a percentage of protein that compares favourably with maize, beans, squashes and tubers; they also contain nutritionally significant levels of iron, vitamin A, riboflavin, niacin and ascorbic acid.

In the *miombo* woodlands of Tanzania, a few tree species represent a secure and a continuous source of food for the Sandawe people (Newman, 1975). Nuts and insects, for example, can be easily stored for consumption when needed. Finally, forest foods can play a particular role when labour is scarce. Among fruit-tree species, in West and Central Africa women value the forest tree *safú* (*Dacroydes edulis*) for its food value: it can be boiled or roasted and consumed with cassava, thus providing an easy meal that is quick to prepare when scarce labour resources have to be devoted to agricultural activities (Schreckenberg et al, 2002; see also Chapter 10, this volume, for discussion of the labour-scarcity implications of household illness).

NUTRITIONAL QUALITY

Forest foods increase the nutritional quality of rural diets. Their nutritional value resides mostly in the availability of micronutrients (vitamins and minerals) and phytochemicals, including non-nutrients like phenolics (tannins and flavonoids) and carotenoids, found in forest fruits and vegetables. Benefits from consuming the latter include antioxidant and hypoglycaemic functions (McCune et al, 2005) and metabolic roles in modifying the activation and detoxication of potential carcinogens (Wargovich, 2000). Nuts, whole grains, fruits and vegetables contain an abundance of phenolic compounds, terpenoids and other natural antioxidants (including vitamins A, C and E) that have been associated with protection from and/or treatment of chronic medical conditions such as heart disease, cancer, diabetes and hypertension (Bloch and Thomson, 1995). Craig and Beck (1999) highlight how an analysis of separate nutrients and food components does not lead to the identification of dietary factors that can explain these protective effects, indicating that phytochemicals act synergistically to exert health benefits and cannot be easily isolated and dispensed within supplements.

Nutritional properties have been well documented for various tree species (e.g., legume trees; Brown et al, 1985; Balogun and Fetuga, 1986) and for other products extracted from woodland and bush lands (e.g., tubers, bush onion, *taro*; Cherikoff et al, 1985). However, the nutrient composition of most wild species is poorly studied (Burlingame, 2000) compared with that of domesticated species, and information on nutritional properties of wild species is not systematically collected. Fruit and nut trees are the best studied with regard to nutritional properties. A compilation of the nutritional properties of edible fruits of indigenous wild trees in Malawi is provided by Saka and Msonthi (1994) (Table 4.1). A study by Kengni et al (2004) illustrates the biochemical composition of forest foods from West and Central Africa (Table 4.2) and shows the macronutrients (carbohydrates, lipids and proteins), micronutrients (vitamins and minerals) and fibre content of some tree products (Ejoh et al, 1996; Tchiegang et al, 1998; Kengni, 2003). Most notably, fruit pulp from *Dacryodes edulis* (*safou*, or African plum) and kernel fats from *Irvingia gabonensis* (*dikanut*, bush mango) and *Ricinodendron heudelotii* are very rich in energy, proteins, iron and unsaturated fatty acids (Tchiegang et al, 1998; Kengni, 2003). In addition, *Irvingia gabonensis* and *Dacryodes edulis* are particularly rich in vitamin A, iron, copper, manganese and zinc, therefore addressing nutritional needs and providing health benefits for vulnerable rural communities.

The value of forest foods has to do with not only the timing of supply but also the contribution of particular nutrients to the diet. More important than the quantity of food supplied can be its quality and correspondence to dietary needs. In most instances, a full assessment of the importance of forest foods involves more than just calories and is associated rather with the supply of vitamins, trace minerals and other essential nutrients.

In many traditional subsistence systems that depend on one or more staples, such as cassava, sago, rice or maize, balanced diets are obtained by ingesting small but complementary amounts of animal-source foods and plants as sauces, condiments, snacks and beverages (Johns and Maundu, 2006). In some parts of Africa, diets based on staple grains depend for their supply of essential vitamins largely on tree products, in the form of oil seeds, edible leaves and fruits (Hoskins, 1990; Odgen, 1990). The poorer households in Mali complement millet porridge by mixing it with fruits from *Boscia senegalensis* (Martin, 1985; Toulmin, 1986). In Venezuela and Colombia, *Jessenia* fruits are a source of protein, when the supply of protein from fishing and hunting is not sufficient (Beckerman, 1984).

Moreover, around the world, including in sub-Saharan Africa, many tree nuts and fruits are used for medicinal purposes, and health care can be largely a forest-based service (Chege,

Table 4.1 *Chemical composition of edible wild fruits of 14 indigenous wild tree species in Malawi*

Fruits	*Dry matter (%)*	*Ash (%)*	*Crude protein (%)*	*Fat (%)*	*Fibre (%)*	*Total carbohydrate (%)*	*Energy value (kJ $100g^{-1}$)*	*Mineral ($\mu g\ g^{-1}$)*					
								P	Ca	Mg	Fe	K	Na
Adansonia digitata	86.8	5.0	3.1	4.3	8.3	79.4	1480	450	1156	2090	58	28,364	188
Annona senegalensis	29.8	5.0	15.1	8.4	17.8	53.7	1426	1507	ND	1762	84	21,812	436
Azanza garckeana	52.8	6.3	12.0	1.1	45.3	35.2	810	1476	95	1453	84	26,190	202
Bauhinia thonningii	91.6	2.4	2.6	5.9	25.8	63.3	1276	979	983	1245	ND	22,312	348
Flacourtia indica	19.2	5.7	4.2	3.6	5.7	80.7	1290	1057	354	1380	734	24,281	589
Parinari curatellifolia	27.1	1.8	3.0	1.5	5.5	88.2	1517	339	129	830	103	10,380	252
Strychnos innocua	21.8	3.7	11.5	6.0	17.9	61.0	1390	2106	60	1633	60	28,670	459
Strychnos spinosa	22.1	4.1	5.4	31.2	17.6	42.1	1923	1081	149	430	136	19,683	253
Syzigium guineese	19.8	7.1	10.1	4.0	30.3	48.5	1096	303	227	2247	758	8252	152
Tamarindus indica	73.1	3.4	4.1	1.6	5.9	85.0	1490	1081	171	1282	68	12,269	111
Trichilia emetica	58.1	4.5	17.0	22.9	8.1	47.5	1897	3164	ND	1129	43	13,017	146
Uapaca kirkiana	27.4	2.2	1.8	1.1	8.4	86.5	1456	555	33	1106	431	13,682	365
Vangueria infausta	26.5	3.4	5.7	2.6	10.2	78.1	1445	823	132	1811	283	18,208	245
Vitex doniana	27.0	4.8	2.6	0.7	5.2	86.7	1459	2837	926	700	93	21,511	278
Ximenia caffra	17.2	11.0	7.6	5.2	2.3	78.8	1506	1674	29	459	366	41,791	198
Ziziphus mauritania	14.8	10.1	4.1	9.5	3.4	73.0	1588	2162	135	507	ND	17,318	426

Source: Saka and Msonthi, 1994, p4

Note: P = phosphorus; Ca = calcium; Mg = magnesium; Fe = iron; K = potassium; Na = sodium

Table 4.2 *Chemical composition and nutritional value of major forest foods from West and Central Africa*

Constituents	*Irvingia gabonensis*[1]	*Dacryodes edulis*[2]	*Ricinodendron heudelotii*[3]*	*Gnetum africaunum*
Moisture	6.3–29.2	11.0–59.1	3.1–5.6	16.3–37.4
Ash	2.5–3.7	2.2–4.7	10.5–16.0	4.7–7.0
Lipids	44.3–68.4	56.2–67.4	37.7–55.5	5.9–14.2
Carbohydrates	7.4–13.5	4.5–8.7	0.8–5.6	33.5–70.6
Proteins	6.2–13.5	6.2–14.7	16.3–24.6	10.2–16.5
Available energy	444–668.6	434.3–629.6	441.0–620.3	200.8–476.2
Dietary fibres	14.1–24.1	6.4–27.0	5.4–9.8	–
Vitamin A	490–780	570–740	340–374	–
Vitamin C	4.4–66.4	22.0–32.1	2.1–7.5	–
Calcium (Ca)	0.14–232	0.73–730	–	0.28–830
Magnesium (Mg)	0.26–160	0.73–450	–	0.41–395
Iron (Fe)	0.19–47	28–181	–	0.05–0.16
Zinc (Zn)	0.04–12	6–36	–	0.05–0.08
Copper (Cu)	0.01–3.1	6.0–5.3	–	0.01–0.04
Iodine value	4.2–5.2	59.6–75.2	150.5–160.0	–

Source: Kengni et al, 2004

Moisture is in g/100g; Zn and Ca are in parts per million (ppm (mg/kg); Cu, Mg, Fe and vitamin C are in mg/100g; vitamin A in µg retinol/100g; energy in kcal/100g; ash in g/100g; iodine value in g iodine/100g fat; dietary fibres in g/100g and macronutrients (carbohydrates, lipids and proteins) in g/100g.

1 Kengni, 2003;

2 Tchiegang et al, 1998;

3 Mialoundama, 1993;

*now called *Schinziophyton rautanenii*

1994; see Chapter 3, this volume). While targeting specific diseases, medicinal plants may at the same time more broadly support the functioning of the immune system, other systems and overall health (Kinyuy, 2001). A particularly interesting example of the contribution of wild plants to balanced diets is the dietary practices of the Maasai people (Box 4.1).

BOX 4.1 FOOD PLANTS USED BY THE MAASAI

The Maasai, a cattle-herding pastoral people of Kenya and Tanzania, have been described as having the 'worst diet' in the world because they consume primarily milk products and meat and obtain up to two-thirds of their calories from animal fat. However, they do not suffer from the diet-related health problems or diseases typically associated with high consumption of fat, apparently because of the wild plants that they add regularly to their soups and stews and the plants they chew as gums and resins or consume as medicinal tonics (Johns and Chapman, 1995; Johns, 1999; Johns et al, 2000). Among the most favoured products is acacia tree bark, added to soups, which has been shown to contain biochemicals that remove cholesterol from the body or mediate its affects through an antioxidant role.

A study undertaken near Loita, located between the Mara and Serengeti plains and the forests of the western escarpment of the Rift Valley, showed the breadth of food plants that the Maasai utilize from vegetation types ranging along a gradient from open or wooded grasslands to closed-canopy forest (Table 4.3). Regrettably, the Maasai society, along with its system of health care, is under threat from continuing urbanization, and there will be a consequent loss of the traditional knowledge of ethno-botanical medicine (cf. Chapters 3, 11 and 13).

Forests supply food and satisfy the nutritional needs of groups with different lifestyles, different strategies for obtaining food from the surrounding environment and different food cultures. The linkages between forest biodiversity and dietary diversity need to be better documented, but the erosion of traditional knowledge about forest biodiversity has been observed to affect food choices considerably and to lead to dietary simplification and negative repercussions on human health (Johns and Sthapit, 2004; Johns and Maundu, 2006).

Moving to urban environments may not always have as an immediate consequence a loss of knowledge about traditional food systems and indigenous technologies for food processing. Urban agricultural activities are undertaken (Ruel et al, 1998) despite their associated health hazards, such as the increase of vector-borne infectious diseases and exposure to contamination from pesticides, heavy metals in the soil and untreated wastewater used for irrigation (McMichael, 2000). However, the delocalization of food supply (Kuhnlein and Receveur, 1996) and the increased role played by supermarkets in food distribution are inevitably leading to a decreased use of traditional food and food systems, and particularly to reduced opportunities for local producers to sell wild foods collected from forests (Reardon et al, 2003).

Table 4.3 *Some food plants used by the Maasai*

Products obtained (or extracted)	*Species used and preparation*
Edible inner bark	The thin, inner, light-coloured layer of some trees, particularly *Acacia* spp, is removed and chewed for its sweet taste and water content, and as an exercise for the jaws.
Infusions	Bark and stems of several species are used to prepare a brown, tea-like infusion, sometimes also using leaves and fruits (e.g., *Zanthoxylum usambarense*). These infusions are common drinks and are also medicinal, curative or preventive.
Gums and resins	Exudates are extracted. Often tasteless, they are used simply to exercise the mouth and to pass time (Johns et al, 2000). Resins are mainly obtained from *Commiphora* spp and gums from *Acacia* spp. The latex from *Acokanthera* spp (otherwise known as the source of arrow poison) and *Carissa edulis* can be made into chewing gum for children.
Fermentation of beer	The roots of various aloe species (*osuguroi*) are dug up, cleaned and used to ferment and flavour honey beer; Bantu communities use instead the fruits of the sausage tree, *Kigelia africana* (Bignoniaceae).
Edible fruits	The five most preferred fruits are *Carissa edulis* (*olamuriaki*), *Vangueria apiculata* (*olgum*), *Pappea capensis (oltimigomi, orkisikong'o), Syzygium cordatum* (*olairagai*) and *Flacourtia indica* (*oldongururwo*). They are consumed as snacks but constitute the major part of the food ingested by children and women.
Edible galls	The galls of *Acacia drepanolobium* (*eluai*) are fleshy, hollow, up to 4cm in diameter and usually inhabited by black or brown ants as they dry. The fresh, soft galls are edible. They have a sweet and somewhat bitter taste.
Edible roots and tubers	Some plants, especially in the genus *Ipomoea* (Convolvulaceae) and several genera in the Asclepiadaceae, have edible tubers. All are characterized by a slight sweet taste and a juicy consistency and are also preferred for their water.
Edible stems	Soft stems of some species are chewed mainly for their taste, usually sour. In most cases only the juice is swallowed. Common species include *Rhus* spp (Anacardiaceae) and *Rhoicissus tridentate* (Vitaceae). Some of the species may be used as a toothbrush at the same time.
Plant extracts for soup	Probably the most important wild-plant elements in the Maasai diet are the plant extracts used in soups. Plants are added to improve the taste and to prevent and treat diseases. In most cases, roots and bark or parts of stems are used. The most frequently used soup species is *Acacia nilotica* (*olkiloriti*), although *Pappea capensis* (*oltimigomi*), *Carissa edulis* (*olamuriaki*), *Scutia myrtina* (*osananguruti*) and others are commonly used.

Source: Maundu et al, 2001

In addition, some forest foods are considered old-fashioned and backwards, despite their potential beneficial effect on diets (see Chapter 13, this volume, for an Indonesian example). When this is the case, public awareness campaigns can influence food consumption behaviour and win back people's attention to resources that have disappeared from the markets and diets of local people. A successful example is the action undertaken by Bioversity International to promote the production, utilization and maintenance of diversity in the undervalued African leafy vegetables that have been found beneficial for the nutritional status of households and the incomes of women farmers (Chweya and Eyzaguirre, 1999).

BOX 4.2 YEARLY HOUSEHOLD CONSUMPTION OF FOREST FOODS IN NORTHEAST THAILAND

A study that recorded a year's worth of consumption of forest foods by rural villagers in three communities in northeast Thailand provides an idea of the wide range of forest foods and shows the importance of this food reserve in particular seasons. Wild foods gathered from forests and field margins make up half of all food eaten by villagers during the rainy seasons. Forest foods are collected in different environments: remote forest areas, on hillsides or hilltops, or in the shrubland areas. Some products are gathered from trees, but a large amount comes from ponds or streams. A total of 126 kinds of forest food are gathered: 49 species of animals (39 per cent), 16 species of mushrooms (13 per cent), 6 species of bamboo shoots (5 per cent), 43 species of other vegetables (34 per cent) and 12 species of fruits (9 per cent) (Figure 4.1).

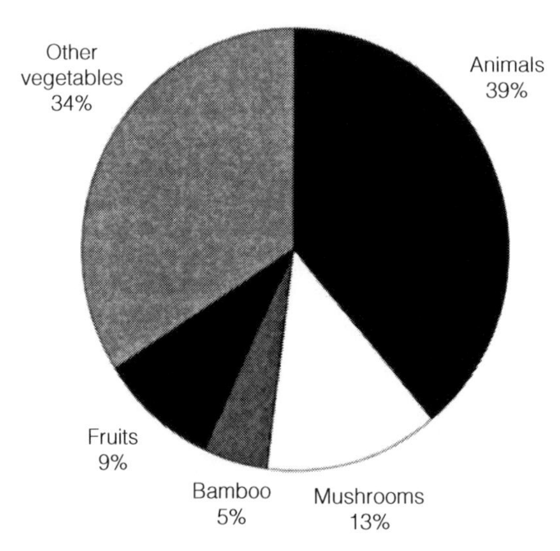

Figure 4.1 *Annual percentage of kinds of forest food gathered for consumption by the households studied*

Table 4.4 *Mean (± standard error) amounts used per household*

Resource	*Units used*	*Amount*	*Range*	*Number of villages in sample*
Wild spinaches	kg/yr	58.2 ± 26.3	12.8–198.4	7
Fuelwood	kg/day	14.5 ± 1.6	8.2–23.2	10
Grass hand-brushes	no/yr	4.5 ± 0.5	3.3–8.6	10
Wild fruits	kg/yr	104.2 ± 15.6	19.4–165.1	10
Twig hand-brushes	no/yr	4.6 ± 0.3	4.0–5.6	6
Wooden poles for fences and *kraals* (excluding brushwood)	no/hh	143.1 ± 31.3	33.1–273.0	10
Wooden poles for housing (excluding laths and brushwood)	no/hh	43.2 ± 11.8	0–113.3	10

Source: Shackleton and Shackleton, 2003

Note: kg = kilogram; yr = year; no = number; hh = household

Given their different degrees of complexity and richness of species (e.g., tropical moist forest, tropical dry forest, dry woodland, mangrove), forestlands play a critical role as reservoirs of food plants for communities with differing lifestyles, from forest-dwelling hunter-gatherers (Malaisse and Parent, 1985; May et al, 1985; Maundu, 1987; Davies and Richards, 1991; Chapters 11, 12, 13 and 16, this volume) to communities of pastoralists (Dei et al, 1989), and from communities living at the margin of the forest (Dei et al, 1989) to agricultural communities (Hladik et al, 1990).

The range of products extracted from forests and woodlands is wide and rich (Box 4.2). The use of non-timber forest products differs between households and communities in response to different local and external contextual conditions, such as resource endowment, availability of substitutes, availability of labour to gather products, education and disposable income. Shackleton and Shackleton (2003) described the non-timber forest products used by South African rural households. They reported that 85 per cent or more of households were collecting non-timber forest products, and more than half the households surveyed made use of edible insects, wood for construction, bushmeat, wild honey and reeds for weaving (Table 4.4). In the Bushbuckridge low-veld, Shackleton and colleagues described the use of dozens of species by individual households; as many as 20 edible fruit species (Shackleton et al, 2000) and 21 edible herb species (Shackleton et al, 1998).

KINDS OF FOREST FOODS

The case studies that follow provide an illustrative overview of the different categories of forest foods used by rural communities, and consumed directly or indirectly (e.g., fodder to feed livestock).

Leaves

Wild leaves, either fresh or dried, are one of the most widely consumed forest foods. As the base for soups, stews and relishes, they add flavour to otherwise bland staples such as rice or maize, making them more palatable and thus encouraging consumption. They can also be preserved, dried and powdered, or fermented; they can later be made into a paste used in stews and soups as a meat substitute. The beta-carotene, vitamin C, calcium and iron content of leaves varies greatly. In Africa, among the leaves of several tree species used for food, the well-known baobab (*Adansonia* spp) occurs in seasonally arid areas. Leaves are commonly used as a vegetable, eaten both fresh and in the form of a dry powder, throughout the dryland areas of Africa.

Another well-documented example is the genus *Gnetum*, which occurs throughout the tropics in Asia, South America (Mialoundama and Paulet, 1986) and in Central Africa (Watt and Breyer-Brandwijk, 1962). Two evergreen understorey lianas belonging to this genus (*G. africanum* and *G. buchholzianum*) and found in humid tropical forests in Africa, from Nigeria through Cameroon, Central African Republic, Gabon, Democratic Republic of Congo to Angola (Mialoundama, 1993), have significant value to many forest-based communities for their contribution to diets and their medicinal use. Their leaves, either eaten raw or finely shredded and added to soups and stews (Burkill, 1994), have a very high nutritional value (Table 4.5) as a source of protein, essential amino acids and mineral elements (Busson, 1965; Ouabonzi et al, 1983; Fokou and Domngang, 1989; Mialoundama, 1993). Some local tribes in east Cameroon and the Congo eat the root tubers or roots of *Gnetum* as wild yams (Bahuchet, 1990; Chapter 13, this volume). However, market demand and the volume of export trade in these leafy vegetables have significantly increased in recent years, particularly in Cameroon, such that *Gnetum* is becoming scarcer in increasingly remote parts. Intensive harvesting, including destructive methods such as uprooting, is contributing to an increasing threat of genetic erosion of such species (see also Chapter 3).

The range of the main traditional vegetables consumed by forest-based communities in Nigeria is shown in Table 4.5. In a diet dominated by starchy staple foods, these plants are essential sources of proteins, vitamins, minerals and amino acids. Several species, including *Gnetum* spp, *Heinsia crinata* and *Lasianthera africana*, are still harvested only from the wild (Okafor, 1983) and are available during the dry season, when conventional cultivated vegetables are scarce.

Fruits and nuts

Fruits and nuts play significant roles in human nutrition. As mentioned above, fruit and nut trees are the best-studied species with regard to nutritional properties (see examples in Tables 4.1 and 4.2). Most fruits are rich in fructose and other sugars, vitamins, minerals and dietary fibre (Quebedeaux and Bliss, 1988; Quebedeaux and Eisa, 1990; Craig and Beck, 1999; Wargovich, 2000). Within their pulp, arils or seeds, fruits of some species contain protein and fat rather than sugars, whereas other fruits and many large seeds are rich in starch. Fruits are commonly rich in ascorbic acid (vitamin C), and orange-fleshed

Table 4.5 *Chemical analysis of some leafy vegetables from southeastern Nigeria*

Scientific name	*Moisture (% fresh)*	*Ash (% dry)*	*Oil (%)*	*Protein (%)*	*Na (ppm)*	*K (ppm)*	*Ca (ppm)*	*Mg (ppm)*	*Fe (ppm)*	*Zn (ppm)*	*Cu (ppm)*	*Mn*	*P*
Gnetum africanum	37.39	4.72	14.2	10.18	26	126	28.35	14.75	5.23	0.49	0.06	–	–
Gongronema latifolium	71.14	10.94	18.77	62.66	58	336	20.75	56	8.17	0.9	0.12	–	–
Myrianthus arboreus	75.52	9.83	1.34†	20.01	–	–	1.132‡	–	0.036‡	–	–	2.134‡	2.543‡
Piper guineense	78.58	15.56	9.3	18.54	35	320	7.11	147	6.12	2.05	0.15	–	–
Pterocarpus soyauxii	74.69	2.83	3.6	3.84	1.8	11.1	19.37	35.5	3.79	0.84	0.21	–	–
Vernonia amygdalina	79.67	10.13	4.5	23.24	46	304	40	58	4.72	0.81	0.1	–	–
Vitex doniana	84.05	7.61	1.10†	22.07	–	–	0.185‡	–	0.045‡	–	–	0.444‡	6.136‡

Source: Okafor, 1997

Notes: † = fat

‡ Determined as a percentage of dry matter (not as ppm)

Na = sodium; K = potassium ; Ca = calcium; Mg = magnesium; Fe = iron; Zn = zinc; Cu = copper; Mn = manganese; P = phosphorus fruits such as mangoes or papaya contain large amounts of carotene (pro-vitamin A). Many less widely distributed species undoubtedly also contain important quantities of this vitamin, which is often limited in tropical diets. For example, Kenyan researchers have recently demonstrated that some wild fruits, especially *Grewia tenax* and *Cordia sinensis*, consumed seasonally by the Maasai of Kajiado District (Oiye et al, unpublished results), have higher levels of beta-carotene by weight than either mangoes or papaya. Although these fruits are small, they are gathered in abundance and can provide an important buffer against malnutrition for people whose diets are predominately milk and maize meal. Other vitamins found in significant amounts in some fruits include thiamine (B1), niacin (B3), pyridoxine (B6), folate (B9) and α-tocopherol (E) (Quebedeaux and Bliss, 1988; Quebedeaux and Eisa, 1990; Wargovich, 2000).

Tree nuts such as almond, filbert, pecan, pistachio and walnut, which are rich in essential amino acids and therefore of high-quality protein, make important dietary contributions. Nuts are also a good source of fibre, vitamin E, minerals and essential fatty acids. Some have been identified as possessing a high content of specific oils, such as omega-3 (walnuts) and mono-unsaturated (almonds, macadamia, pistachio, hazelnuts) fatty acids, which reduce the risk of cardiovascular and other diseases (Maguire et al, 2004). Some lesser-known nuts, such as argan, offer similar benefits (Drissi et al, 2004), but the specific fatty acid composition of most forest species has not been determined.

Examples of nutritious foods collected in African agroforestry parklands and fallows include the fruits of the tamarind tree (*Tamarindus indica*) and the pods of the locust bean tree (*Parkia biglobosa*), both high in vitamins and used in many preparations and recipes. The seeds of *Parkia* are an important element in the diet of most parts of the Sahel. They are fermented to improve their digestibility and to increase the vitamin content, thus providing a nutritious, protein and fat-rich food known as *dawadawa*. *Parkia* is an important ingredient in side dishes, soups and stews made to accompany porridges in northern and western Africa. Other important forest products include the *drupes* (one-seeded fruits) of *Spondias mombin*, and the fruits of *Detarium* spp. The dry pulp of baobab fruits, minus the seeds and fibres, is eaten directly or mixed into porridge or milk. The seeds are mostly used as a thickener for soups but may also be fermented into a seasoning, roasted for direct consumption, or pounded to extract vegetable oil.

Indigenous fruits of the miombo (*Uapaca kirkiana*, *Parinari curatellifolia*, *Strychnos cocculoides*, *Flacourtia indica*) are an important source of food for many rural communities in Malawi. Fruits have long been used to complement or supplement diets because they contain vital nutrients and essential vitamins. Nutritional studies have shown that *Parinari curatellifolia*, *Strychnos cocculoides* and *Azanza garkeana* contain more than 30 per cent fat and about 45 per cent crude fibre and total carbohydrates, and *Trichilia emetica* and *Annona senegalensis* are important sources of protein (Table 4.1).

Sclerocarya birrea (marula) is an integral part of the diet, tradition and culture of communities in southern Africa (Wynberg et al, 2003). Marula fruits and juice are high in vitamin C, providing about 2mg per gram of fresh juice, an amount approximately four times that found in orange juice (Fox and Stone, 1938; Shone, 1979). The fruit and the kernel are rich in protein and minerals and are consumed with sorghum, millet or maize porridge. In Botswana the species is protected and preserved by local people in areas of natural occurrence. Local farmers always select and retain the species when clearing the woodland for arable agriculture.

A study in the humid primary forest region in the eastern part of Madagascar has identified 150 wild fruit species used by the rural people, who possess an intimate knowledge of indigenous plant resources in a region where such information is scarce and infrequently mentioned in botanical works and inventories (Abadie, 1953; Turk, 1995). In the region studied, people regularly consume wild fruits while travelling through the forest or working in the fields; they also collect roots and tubers. Wild fruits and nuts supplement the daily diet, which is high in carbohydrates and low in important vitamins and micronutrients (Hardenbergh, 1997). Wild fruits substitute for exotic fruits, especially in periods of food shortage, and are appreciated by children (Ferraro, 1994; Isaia, 1995), who are by far the main consumers.

Another example of a non-timber forest product with important nutritional properties is camu-camu (*Myrciaria dubia*), a shrubby species native to the floodplains of Amazonia (McVaugh, 1969) whose range extends from the centre of Pará (Brazil) along the mid and upper Amazon River to the eastern part of Peru; in the north it appears in the Casiquiare and the upper and middle Orinoco River basin. The major potential use of camu-camu is as a source of organic vitamin C, since this may reach 2.4 to 3g per 100g of fresh fruit pulp (Roca, 1965; Justi et al, 2000). Mendoza et al (1989) affirm that this value is 60 times that of the lemon, which contains only 44mg per 100g. Vitamin C is present in the pulp and the fruit rind, both commonly used (Calzada, 1980).

Palms

Non-timber forest products from palms are of local importance as food (see review by Nepstad and Schwartzman, 1992). Several palm species are particularly appreciated for their nutritional roles in the diets of forest-dwellers, especially in the humid tropics (Table 4.6). Mesocarps of some palms are rich in carotenes (pro-vitamin A) (Aguiar et al, 1980; Arkcoll and Aguiar, 1984; Rodriguez-Amaya, 1999), with significant genetic variation in vitamin

Table 4.6 *Amazonian palms that are important as food sources*

Species	*Product*	*Distribution*
Bactris gasipaes	Heart, fruit, oil	Throughout Amazon
Elaeis oleifera	Oil, genetic improvement	Throughout Amazon
Euterpe oleracea	Heart, fruit	East Amazon
Euterpe precatoria	Heart, fruit	Central and east Amazon
Jessenia bataua	Fruit, oil	Throughout Amazon
Mauritia flexuosa	Fruit, starch, *Coleoptera* larvae	Throughout Amazon
Orbignya phalerata	Oil, starch	South and east Amazon

Source: Adapted from Kahn, 1993

content. The nutritional composition of several palm species is reported by FAO (1995a) in a compilation of species from different regions of the world.

For centuries, many palm species have been tapped throughout the tropical world to produce fresh juice (sweet toddy), fermented drinks (toddy, wine, *arak*), syrup ('honey'), brown sugar (*jaggery*) or refined sugar. Cunningham and Wehmeyer (1988) describe the nutritional value of palm wine from *Hyphaene coriacea* and *Phoenix reclinata.* In some species, several parts of the plant are extracted for consumption. For example, the fruit pulp of *P. reclinata* is highly nutritious, quite palatable (van Wyk and Gericke, 2000) and eaten raw or dried for storage. The heart of the crown and the young leaf bases are eaten as a vegetable (Pooley, 1993). The roots yield an edible gum (Roodt, 1998), mainly eaten by children.

A species that has been recognized as having particularly high potential, both for reforestation and for non-timber forest products, is *Orbignya phalerata,* locally known as babassu palm. It occurs widely in Brazil along the transition belt between the central savannahs and the Amazon forest (May et al, 1985). Babassu kernels are good sources of oil and protein, which are often not readily available to the native populations of Amazonia. In pre-colonial Brazil, Tupinambá Indians on the floodplains of Maranhão consumed large quantities of the products from babassu and buriti palms *(Mauritia flexuosa)* in their diets (Steward, 1963). For nomadic groups, these palms were important sources of fuel and fibre and formed buffers against starvation. The Guajá of the eastern Amazon, a small group that constitutes one of the few remaining hunter-gatherer cultures, consume the fleshy mesocarp of the fruit and sometimes grate or grind the seeds, which are mixed with water to prepare babassu 'milk'. Peasants and indigenous groups find babassu a reliable source of food (Anderson and May, 1985). De Beer and McDermott (1989) point out the importance of rattans and edible palm products in Southeast Asia. Falconer and Koppell (1990) document the significance of palms among the forest products in West Africa.

Wild roots and tubers

Starch reserves in stems, roots and tubers usually constitute a major food source in forest areas. The tropical humid forests and woodlands contain a host of plants that produce edible starchy roots and tubers. Forest yams are consumed in Africa, Australia and Asia (e.g., the Philippines, Indonesia, Malaysia and Thailand). In Central Africa, yams constitute a large group of edible species (*Dioscorea* spp; Hamon et al, 1995; see Chapter 13, this volume), which vary greatly in water, protein and starch content (Hladik et al, 1984). Of particular importance from a health and nutrition perspective are condiments made with ginger species, which have major impacts on health functions (Fuhrman et al, 2000).

Mushrooms

Most cultures have used mushrooms to provide supplementary food, especially during the rainy season (Chipompha, 1985), but the potential of mushrooms found in forest ecosystems and other landscapes is largely neglected in many parts of the developing world.

Recent attempts to fill this gap include a global review of the use and importance to people of wild edible fungi published by FAO (2004).

Edible mushrooms represent a broad spectrum of species and are typically consumed only to a limited extent and according to personal preference by local people. Various studies have documented the utilization of up to 60 species of edible mushrooms sold along roadsides in Malawi, Zimbabwe and neighbouring countries during the rainy season, particularly by women and children (Clarke et al, 1996). The trade of this product depends on mushroom fruiting, but appropriate storage practices can extend the season (Karmann and Lorbach, 1996). Most wild edible mushrooms grow on the dead wood and leaf litter normally found in indigenous woodlands. However, the indigenous forests are declining because of deforestation, and as a result, mushroom yields from the woodlands are also declining, and efforts to cultivate most species have been unsuccessful. Fungi nonetheless remain important in traditional diets and as a source of income for communities with access to forest ecosystems.

Given the popularity of forest mushrooms in local cuisine and their reputed nutritional value, they have been singled out as a potential product for income generation that contributes to livelihood security. Gathering mushrooms together with other non-timber forest products is also widespread outside tropical and subtropical areas. In the present-day Russian Federation, non-timber forest product collection is a central element of the traditional culture: part of a strong relationship with nature, an occasion for recreation and a way of life for the indigenous peoples inhabiting northern and central Siberia. Since the beginning of economic reforms in the Russian Federation in 1992, the number of poor people has increased with the closing of many logging and mining enterprises. A study undertaken in Krasnoyarsk, in central Siberia, examined the role of non-timber forest products in local food security (Vladyshevskiy et al, 2000). The harsh climate severely limits agriculture, and several indigenous tribes (Keto, Nenets, Dolgan) live in widely distributed, sparsely settled communities with low standards of living. Wild game, fish, mushrooms, nuts (seeds) and cones from the Siberian pine are consumed and sometimes sold together with berries and medicinal plants. Historically, gathering was more important under unfavourable socioeconomic conditions. Because of the current socioeconomic collapse in many parts of the former Soviet Union, the use of mushrooms and Siberian pine nuts has increased two- to three-fold, the use of wild onion three- to five-fold, and the collection of berries one-and-a-half- to two-fold. In most areas, where there is no employment non-timber forest products are often the main source of food and income for village populations, representing as much as 30 to 40 per cent of family income.

Insects

The food potential of non-timber forest products is increasingly recognized, but the role of edible insects is less well known, even though, according to a study by Ramos-Elorduy (1991), more than 1300 species are edible. In Central Africa, edible insects are widely available in village markets, and some favourite species reach urban markets and restaurants (de Foliart, 1992; Tabuna, 2000). In Zaire, a national study of animal protein intake revealed that insects furnished 10 per cent of the animal proteins produced annually in the country, compared with 30 per cent for game, 47 per cent for fishing, only 1 per cent for fish culture, 10 per cent for grazing animals, and 2 per cent for poultry (Gomez et al, 1961). However, in some parts of the country, insects furnished between 22 and 64 per cent of the animal protein intake, showing the great importance that insect foods can have for some local communities.

The nutritional value of various caterpillars has been investigated in the Congo Basin region. Results show that caterpillars have higher protein and fat content and provide more energy per unit than meat and fish (Malaisse, 1997). As reported by de Foliart (1992) and Malaisse (1997), respectively, 100g of cooked insects provides more than 100 per cent of daily requirements of minerals and vitamins, and a consumption of 50g of dried caterpillars matches the daily requirements of riboflavin and pantothenic acid.

Insects are rich in several essential amino acids, different minerals (e.g., potassium, calcium, magnesium, zinc, phosphorus and iron) and/or vitamins (e.g., thiamine/B1, riboflavin/B2, pyridoxine/B6, pantothenic acid, niacin). Some groups (especially *Coleoptera* larvae and some caterpillars) contain high levels of fat, generally unsaturated lipids (Dufour, 1981). Ramos-Elorduy et al (1981), studying edible insects in Mexico, found their digestibility to be very high. Flour made from larvae is fed to children to counter malnutrition. Species rich in calcium, protein or iron are given to anaemic people and pregnant and lactating women.

Finally, the market for edible caterpillars is growing and lucrative, generating considerable household income and thus contributing to livelihood security. Caterpillars from the mopane tree (*Colophospermum mopane*) are a cash crop in southern Africa (Menzel and D'Aluisio, 1998; Stack et al, 2003). However, over-extraction of caterpillars is threatening the sustainability of this activity, and efforts are underway to establish mopane farming. Considering the strong market demand for edible caterpillars and the pressure placed on these resources by deforestation and forest degradation, the relationships between caterpillar harvesting and its impact on forests on the one hand, and between forest harvesting on caterpillar populations on the other, are poorly understood and deserve further investigation.

Foods from mangrove forests

Mangrove forests cover an estimated 160,000–170,000km^2 worldwide, with the largest forests in Brazil, Indonesia, Australia, Nigeria and Malaysia. These forests are under increasing threat from human activities. Mangroves harbour large quantities of fish, molluscs, crabs and shrimp for a total annual catch estimated at around 1 million tonnes, slightly over 1 per cent of the total world fish catch (FAO, 1992). Mangrove forests also supply other foods, from honey and edible algae to fruits and leaves for animal fodder, and a variety of traditional medicines (Figure 4.2).

The tender leaves of *Acrostichum* spp and the hypocotyls of *Bruguiera* spp (mangrove bean) are staple foods of some Papua New Guineans. Fruits of *Avicennia marina* are universally used as vegetables. The fruits of *Kandelia candel* and *Bruguiera gymnorrhiza*

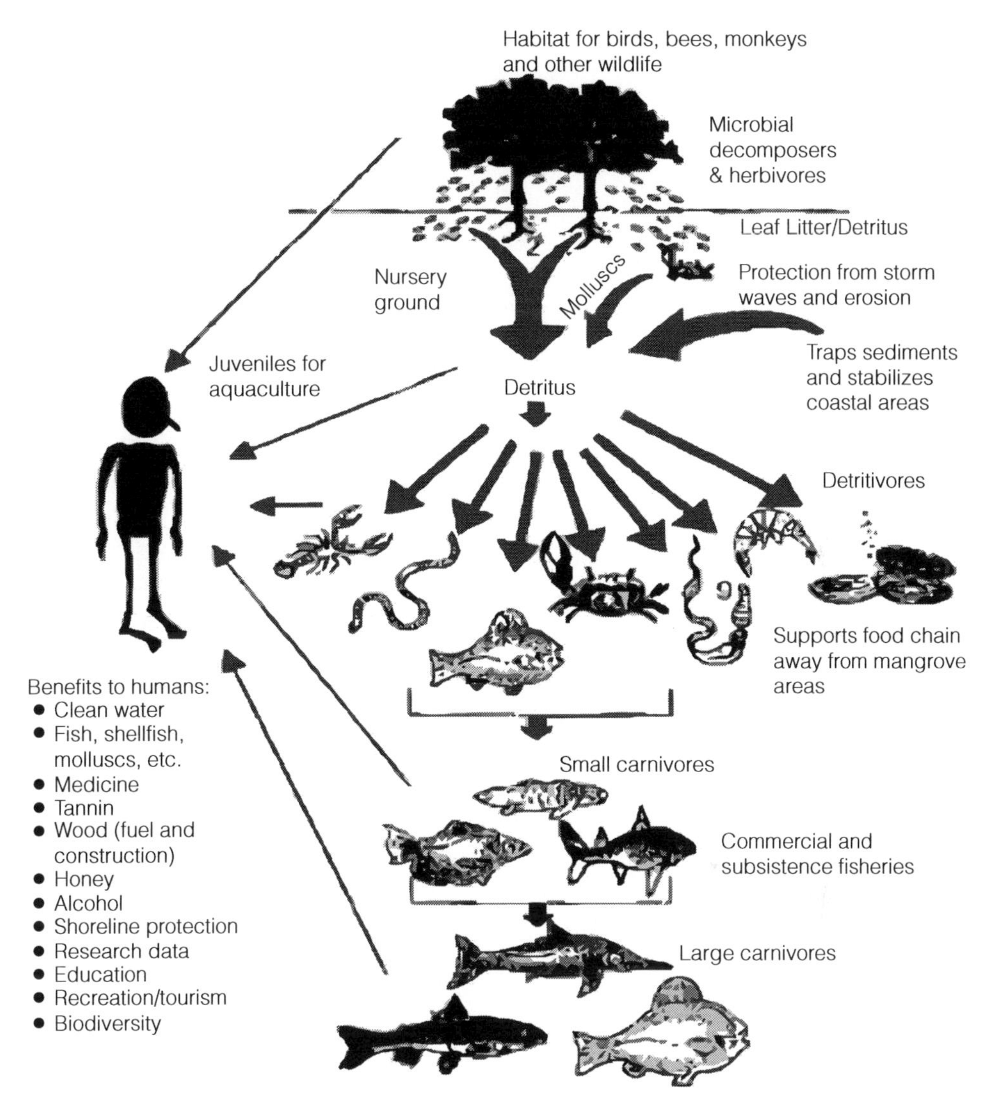

Figure 4.2 *Mangroves and their ecological and economic benefits*

Source: From Berjak et al, 1977

contain starch and can make excellent cakes or sweetened stuffing for pastry if sliced, soaked in water to rinse out tannins and then ground to a paste. 'Sagu' is taken from the mangrove palm tree *Metroxylon sagu*, found in Southeast Asia, where the hypocotyls of *Bruguiera* are also an accepted food item. Intoxicating drinks are made from the sap of the 'coconut' of *Nypa* and *Borassus*. Extracts of the heartwood of *Avicennia alba* and *A. officinalis* have tonic properties. It is reported that some mangrove plants and extracts are used as incense, perfumes, hair products, condiments and aphrodisiacs. Edible jelly and a kind of salt are made from the ashed leaflets (Bandaranayake, 1999).

Fodder for livestock

An adequate supply of livestock fodder is a crucial part of food production for millions of people, yet the role of forests in livestock raising has been largely ignored. Animal-source foods generally provide more available iron, vitamin A and protein than plants (Murphy and Allen, 2003), and even small increases in intake offer real benefits to the majority of the world's malnourished. Different examples of the use of trees and shrubs as fodder species in various regions of the world are presented in a review published by FAO (1991). Fodder from trees and shrubs is particularly important during periods when the nutritional quality of grasses falls. Dry grasses contain only minimal levels of energy and protein and lack essential minerals, especially phosphorus, so pastoralists keep their herds alive on arid and semi-arid land by supplying them with twigs, leaves, small branches, seed-pods and fruit from trees and shrubs. There is evidence that forest ecosystems can host a large number of such forage species. A recent study conducted in Quintana Roo (Mexico) revealed that a total of 18 forage species are derived from the tropical forest (Dalle, 2006).

In parts of northern Senegal, to keep their herds alive and healthy, pastoralists rely on high-quality supplements of browse leaves, fruits and seed-pods from trees and shrubs for at least six months a year. Unfortunately, supplies are increasingly threatened because over-browsing is hindering the regeneration of trees and shrubs.

ADDRESSING NUTRITIONAL AND HEALTH PROBLEMS WITH FOREST FOODS

Even though forests offer many contributions to food supply and dietary quality, most diseases related to micronutrient deficiency in developing countries continue to be addressed with expensive, single-nutrient-based interventions. Instead, non-timber forest products (wild foods or forest foods) could supplement staple foods with the nutrients needed by rural people and could be more effectively used for disease mitigation (Table 4.7). Complicating the problem, policies and programmes for natural resource management and environmental protection have traditionally been perceived, managed, implemented and financed independently of those for food security and poverty alleviation.

To fully realize the contribution of forest foods to diets, better use and sustainable management of forest resources should be promoted, based on an improved understanding of forest food characteristics and their potential contribution to nutrition. An approach to health based on the quality of food supplies requires the support of a considerable amount of information on the nutritional properties of forest foods.

Most of the studies carried out thus far indicate how difficult it is to assess precisely the nutritional contribution of forest foods, since it is hard to determine how frequently and in what quantity they are consumed. However, measures of the contribution of forest foods to diets should be not merely quantitative but also qualitative, and aimed at identifying thresholds in the amounts consumed that make their positive impact on diets significant. Suitable methods to determine the nutritional role of forest foods and their impact on health are based on food consumption studies associated with nutritional analyses and the assessment of the role of micronutrients and functional phytochemicals.

Table 4.7 *Common nutritional deficiencies and the potential role of forest food*

Nutrient-related problems	*Forest food with potential for combating deficiencies*
Protein and energy malnutrition due to inadequate food consumption: reduced growth, susceptibility to infection, and changes in skin, hair and mental facility	Energy-rich food available during seasonal or emergency food shortages, especially nuts, seeds, oil-rich fruit and tubers; e.g., the seeds of *Geoffroea decorticans*, *Ricinodendron rautanenil* and *Parkia* sp; oil of *Elaeus guineensis*, babassu, palmyra and coconut palms; protein-rich leaves such as baobab (*Adansonia digitata*) and wild animals, including snails, insects and larvae
Vitamin A deficiency: impaired vision and immune function, and in extreme cases blindness and death; responsible for blindness of 250,000 children per year	Forest leaves and fruit are often good sources of vitamin A: e.g., leaves of *Pterocarpus* sp, *Moringa oleifera*, *Adansonia digitata*, the gum of *Sterculia* sp, palm oil of *Elaeus guineensis*, bee larvae and other animal food
Zinc deficiency: retarded growth and development, increased incidence of pregnancy complications, suppressed immunity and impairments in neuro-psychological functions	Animal-source foods, particularly red meat, are excellent sources of zinc, although many nuts (e.g., pine nuts, coconut meat, pecans, brazil nuts) have comparable values on a per-weight basis, and mushrooms can make significant contributions
Iron deficiency: anaemia, weakness and increased susceptibility to disease (especially in women and children), impaired cognitive development	Wild animals (including insects such as tree ants), mushrooms (often consumed as meat substitutes) and forest leaves (e.g., *Leptadenia hastata*, *Adansonia digitata*); baobab fruit pulp is also an important source of iron
Folate deficiency: anaemia, neural tube defects	Leafy and other vegetables, and many fruits
Niacin deficiency: dementia, diarrhoea and dermatitis; common in areas with maize as staple diet	Forest fruit and leaves rich in niacin, such as *Adansonia digitata*, fruit of *Boscia senegalensis* and *Momordica balsamina*, seeds of *Parkia* sp, *Irvingia gabonensis* and *Acacia albida*
Riboflavin deficiency: skin problems; common throughout Southeast Asia among those with rice diets	Forest leaves, notably *Anacardium* sp, *Sesbania grandiflora* and *Cassia obtusifolia*, are especially high in riboflavin, as are animal-source foods, especially insects
Vitamin C deficiency: increased susceptibility to disease and impaired iron status (leading to anaemia); common to those consuming monotonous diets	Forest fruits, particularly the fruit of *Ziziphus mauritiana*, *Adansonia digitata* and *Sclerocarya caffra*; and leaves (such as *Cassia obtusifolia*) and the gum of *Sterculia* sp

Source: Adapted from FAO, 1995b

Methodologies are increasingly being developed to include nutrition and household food security concerns in forestry planning (Odgen, 1990; Egal et al, 2000). They need to be supported by more evidence and combined with documentation of the nutritional characteristics and patterns of use of forest foods. In addition, investigations into traditional management can help track exploitation trends and address any over-harvesting.

Trees with nutritionally important traits, once selected from the natural stands, may be promoted for wider cultivation by farmers in their various farming systems (Leakey et al, 2002; Kengni, 2003), but cultivated sources represent a small number of individuals and lead to a shrinking genetic pool. Furthermore, some cultivated food plants do not fully retain the characteristics for which the wild sources are extremely valued by local people. The observed variation in the biochemical composition found between varieties and among individuals of the same variety depends on environment (climate and soil), genetic make-up, the age of the individuals and the parts of the plant harvested. More understanding of this variation in desirable characteristics and their causes is needed.

To identify the potential of important species for cultivation in home gardens and use in agroforestry systems, selection and development should be supported by investigations that attempt to address the following questions:

- Are cultivation and domestication generally desirable options to reduce over-exploitation?
- What are the biological limitations of removing trees from the forest and retaining forest food production (e.g., sustaining pollinators, agents of seed dispersal)?
- What about the long-term viability of small populations of trees cultivated outside the forest, and the long-term maintenance of desirable traits and nutritional properties?
- Can we afford to lose the wild gene-pool of those species found to play fundamental roles in food security and nutrition?
- Are the species or populations that provide critical forest foods adequately conserved in collections?

The productivity and biochemical composition of wild and semi-domesticated tree species are far less well known than for cultivated species. FAO's food composition tables allow comparisons between wild and cultivated species, but do not account for the variation in biochemical composition due, for instance, to age and geographic origin. It has been observed, however, that domesticated plant sources tend to be more energy-dense than wild sources, as a result of the domestication and breeding that favour higher concentrations of fats and carbohydrates, as reported by Dounias et al (2004). However, many indigenous wild fruit tree species have been found to be higher in vitamins and other important nutrients than domesticated varieties. The vitamin C content of an orange is famously high at 57mg per100g, but the fruit of the baobab tree has 360mg per100g, and *Ziziphus jujube* var. *spinosa* 1000mg per 100g (FAO, 1992). Similarly, by weight, wild leaf vegetables contain more riboflavin than eggs, milk, nuts and fish.

More research is needed to understand the patterns of content variation of micronutrients and the factors responsible for the observed variations. For products other than edible fruits, very limited research has been undertaken to compare the nutritional value of wild versus cultivated varieties. Yam species, whose wild gene-pool is of great importance for breeding, provide an interesting example. Biochemical studies on the phylogeny of cultivated African yam varieties (Terauchi et al, 1992, reported by Hladik and Dounias, 1993), provide evidence that some forest yams are ancestors of cultivated varieties and that wild forest species comprise a very valuable genetic pool from which useful characteristics can be drawn to improve cultivated yam varieties. The data presented above seem to indicate a great affinity between some wild (Table 4.8) and cultivated varieties (Table 4.9), and lead to the conclusion that the wild species described are the genetic sources of the cultivated forms.

The study also indicates that empirical selection by local communities has generated cultivated varieties that have comparable and even higher amounts of some macronutrients (Tables 4.8 and 4.9). This type of study should be extended to other critical forest-food species to understand the impact of human selection on some non-observable traits, such as the content of micronutrients, in tree species that are introduced in cultivated systems. The following research questions should be answered:

- What is the range of variation in nutritional properties of forest foods in these species?
- How can we identify and control the drivers of variation in nutritional properties of food species within forest ecosystems while putting trees under cultivation?
- Can we take trees out of the forest for cultivation and retain unaltered their characteristics and nutritional properties?
- Are there examples of concerns about the long-term maintenance of particular nutritional properties and preferred traits in the practices developed by local communities to conserve and use forest foods (wild collection and cultivation)?
- How does empirical selection affect the nutritional content of cultivated varieties? Are the traits preferred by farmers retained in cultivated trees? Can we identify general predictable patterns?

In addition, the taste of forest foods is often considered superior by local populations. Recently, it has become increasingly evident that urban-focused commercialization of non-timber forest products involves the sale of 'wild' products, which from a cultural point of view may be held in high esteem (Cocks and Wiersum, 2003; Chapter 3, this volume). Market prices often reflect the higher value attached to wild-harvested products. Wild foods may have a cultural value and be consumed during special feasts. Some forest sites and species may also have a sacred value, and there are social and spiritual aspects to gathering forest foods in the wild. For such reasons, cultivated systems cannot replace entirely forest ecosystems.

In West Africa, people still seek to obtain the bulk of micronutrients in their diet from shrub leaves, and prefer pulp and nuts from wild fruits (Okafor et al, 1996). Subsistence farmers in the region produce exotic fruits, such as citrus, but largely for market purposes (Kengni et al, 2004). A study in Lushoto, Tanzania, revealed that people who consumed wild leaf relishes favoured the taste of wild leaves over introduced cultivated vegetables (Fleuret, 1979).

Table 4.8 *Composition of African (Central African Republic and Gabon) wild yams (*Dioscorea *spp) tuber, raw*

Species	*Water %*	*Percentage dry weight*							
		Protein	*Lipids*	*Reducing sugars*	*Hemi-celluloses*	*Cellulose*	*Lignin*	*Starch*	*Minerals*
EDIBLE FOREST YAMS									
D. semperflorens fruiting phase	75	5.5	–	3.3	4.7	2.6	0.4	78.8	2.1
D. semperflorens sterile phase	65	5.3	0.1	1.2	5.0	1.6	0.3	81.3	1.7
D. mangenotiana	68	9.0	–	1.0	6.2	2.0	0.3	75.9	3.5
D. praehensilis	–	7.1	0.6	5.8	11.9	7.6	2.7	58.3	3.1
D. burkilliana sterile phase	67	6.8	0.2	1.7	12.8	1.8	0.5	69.9	2.5
D. burkilliana flowering phase	55	5.6	0.2	0.7	7.0	1.6	0.2	78.2	2.5
D. minutiflora	69	4.6	–	3.4	11.4	4.6	0.3	73.4	2.3
OPEN AREA YAMS									
D. dumetorum	–	9.1	1.0	2.9	5.9	6.6	1.5	68.2	2.3
D. preussii	82	9.4	0.2	4.6	21.6	5.5	2.2	48.4	6.1
D. bulbifera	68	5.8	0.5	3.1	7.7	6.7	2.7	57.6	2.9

Source: From Hladik and Dounias, 1993

Table 4.9 *Composition of African cultivated yams (*Dioscorea *spp), both present and absent as indigenous wild species, collected in Central African Republic (CAR) and Gabon (G)*

Species		Water %	Percentage of dry weight							
			Protein	Lipids	Reducing sugars	Hemi-celluloses	Cellulose	Lignin	Starch	Minerals
D. dumetorum	*lobo* (G),	73	13.2	0.6	4.8	7.0	4.1	0.2	63.6	3.4
	moyo (G)	73	12.4	0.6	1.9	7.1	3.7	0.4	67.8	2.7
D. complex	*nzo* (G)	73	12.2	0.2	0.6	25.2	2.3	0.3	53.6	3.1
D. cayenesis-rotundada	*engendi* (CAR)	62	11.3	0.2	1.3	5.1	3.0	0.4	73.1	3.5
	botoko (CAR)	61	10.1	–	1.3	5.9	1.2	4.1	72.1	2.9
	ekata (CAR)	58	9.8	–	1.8	4.4	3.1	0.4	75.4	2.5
	ngbongbo (CAR)	62	8.7	0.3	1.9	5.7	3.0	0.5	74.8	2.7
	kambele (CAR)	–	8.2	–	2.4	7.6	3.3	0.8	72.9	2.7
D.alata	*mokondo* (CAR)	74	8.7	–	1.2	6.4	2.5	0.2	73.6	5.2
	mbondo (CAR)	73	8.5	0.3	2.9	3.9	2.3	0.3	76.7	3.0
	molobilo (CAR)	79	8.4	0.3	3.4	6.3	3.8	0.8	70.7	3.9
	mongandi (CAR)	69	7.7	–	1.3	6.1	2.2	0.5	78.5	4.2
D. burkilliana	*nji* (G)	74	7.3	0.5	1.8	8.9	2.0	0.6	71.6	2.7
	kobo (CAR)	67	6.6	0.1	7.3	6.4	3.5	0.6	71.2	3.4

Source: Adapted from Hladik et al, 1984

Note: a dash indicates 'not measured'

A preference for wild products also characterizes the use of medicinal plants. For example, ginseng, a deciduous perennial whose roots are highly prized for medicinal purposes, has been an important component of Chinese folk medicine for more than 4000 years and now occurs in the USA, east of the Mississippi River; from here, roots have been traded to China since the Revolutionary War. Cultivated ginseng can grow in artificial shade, but the preferred 'wild' ginseng requires natural forest (Hankins, 1998). Although the production is much higher, cultivated roots usually command a lower price. Chinese buyers prefer wild ginseng over cultivated because it more closely resembles the revered wild Asian ginseng. Research in different social and environmental contexts is needed to determine whether, from a cultural perspective, wild forest foods can be replaced by cultivated varieties.

Despite the often higher appreciation of wild-harvested forest foods, in some circumstances their use in local diets is declining because of several factors:

- a loss of knowledge about which local plants and other organisms are edible (Robson, 1976; Tabuti et al, 2004);
- changes in livelihood patterns that limit the amount of time spent in the forest, as when roaming people become settled (see Chapters 12 and 13) and women must devote time to other activities (Arnold and Ruiz Pérez, 1998; Brown and Lapuyade, 2002); and
- the general shift from gathered to purchased foods as rural communities join the market economy and non-timber forest products lose ground to other food sources (Godoy et al, 1995; Arnold and Ruiz Pérez, 1998; Byron and Arnold, 1999).

Those factors, combined with the introduction of exotic foods with lower nutritional values (Udosen, 1995; Okafor, 1997), the contraction of food species in local farming systems (Future Harvest, 2001) – both associated with a change of taste – and the disappearance of local knowledge on traditional products, are leading to a general simplification of diets.

Thus, it is fundamental to understand trends in consumption and patterns of change. New studies are warranted in places where previous investigations have been undertaken to compare past and current food consumption behaviour and quality of the diet. This would also help identify policies to promote the conservation of forests as a means to address nutritional and health issues. Comparative studies would look for answers to the following questions:

- What are the determinants of consumer behaviour vis-à-vis forest foods?
- How can consumption be influenced through public awareness initiatives?

Some studies have highlighted an inverse correlation between the nutritional status of certain communities and their distance from forests (Fleuret, 1979; Somnasang et al, 1988). However, more recent data indicate a large incidence of malnutrition in rural areas, sometimes higher than in urban areas. For instance, in some parts of Brazil, according to FAO (2000), malnutrition is higher among the rural than the urban population. According to UNICEF (2006), a study of children under five in Penambuco State showed a 56 per cent incidence of iron deficiency anaemia in rural areas, compared with 41 per cent in urban areas. According to the same source, in Peru, maternal iron deficiency anaemia rates are high in rural areas (41 per cent) and the mountains (42 per cent). These patterns are presented also in a report by Ruel et al (1998), who provide an interesting review of the literature on urban–rural comparisons in selected health and nutrition indicators. The high incidence of malnutrition in rural areas could be linked to the conditions of rural environments, in particular to cultural erosion, poverty and the degradation and contraction of forest ecosystems (Johns and Maundu, 2006). Thus, more research is needed to understand the particular role played by forest foods in communities in different states of conservation and with different food systems. The range of products extends from food to medicinal plants, with the two groups sometimes not clearly distinguishable. Because dietary quality has been associated partly but not exclusively with nutrient content and has been shown to improve with greater food diversity (Shimbo et al, 1994; Slattery et al, 1997; Hatløy et al, 1998), the disruption in environmental integrity would affect patterns of human health, disease and nutritional status (Johns and Eyzaguirre, 2000). Efforts should be devoted to understanding the impact of limited access to forest biodiversity and the repercussions of forest degradation on the diets of forest-based communities. More research is needed to understand the nutritional, health and social implications of lack of access to forest biodiversity, tackling such questions as these:

- Are nutritional and health gradients associated with levels of degradation of forests and woodlands, or with increasing distance from and lack of access to the forest?
- Can market distribution channels mitigate problems related to limited or total lack of access to wild forest foods?

Forest foods also generate income and thus have an indirect effect on people's diet and health. A study by Gockowski et al (2003) in sub-Saharan Africa describes resource-poor households' small-scale marketing of traditional leafy vegetables (including *Gnetum africanum* harvested from the forest). However, it seems that people's reliance on the forest as a safety net is not uniform (McSweeney, 2003), and even the most remote rural communities are becoming more involved in off-farm activities (e.g., in Latin America; Reardon et al, 2001). Thus, the sale of forest products, as much as their subsistence use, forms just part of the portfolio of coping strategies adopted by smallholders. Finally, generating income from the sale of forest products does not necessarily slow the conversion of forests to agriculture (Ellis, 1998). Therefore, the relationships between dependence on forest foods and forest conservation need to be better understood to help policy-makers ensure the continued use of forest foods, especially by vulnerable groups – by, for example, supporting diversity-rich production options, conservation actions or marketing initiatives.

CONCLUSIONS

Forests are a very important source of food and particularly crucial for vulnerable groups. Forest products complement seasonal crops and become vital during famine, wars and droughts. The studies reported above illustrate the wealth of non-timber forest products that provide food, whether directly or indirectly, and their contributions to human nutrition. The range of products consumed extends from leaves to fruits and nuts, from mushrooms to insects, and supports the livelihood of many rural communities.

The potential of forest foods to mitigate food insecurity and malnutrition in developing countries is only partly understood. Considerable efforts should be made to better document the nutritional characteristics of forest foods and develop sustainable use and conservation practices in ecosystems that harbour species with particularly valuable nutritional characteristics. These interventions need to be supported by research on how forest management can promote nutritional and food security for rural communities. The main research questions posed in this chapter address the variation in nutritional properties of non-timber forest products, the impact of cultivation on nutritional properties of forest foods now extracted from wild sources, and the effects of limited access to forest biodiversity on nutrition and health. Filling the gaps in our knowledge about the contributions, both quantitative and qualitative, of forest foods in local diets could also contribute to a greater appreciation of the value of forest ecosystems per se.

REFERENCES

Abadie, M.C. (1953) 'Inventaire des espèces fruitières comestibles à Madagascar', *Bulletin de l'Académie Malgache* 30: 185–204

Aguiar, J.P.L., Marinho, H.A., Rebelo, R. and Shrimpton, R. (1980) 'Aspectos nutritivos de alguns frutos da Amazônia', *Acta Amazônica* 13: 953–954

Anderson, A. and May, P. (1985) 'A palmeira de muitas vidas', *Ciência Hoje* 4(20): 58–64

Arkcoll, D.B. and Aguiar, J.P.L. (1984) 'Peach palm (*Bactris gasipaes* HBK), a new source of vegetable oil for the wet tropics', *Journal of the Science of Food and Agriculture* 35: 520–526

Arnold, J.E.M. and Ruiz Pérez, M. (1998) 'The role of non-timber forest products in conservation and development', in E. Wollenberg and A. Ingles (eds) *Incomes from the Forest: Methods for the Development and Conservation of Forest Products for Local Communities*, CIFOR, Bogor

Arnold, M. and Townson, I. (1998) *Assessing the Potential of Forest Product Activities to Contribute to Rural Incomes in Africa*, Overseas Development Institute, London

Asibey, E.O.A. and Beeko, C.Y.A. (1998) 'Forest and food security', paper presented at the Symposium on Ghana's Forest Policy, 3–7 April, Grenhill, Accra, Ghana

Bahuchet, S. (1990) 'The Akwa Pygmies: Hunting and gathering in the Lobaye Forest', in *Food and Nutrition in the African Rain Forest*, Food Anthropology Unit 263, UNESCO, Paris

Balogun, A.M. and Fetuga, B.L. (1986) 'Chemical composition of some under-exploited leguminous crop seeds in Nigeria', *Journal of Agricultural and Food Chemistry* 34: 189–192

Bandaranayake, W.M. (1999) 'Economic, traditional and medicinal uses of mangroves', AIMS Report 28, Australian Institute of Marine Science, Townsville, Australia

Beckerman, S. (1984) 'Swidden in Amazonia and the Amazon Rim', in B.L. Turner and S.B. Brush (eds) *Comparative Farming Systems*, Guildford Press, New York, NY, pp55–94

Berjak, P., Campbell, G.K., Huckett, B.I. and Pammenter, N.W. (1977) *In the Mangroves of Southern Africa*, Wildlife Society of South Africa, Durban

Bloch, A. and Thomson, C.A. (1995) 'Position of the American Dietetic Association: Phytochemicals and functional foods', *Journal of the American Dietetic Association* 95: 493–496

Brown, A.J.P., Robert, D.C.K. and Cherikoff, V. (1985) 'Fatty acids in indigenous Australian foods', *Proceedings of the Nutrition Society of Australia* 10: 209–212

Brown, K. and Lapuyade, S. (2002) 'Changing gender relationships and forest use: A case study from Komassi, Cameroon', in C.J.P. Colfer and Y. Byron (eds) *People Managing Forests: The Link Between Human Well Being and Sustainability*, RFF, Washington, DC/CIFOR, Bogor, Indonesia, pp90–110

Burkill, H.M. (1994) *The Useful Plants of West Tropical Africa, Vol. 2: Families E-I*, Kew Royal Botanic Gardens, Kew, UK

Burlingame, B. (2000) 'Wild nutrition', *Journal of Food Composition and Analysis* 13: 99–100

Busson, F. (1965) *Plantes alimentaires de l'Ouest Africain: Etude botanique, biologique et chimique*, Ministere de la Cooperation, Marseilles

Byron, R.N. and Arnold, J.E.M. (1999) 'What futures for the people of the tropical forests?' *World Development* 27(5): 189–805

Calzada B., José (1980) *Ciento cuarenta y tres frutales nativos*, Distributor Librería El Estudiante, Lima, Peru

Campbell-Platt, G. (1980) 'African locust bean and its West African fermented products: Dadawa', *Ecology of Food and Nutrition* 9: 123–132

Chege, N. (1994) 'Africa's non-timber forest economy', *World Watch* 7(4):19–24

Cherikoff, V., Brand, J.C. and Truswell, A.S. (1985) 'Australian aboriginal bushfoods: Corms, roots, tubers and yams', *Proceedings of the Nutrition Society of Australia* 10: 183

Chipompha, N.W.S. (1985) 'Some mushrooms of Malawi', Forestry Research Record No. 63, Forestry Research Institute of Malawi (FRIM), Malawi

Chweya, J.A. and Eyzaguirre, P.B. (eds) (1999) *The Biodiversity of Traditional Leafy Vegetables*, IPGRI, Rome

Clarke, J., Cavendish, W. and Coote, C. (1996) 'Rural households and miombo woodlands: Use, value and management', in B. Campbell (ed) *The Miombo in Transition: Woodlands and Welfare in Africa*, CIFOR, Bogor

Cocks, M.L. and Wiersum, K.F. (2003) 'The significance of plant diversity to rural households in Eastern Cape province of South Africa', *Forest, Trees and Livelihoods* 13: 39–58

Craig, W. and Beck, L. (1999) 'Phytochemicals: Health protective effects', *Canadian Journal of Dietetic Practice and Research* 60: 78–84

Cunningham, A.B. and Wehmeyer, A.S. (1988) 'Nutritional value of palm wine from *Hyphaene coriacea* and *Phoenix reclinata* (Arecaceae)', *Economic Botany* 42(3): 301–306

Dalle, S.P. (2006) 'Landscape dynamics and management of wild plant resources in shifting cultivation systems: A case study from a forest ejido in the Maya Zone of Quintana Roo, Mexico', PhD thesis, Department of Plant Science, McGill University, Montreal

Davies, A.G. and Richards, P. (1991) 'Rain forest in Mende life: Resources and subsistence strategies in rural communities around Gola North Forest Reserve (Sierra Leone)', report to the Economic and Social Committee on Overseas Research (ESCOR), Overseas Development Administration, London

de Beer, J.H. and McDermott, M.J.M. (1989) *Economic Value of Non-Timber Forest Products in Southeast Asia*, Council for IUCN, Amsterdam

de Foliart, G. (1992) 'Insects as human food', *Crop Protection* 5(11): 395–399

Dei, G.J.S., Sedgley, M. and Gardner, J.A. (1989) 'Hunting and gathering in a Ghanaian rain forest community', *Ecology of Food and Nutrition* 22: 225–243

Dounias, E., Kishi, M., Selzner, A., Kurniawan, I. and Levang, P. (2004) 'No longer nomadic, changing Punan Tubu lifestyle requires new health strategies', *Cultural Survival Quarterly* 28(2): 15–20

Drissi, A., Girona, J., Cherki, M., Godas, G., Derouiche, A., El Messal, M., Saile, R., Kettani, A., Sola, R., Masana, L. and Adlouni, A. (2004) 'Evidence of hypolipemiant and antioxidant properties of argan oil derived from the argan tree (*Argania spinosa*)', *Clinical Nutrition* 23(5): 1159–1166

Dufour, P.A. (1981) 'Insects: A nutritional alternative', mimeograph, Department of Medicine and Public Affairs, The George Washington University Medical Center, Washington, DC

Egal, F., Ngom, A. and Ndione, P. (2000) 'Integration of food security and nutrition into forestry planning: The role of participatory approaches', *Unasylva* 51(3): 19–23

Ejoh, R., Tchouanguep Mbiapo, F. and Fokou, E. (1996) 'Nutrient composition of the leaves and flowers of *Colocasia esculenta* and the fruits of *Solanum melongena*', *Plant Foods for Human Nutrition* 49: 107–112

Ellis, F. (1998) 'Household strategies and rural livelihood diversification', *The Journal of Development Studies* 35(1): 1–38

Falconer, J. and Arnold, J.E.M. (1988) 'Forests, trees and household food security', Social Forestry Network Paper 7a, Overseas Development Institute, London

Falconer, J. (1990) 'Hungry season food from forests', *Unasylva* 41: 14–19

Falconer, J. and Koppell, C.R.S. (1990) 'The major significance of "minor" forest products: The local use and value of forests in the West African humid forest zone', Community Forestry Note No. 6, FAO, Rome

FAO (1989) 'Forestry and food security', Forestry Paper No. 90, FAO, Rome

FAO (1991) 'Legume trees and other fodder trees as protein sources for livestock', Animal Production and Health Paper 102, FAO, Rome

FAO (1992) 'Forests, trees and food', miscellaneous paper, FAO, Rome

FAO (1995a) 'Tropical palms', Non-timber Forest Products 10, FAO, Rome

FAO (1995b) 'Report of the international expert consultation on non-timber forest products', Non-timber Forest Products 3, FAO, Rome

FAO (2000) 'Nutrition country profiles – Brazil', FAO-ESNA, Rome

FAO (2004) 'Wild edible fungi: A global overview of their use and importance to people', Non-timber Forest Products 17, FAO, Rome

Ferraro, P.J. (1994) 'Natural resource use in the southeastern rain forests of Madagascar and the local impacts of establishing the Ranomafana National Park', MSc thesis, Duke University, Durham, NC

Fleuret, A. (1979) 'The role of wild foliage in the diet: A case study from Lushoto, Tanzania', *Ecology of Food and Nutrition* 8(2): 87–93

Fleuret, A. (1986) 'Indigenous responses to drought in sub-Saharan Africa', *Disasters* 10(3): 224–229

Fokou, E. and Domngang, F. (1989) 'In vivo assessment of the nutritive value of proteins in situ in the leaves of *Solanum nigrum* L., *Xanthosoma* spp and *Gnetum africanum* L.', *Indian Journal of Nutrition and Dietetics* 26(12): 366–373

Fox, F.W. and Stone, W. (1938) 'The anti-scorbutic value of Kaffir beer', *South African Journal of Medical Sciences* 3: 7–14

Fuhrman, B., Rosenblat, M., Hayek, T., Coleman, R. and Aviram, M. (2000) 'Ginger extract consumption reduces plasma cholesterol, inhibits LDL oxidation, and attenuates development of atherosclerosis in atherosclerotic, apolipoprotein E-deficient mice', *Journal of Nutrition* 130(5): 1124–1131

Future Harvest (2001) 'With time running out scientists attempt rescue of African vegetable crops', news feature, 13 November, Future Harvest, Washington, DC

Gockowski, J., Mbazo'o, J., Mba, G. and Moulende, T.F. (2003) 'African traditional leafy vegetables and the urban and peri-urban poor', *Food Policy* 28: 221–235

Godoy, R., Brokaw, N. and Wilkie, D. (1995) 'The effect of income on the extraction of non-timber forest products: Model, hypotheses, and preliminary findings from the Sumu Indians of Nicaragua', *Human Ecology* 23(1): 29–52

Gomez, P.A., Halut, R. and Collin, A. (1961) 'Production de proteines animales au Congo', *Bulletin agricole du Congo* 52(4): 689–815

Hamon, P., Dumont, R., Zoundjihékpon, J., Tio-Touré, B. and Hamon, S. (1995) 'Les Ignames sauvages d'Afrique de l'Ouest: Caractères morphologiques', ORSTOM, Paris

Hankins, A. (1998) 'Producing and marketing wild simulated ginseng in forest and agroforestry systems', paper presented at the North American Conference on Enterprise Development through Agroforestry: Farming the Agroforest for Specialty Products, 4–7 October, Minneapolis, MN

Hardenbergh, S.H.B. (1997) 'Why are boys so small? Child growth, diet and gender near Ranomafana, Madagascar', *Social Science and Medicine* 44(11): 1725–1738

Hartley, C.W.S. (1988) *The Oil Palm*, Longman Scientific and Technical, Harlow, UK

Hatløy, A., Torheim, L.E. and Oshaug, A. (1998) 'Food variety – a good indicator of nutritional adequacy of the diet? A case study from an urban area in Mali, West Africa', *European Journal of Clinical Nutrition* 52: 891–898

Hladik, A., Bahuchet, S., Ducatillion, C. and Hladik, C.M. (1984) 'Les plantes à tubercules de la forêt dense d'Afrique centrale', *Revue d'Ecologie (Terre et Vie)* 39: 249–290

Hladik, A. and Dounias, E. (1993) 'Wild yams in the African forest as potential food resources', in C.M. Hladik, A. Hladik, O.F. Linares, H. Pagezy, A. Semple and M. Hadley (eds) *Tropical Forests, People and Food: Biocultural Interactions and Applications to Development*, UNESCO, Paris/Parthenon Publishing, Carnforth, UK, pp163–176

Hladik, C.M., Bahuchet, S. and de Garine, I. (1990) *Food and Nutrition in the African Rain Forest*, UNESCO/MAB, Paris

Hoskins, M. (1990) 'The contribution of forestry to food security', *Unasylva* 41(1): 3–13

Isaia, R. (1995) 'Rapport de mission dans le cadre du programme sur l'utilisation de la forêt à Ambanizana', in Projet Masoala (ed) *Proposition des limites du Parc National Masoala, Soumise à la Direction des Eaux et Forêts*, Care International Madagascar/Wildlife Conservation Society/Perigrine Fund Projet Masoala, Antananarivo, Madagascar

Jodha, N.S. (1990) 'Rural common property resources: Contributions and crises', *Economic and Political Weekly* 25: A65–A78

Johns, T. (1999) 'Plant constituents and the nutrition and health of indigenous peoples', in V.D. Nazarea (ed) *Ethnoecology: Situated Knowledge/Located Lives*, University of Arizona Press, Tucson, AZ

Johns, T. and Chapman, L. (1995) 'Phytochemicals ingested in traditional diets and medicines as modulators of energy metabolism', in J.T. Arnason and R. Mata (eds) *Phytochemistry of Medicinal Plants: Recent Advances in Phytochemistry 29*, Plenum Press, New York, NY, pp161–188

Johns, T. and Eyzaguirre, P.B. (2000) 'Nutrition for sustainable environments', *SCN News* 21: 24–29

Johns, T. and Maundu, P. (2006) 'Forest biodiversity, nutrition and population health in market-oriented food systems', *Unasylva* 224(57): 34–40

Johns, T., Nagarajan, M., Parkipuny, M.L. and Jones, P.J.H. (2000) 'Maasai gummivory: Implications for paleolithic diets and contemporary health', *Current Anthropology* 41(3): 453–459

Johns, T. and Sthapit, B.R. (2004) 'Biocultural diversity in the sustainability of developing country food systems', *Food and Nutrition Bulletin* 25: 143–155

Justi, K.C., Visentainer, J.V., Evelazio de Souza, N. and Matsushita, M. (2000) 'Nutritional composition and vitamin C stability in stored camu-camu (*Myrciaria dubia*) pulp', *Archivos latinoamericanos de nutrición* 50(4): 405–408

Kahn, F. (1993) 'Amazonian palms: Food resources for the management of forest ecosystems', in C.M. Hladik, A. Hladik, O.F. Linares, H. Pagezy, A. Semple and M. Hadley (eds) *Tropical Forests, People and Food*, Man and the Biosphere Series (Vol. 13), UNESCO, Paris, pp153–176

Karmann, M. and Lorbach, I. (1996) 'Utilization of non-timber tree products in dryland areas: Examples from southern and eastern Africa', in 'Domestication and commercialization', Non-timber Forest Products 9, FAO, Rome

Kengni, E. (2003) 'Food value of fruits from indigenous fruit trees of the lowland humid tropics of West and Central Africa: A case study of *Irvingia gabonensis* and *Dacryodes edulis* in Cameroon', PhD thesis, University of Yaounde, Cameroon

Kengni, E., Mbofund, C.M.F., Tchouanguep, M.F. and Tchoundjeu, Z. (2004) 'The nutritional role of indigenous foods in mitigating the HIV/AIDS crisis in West and Central Africa', *International Forestry Review* 6(2): 149–160

Kinyuy, W. (2001) 'Cameroon: Traditional food preparations improve immune system of HIV/AIDS patients', World Bank, Washington, DC

Kuhnlein, H.V. and Receveur, O. (1996) 'Dietary change and traditional food systems of indigenous peoples', *Annual Review of Nutrition* 16: 417–442

Leakey, R.R.B., Atangana, A.R., Kengni, E., Waruhiu, A.N., Usuro, C., Anegbeh, P.O. and Tchoundjeu, Z. (2002) 'Domestication of *Dacryodes edulis* in West and Central Africa: Characterisation of genetic variation', *Forests, Trees and Livelihoods* 12: 57–72

Lee, R.B. (1973) 'Mongongo: The ethnography of a major wild food resource', *Ecology of Food and Nutrition* 2: 307–321

McCune, L.M., Owen, P. and Johns, T. (2005) 'Flavonoids, xanthones and other antioxidant polyphenols', in A. Soumyanath (ed) *Medicines for Modern Times: Antidiabetic Plants*, CRC Press, Boca Raton, FL

McMichael, A.J. (2000) 'The urban environment and health in a world of increasing globalization: Issues for developing countries', *Bulletin of the World Health Organization* 78(9): 1117–1126

McSweeney, K. (2003) 'Tropical forests as safety nets? The relative importance of forest product sale as smallholder insurance, eastern Honduras', proceedings of CIFOR International Conference on Rural Livelihoods, Forests and Biodiversity, 19–23 May, Bonn, Germany

McVaugh, R. (1969) 'Botany of the Guyana highland', *Memoirs of the New York Botanical Garden* 18(2): 55–286

Maguire, L.S, O'Sullivan, S.M., Galvin, K., O'Connor, T.P. and O'Brien, N.M. (2004) 'Fatty acid profile, tocopherol, squalene and phytosterol content of walnuts, almonds, peanuts, hazelnuts and the macadamia nut', *International Journal of Food Sciences and Nutrition* 55(3): 171–178

Malaisse, F. (1997) *Se nourrir en forêt claire africaine, Approche écologique et nutritionelle*, Les Presses Agronomiques de Gembloux, Gembloux, Belgium

Malaisse, F. and Parent, G. (1985) 'Edible wild vegetable products in the Zambian woodland area: A nutritional and ecological approach', *Ecology of Food and Nutrition* 18: 43–82

Martin, M. (1985) 'Design of a food intake study in two Bambara villages in the Segou region of Mali with preliminary findings', in A. Hill (ed) *Population, Health and Nutrition in the Sahel*, KPI Limited, London

Maundu, P. (1987) 'The importance of gathered fruits and medicinal plants in Kakuyuni and Kathama areas of Machakos', in K.K. Wachiira (ed) *Women's Use of Off-farm and Boundary Lands: Agroforestry Potentials*, ICRAF, Nairobi, Kenya, pp56–60

Maundu, P., Berger, D.J., ole Saitabau, C., Nasieku, J., Kipelian, M., Mathenge, S.G., Morimoto, Y. and Höft, R. (2001) 'Ethnobotany of the Loita Maasai', People and Plants Working Paper 8, UNESCO, Paris

May, P.H., Anderson, A.B., Frazao, J.M.F. and Balick, M.J. (1985) 'Babassu palm in the agroforestry systems in Brazil's mid-north region', *Agroforestry Systems* 3(39): 275–295

Mendoza, O., Picón, C., Gonzáles, T.J., Cárdanas, M.R., Padilla, T.C., Mediávilla, G.M., Lleras, E. and de la Delgado, F.F. (1989) 'Informe de la expedición de recolección de germoplasma de camu-camu (*Myrciaria dubia*) en la amazonía peruana', Informe Técnico no. 11, Programa de Investigación en Cultivos Tropicales, INIA, Lima

Menzel, P. and D'Aluisio, F. (1998) *Man Eating Bugs: The Art and Science of Eating Insects*, Ten Speed Press, Berkeley, CA

Mialoundama, F. (1993) 'Nutritional and socio-economic value of *Gnetum* leaves in Central African forest', in C.M. Hladik et al (eds) *Tropical Forests, People and Food: Biocultural Interactions and Applications to Development*, Parthenon Publishing Group, Carnforth, UK

Mialoundama, F. and Paulet, P. (1986) 'Regulation of vascular differentiation in leaf primordia during the rhythmic growth of *Gnetum africanum*', *Canadian Journal of Botany* 64(1): 208–213

Murphy, S.P. and Allen, L.H. (2003) 'Nutritional importance of animal source foods', *Journal of Nutrition* 133: 3932S–3935S

Nepstad, D.C. and Schwartzman, S. (eds) (1992) 'Non-timber forest products from tropical forests: Evaluation of a conservation and development strategy', *Advances in Economic Botany* 9, The New York Botanical Garden, New York

Newman, J. (1975) 'Dimensions of the Sandawe diet', *Ecology of Food and Nutrition* 4(1): 33–39

Odgen, C. (1990) 'Building nutritional considerations into forestry development efforts', *Unasylva* 41(1): 20–28

Oiye, S., ole Simel, J., Oniang'o, R. and Johns, T. 'Indigenous peoples' food systems, food security and health: Maasai', unpublished results

Okafor, J.C. (1983) 'Horticulturally promising indigenous wild plant species of the Nigerian forest zone', *Acta Horticulturae* 123: 165–176

Okafor, J.C. (1997) 'Conservation and use of traditional vegetables from woody forest species in southeastern Nigeria', in L. Guarino (ed) *Traditional African Vegetables: Proceedings of the IPGRI International Workshop on Genetic Resources of Traditional Vegetables in Africa, Conservation and Use*, ICRAF-HQ, Nairobi/Institute of Plant Genetic and Crop Plant Research, Rome, pp31–38

Okafor, J.C., Okolo, H.C. and Ejifor, M.A.N. (1996) 'Strategies for enhancement of utilisation potential of edible woody forest species of south-eastern Nigeria', in L.J.G. van der Maesen, X.M. Van Der Burgt, and J.M. van Medenbach de Rooy (eds) *The Biodiversity of African Plants: Proceedings of the XIVth AETFAT Congress*, Kluwer, Wageningen, The Netherlands, pp684–695

Ouabonzi, A., Bouillant, M.L. and Hapin, J. (1983) 'C-Glycosyflvones from *Gnetum buchholzianum* and *Gnetum africanum*', *Phytochemistry* 22(11): 2632–2633

Peters, C.M. and Pardo-Tejeda, E. (1982) '*Brosimum alicastrum* (Moraceae) uses and potential in Mexico', *Economic Botany* 36(2): 166–175

Pooley, E. (1993) *The Complete Field Guide to Trees of Natal, Zululand and Transkei*, Natal Flora Publications Trust, Durban, South Africa

Pralhad, R. (1989) 'Diet and nutrition during drought – an Indian experience', *Disasters* 13: 58–72

Puleston, D.E. (1968) '*Brosimum alicastrum* as a subsistence alternative for the classic Maya of the central southern lowlands', thesis, Department of Anthropology, University of Pennsylvania, Philadelphia, PA

Puleston, D.E. (1982) *The Role of Ramón in Maya Subsistence*, Academic Press, New York, NY

Quebedeaux, B. and Bliss, F.A. (1988) *Horticulture and Human Health, Contributions of Fruits and Vegetables: Proceedings of First International Symposium on Horticulture and Human Health*, Prentice Hall, Englewood, NJ

Quebedeaux, B. and Eisa, H.M. (1990) 'Horticulture and human health, contributions of fruits and vegetables: Proceedings of Second International Symposium on Horticulture and Human Health', *HortScience* 25: 1473–1532

Ramos-Elorduy, J. (1991) 'Insects as tropical forest peoples' food', paper presented to the International Symposium on Food and Nutrition in the Tropical Forest: Biocultural Interactions and Applications to Development, 10–13 September, UNESCO, Paris

Ramos-Elorduy, J., Pino, J.M. and Gonzales, M.O. (1981) 'Digestibilidad in vitro de algunos insectos comestibles de Mexico II', *Folia Entomologica Mexicana* 49: 141–154

Reardon, T., Berdegué, J. and Escobar, G. (2001) 'Rural nonfarm employment and incomes in Latin America: Overview and policy implications', *World Development* 29(3): 395–409

Reardon, T., Timmer, C.P., Barrett, C.B. and Berdegue, J. (2003) 'The rise of supermarkets in Africa, Asia, and Latin America', *American Journal of Agricultural Economics* 85(5): 1140–1146

Reddy, S.R.C. and Chakravarty, S.P. (1999) 'Forest dependence and income distribution in a subsistence economy: Evidence from India', *World Development* 27(7): 1141–1149

Richards, P. (1986) *Coping with Hunger*, Allen and Unwin, London

Robson, J.R.K. (1976) 'Changing food habits in developing countries', *Ecology of Food and Nutrition* 4: 251–256

Roca, N.A. (1965) 'Estudio químico-bromatológico de la *Myrciaria paraensis*', Berg, Thesis, Faculty of Chemistry, Universidad Nacional Mayor San Marcos. Lima, Peru

Rodriguez-Amaya, D.B. (1999) 'Latin American food sources of carotenoids', *Archivos Latinamericanos de Nutrition* 49: 74S–84S

Roodt, V. (1998) *Trees and Shrubs of the Okavango Delta, Medicinal Uses and Nutritional Value, The Shell Field Guide Series: Part I*, Shell Oil Botswana, Gaborone, Botswana

Ruel, M.T., Garrett, J.L., Morris, S.M., Maxwell, D., Oshaug, A., Engle, P., Purima, M., Slack, A. and Haddad, L. (1998) 'Urban challenges to food and nutrition security: A review of food security, health and caregiving in the cities', IFPRI FCND Discussion Paper No. 51, International Food Policy Research Institute, Washington, DC

Saka, J.D.K. and Msonthi, J.D. (1994) 'Nutritional value of edible fruits of indigenous wild trees in Malawi', *Forest Ecology and Management* 64: 245–248

Schreckenberg, K., Degrande, A., Mbosso, C., Boli Baboule, Z., Boyd, C., Enyong, L., Kanmegne, J. and Ngong, C. (2002) 'The social and economic importance of *Dacryodes edulis* (G Don) H.J. Lam in Southern Cameroon', *Forests, Trees and Livelihoods* 12(1–2): 15–40

Scoones, I., Melnyk, M., and Pretty, J.N. (1992) *The Hidden Harvest: Wild Foods and Agricultural Systems: A Literature Review and Annotated Bibliography*, IIED, London/Swedish International Development Authority and WWF, Gland, Switzerland

Shackleton, C.M. and Shackleton, S.E. (2003) 'Value of non-timber forest products and rural safety nets in South Africa', Proceedings of CIFOR International Conference on Rural Livelihoods, Forests and Biodiversity, 19–23 May, Bonn, Germany

Shackleton, C.M., Dzerefos, C.M., Shackleton, S.E. and Mathabela, F.R. (2000) 'The use of and trade in indigenous edible fruits in the Bushbuckridge savanna region, South Africa', *Ecology of Food and Nutrition* 39: 225–245

Shackleton, S.E., Shackleton, C.M., Dzerefos, C.M. and Mathabela, F.R. (1998) 'Use and trading of wild edible herbs in the central lowveld savanna region, South Africa', *Economic Botany* 52: 251–259

Shimbo, S., Kimura, K., Imai, Y., Yasumoto, K., Yamamoto, K., Kawamura, S., Watanabe, T., Iwami, O., Nakatsuka, H. and Ikeda, M. (1994) 'Number of food items as an indicator of nutrient intake', *Ecology of Food and Nutrition* 32: 197–206

Shone, A.K. (1979) 'Notes on the marula', *Department of Water Affairs and Forestry Bulletin* 58: 1–89

Shriar, A.J. (2002) 'Food security and land use deforestation in northern Guatemala', *Food Policy* 27: 395–414

Slattery, M., Berry, T., Potter, J. and Caan, B. (1997) 'Diet diversity, diet composition and risk of colon cancer', *Cancer Causes and Control* 8: 872–882

Somnasang, P., Rathakette, P. and Rathanapanya, S. (1988) The role of natural foods in northeast Thailand', in G.W. Lovelace et al (eds) *Rapid Rural Appraisal in Northeast Thailand*, KKU-FORD Rural Systems Research Project, Khon Kaen University, Khon Kaen, Thailand

Stack, J., Dorward, A., Gondo, T., Frost, P., Taylor, F. and Kurebgaseka, N. (2003) 'Mopane worm utilization and rural livelihoods in Southern Africa', paper presented to the CIFOR Livelihood Conference, May 19–23, Bonn, Germany

Steward, J. (1963) *Handbook of South American Indians*, US Government Printing Office, Washington, DC

Styger, E., Rakotoarimanana, J.E.M., Rabevohitra, R. and Fernandes, E.C.M. (1999) 'Indigenous fruit trees of Madagascar: Potential components of agroforestry systems to improve human nutrition and restore biological diversity', *Agroforestry Systems* 46(3): 289–310

Tabuna, H. (2000) *Evaluation des échanges des produits forestiers non ligneux entre l'Afrique subsaharienne et l'Europe*, FAO, Rome

Tabuti, J.R.S., Dhillion, S.S. and Lye, K.A. (2004) 'The status of wild food plants in Bulamogi County, Uganda', *International Journal of Food Sciences and Nutrition* 55: 485–498

Tchiegang, C., Kapseu, C. and Mapongmetsem, P.M. (1998) 'Interet nutriotionnel de la domestication d'une Euphorbiaceae: le *Ricinodendron heudelotii* (Bail.)', in B. Duguma and B. Mallet (eds) *Regional Symposium on Agroforestry Research and Development in the Humid Lowlands of West and Central Africa*, Yaounde, Cameroon/CIRAD, Montpellier, France, pp105–112

Terauchi, R., Chikaleke, V.A., Thottappily, G. and Hahn, S.K. (1992) 'Origin and phylogeny of Guinea yams as revealed by RFLP analysis of chloroplast DNA and nuclear ribosomal DNA', *Theoretical and Applied Genetics* 83: 743–751

Thompson, J.E.S. (1930) 'Ethnology of the Mayas of southern and central British Honduras', *Field Museum of Natural History Anthropological Series* 17(2)

Tiwari, R.J. and Banafar, R.N.S. (1995) 'Studies on the nutritive constituents, yield and yield attributing characters in some ber (*Zizyphus jujuba*) genotypes', *Indian Journal of Plant Physiology* 38(1): 88–89

Toulmin, C. (1986) 'Access to food, dry season strategies and household size amongst the Bambara of central Mali', *IDS Bulletin* 17(3): 58–66

Townson, I.M. (1995) 'Forest products and household incomes: A review and annotated bibliography', Tropical Forestry Papers No. 31, CIFOR and Oxford Forestry Institute, Oxford, UK

Turk, D. (1995) *A Guide to Trees of Ranomafana National Park and Central Eastern Madagascar*, USAID/ANGAP, Antananarivo, Madagascar

Udosen, E. (1995) 'Proximate and mineral composition of some Nigerian vegetables', *Discovery and Innovation* 7(4): 383–386

Ulijaszek, S.J. (1983) 'Palm sago (*Metroxylon* sp) as a subsistence crop', *Journal of Plant Foods* 5(3): 115–134

UNICEF (2006) *The State of the World's Children*, available at www.unicef.org/infobycountry/

van Wyk, B.-E. and Gericke, N. (2000) *People's Plants: A Guide to Useful Plants of Southern Africa*, Briza Publications, Pretoria, South Africa

Vladyshevskiy, D.V., Laletin, A.P. and Vladyshevskiy, A.D. (2000) 'Role of wildlife and other non-wood forest products in food security in central Siberia', *Unasylva* 202: 46–52

Wargovich, M.J. (2000) 'Anticancer properties of fruits and vegetables', *HortScience* 35: 573–575

Watt, J.M.A. and Breyer-Brandwijk, M.G. (1962) *The Medicinal and Poisonous Plants of Southern and Eastern Africa*, E. and S. Livingstone, Edinburgh, Scotland

Wynberg, R.P., Cribbins, J., Leakey, R., Lombard, C., Mander, M., Shackleton, S. and Sullivan, C. (2003) 'The knowledge on *Sclerocarya birrea* subsp. Caffra with emphasis on its importance and non-timber forest product in South and southern Africa: A summary, Part 2: Commercial use, tenure and policy, domestication, intellectual property rights and benefit-sharing', *Southern African Forestry Journal* 196: 67–77

5

Wood: The Fuel that Warms You Thrice

Kirk R. Smith

'Wood is the fuel that warms you twice', goes an old New England expression: 'once when chopping and once when burning'. Unfortunately, however, wood seems to have the potential to generate heat a third time, because the smoke from its burning is a major risk factor for respiratory infections and the fever that often accompanies them. It is now believed that the burning of simple household biomass fuels – wood, but also fuels derived from trees, crops, animal dung, shrubs, grasses and root plants – is responsible for some 1.4 million (range: 1 million to 2 million) premature deaths annually, mainly in women and young children of developing countries (Smith et al, 2004). Household use of coal, another solid fuel that produces significant pollution, is responsible for another 200,000 premature deaths a year, mostly in China. This places indoor air pollution from household fuels as the second most important environmental risk factor globally, after poor water and sanitation, being responsible for a health burden well above that from all outdoor air pollution in cities, and tenth among risk factors of all kinds (Ezzati et al, 2002; WHO, 2002; Figure 5.1, overleaf).

When first hearing of the risks from woodsmoke, many people are sceptical because of our long-term association with this natural material. Indeed, it could even be said that the smell of woodsmoke from the hearth is exactly as old as humanity itself, since many anthropologists define the point at which we became 'human' as the moment when our ancestors learned to control fire. 'Natural', however, does not necessarily mean 'benign', and only a rather narrow view of global and historical environments would allow a conclusion that 'natural' is always good for health. Most of humanity has spent most of history trying to protect itself from environmental hazards. Nostalgia triggered by the sight and smell of an open wood fire in the hearth has tricked us into complacency about this source of risk in the past, and it continues to do so today.

Exposure to woodsmoke

Chemically, wood is nearly all carbon, hydrogen and oxygen and – unlike the other major solid fuel, coal – itself contains essentially no toxic materials (Smith, 1987). Thus, in special combustion conditions it can be burned completely to non-toxic carbon dioxide and water. Unfortunately, however, in simple household stoves, combustion is far from complete, and wood releases much of its carbon as products of incomplete combustion, respirable particles, volatile organic chemicals and carbon monoxide.

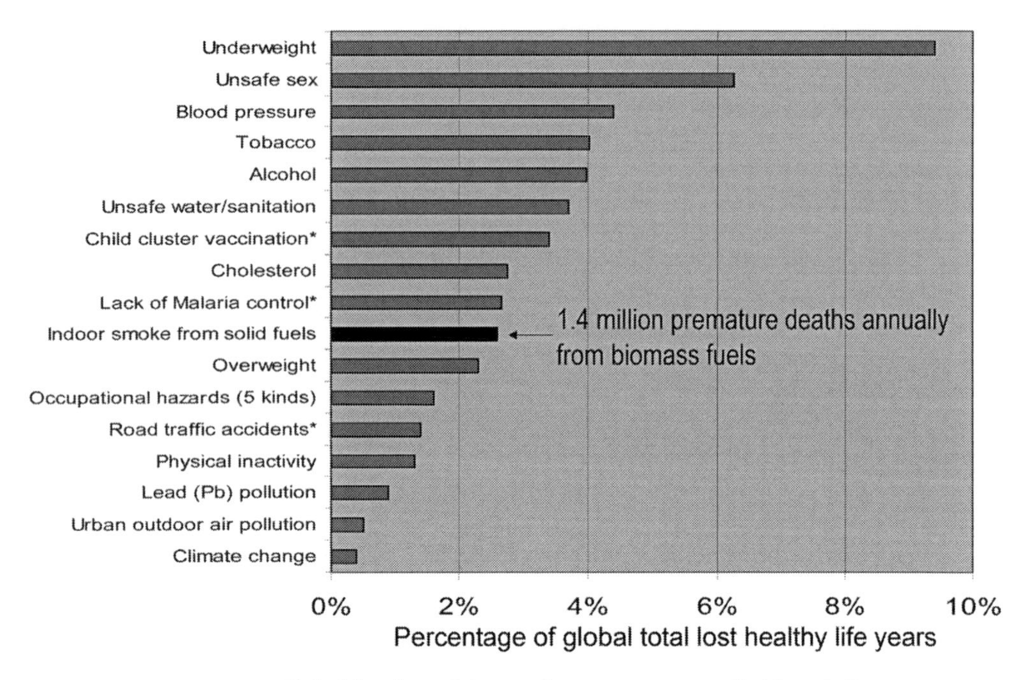

Figure 5.1 *Global burden of disease from major controllable risk factors*

Studies of typical Indian and Chinese biomass stoves, for example, show that 5 to 20 per cent of the carbon is diverted to such products (Smith et al, 2000; Zhang et al, 2000). Because of the complex chemistry created by the highly varying levels of temperature, oxygen and residence times in the flame zone of a simple wood fire, a vast range of compounds are routinely emitted. Thousands have been identified, many dozens of which are known from laboratory, animal or human studies to have toxic effects (Naeher et al, 2007). Box 5.1 lists major categories of materials found in woodsmoke, with a specific example of a major toxic chemical for each. Although there are differences in emissions from different species of wood and other biomass varieties, none burn without significant emissions of these substances in simple, small-scale combustion (Naeher et al, 2007).

The health burden created by a toxic material, however, is due not only to toxicity but also to exposure. No matter how toxic the mixture, if few people actually breathe it, the overall effect will be small. In the words attributed to Paracelsus, the Renaissance pioneer of toxicology and environmental health, 'the dose makes the poison' (Binswanger and Smith, 2000).

Besides containing noxious materials, woodsmoke is created by a process that is perniciously optimal for producing large human exposures. Nearly half the human race is thought still to rely on simple solid fuels for most of their household energy needs. Figure 5.2 shows estimates of the fraction of households using solid fuels in countries around the world, based on a combination of household survey data and econometric modelling (Smith et al, 2004). Only in China does any appreciable fraction (~25 per cent) of

BOX 5.1 MAJOR CHEMICAL CLASSES OF WOODSMOKE CONSTITUENTS

Examples of particular chemicals with known human toxic properties are shown in italics.

- Criteria air pollutants (national standards in most countries):

  small particles (PM2.5, those less than 2.5 microns in diameter), *carbon monoxide, nitrogen dioxide*

- Hydrocarbons

  25+ saturated hydrocarbons such as *n-hexane*
  40+ unsaturated hydrocarbons such as *1,3 butadiene*
  28+ mono-aromatics such as *benzene** and *styrene*
  20+ polycyclic aromatics* such as *benzo(α)pyrene**

- Oxygenated organics

  20+ aldehydes including *formaldehyde** and *acrolein*
  25+ alcohols and acids such as *methanol*
  33+ phenols such as *catechol* and *cresol*
  Many quinones such as *hydroquinone*
  Semiquinone-type and other radicals

- Chlorinated organics such as *methylene chloride* and *dioxin**

*Classified as Group I, 'known human carcinogen' by the International Agency for Research on Cancer.

Source: Naeher et al, 2007

solid-fuel-using households use coal; most households use biomass, with something like 50 to 60 per cent being in the form of wood (logs, branches, twigs) and most of the rest being agricultural residues.

Essentially all this wood fuel is used in simple household stoves, mostly for cooking but also for other tasks such as animal food preparation and, in temperate and highland areas, space heating. An unknown but high percentage of this burning is in unvented stoves – no chimney or hood – that release the smoke directly into the living area of the house. The result is indoor concentrations of many noxious pollutants that are high by any criterion: WHO guidelines, national health-based air pollution standards, urban pollution levels anywhere in the world, and any but the dirtiest industrial workplaces. Small particles, for example, are thought to be the best single indicator of risk for such combustion smokes. Typical long-term levels in households with open cookstoves are between hundreds of micrograms per cubic metre ($\mu g/m^3$) to 1000 or 2000$\mu g/m^3$, with peaks during cooking some five to ten times higher. The WHO's recent revision of its Global Air Quality Guidelines, by comparison, calls for a maximum of 10$\mu g/m^3$ for long-term exposure, and the new US Environmental Protection Agency (EPA) standard is 15$\mu g/m^3$ (WHO, 2005).

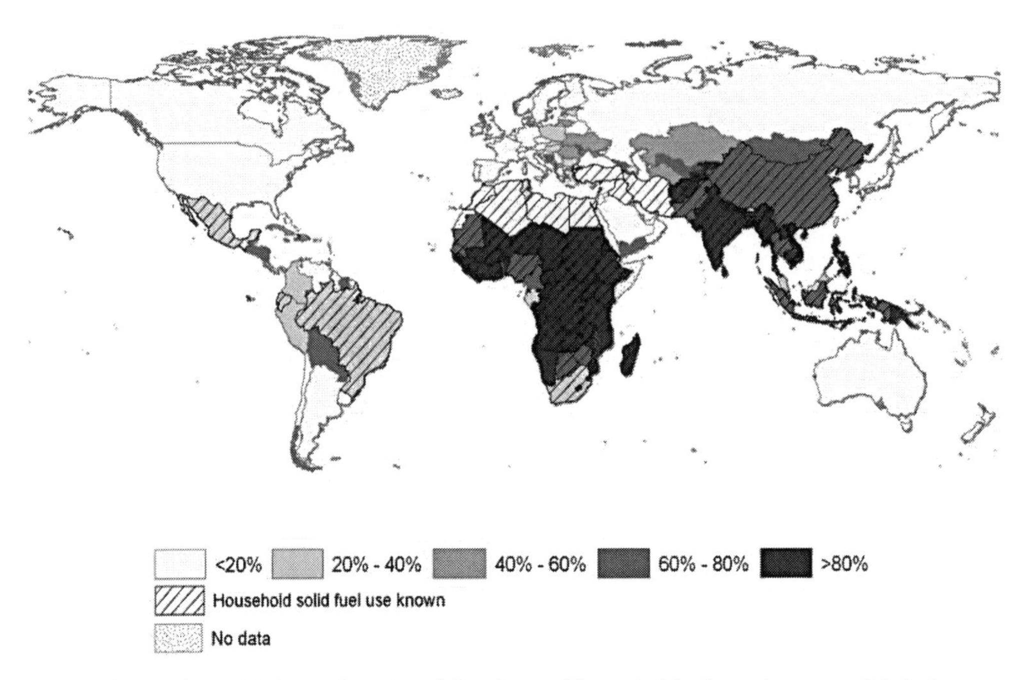

Figure 5.2 *Estimated national fractions of households dependent on solid fuels*

Here, then, is the 'perfect storm' combination for producing exposure: an essential task done two or three times every day in half the world's households releases a toxin-containing mixture directly into the households just at the times when people are present. When one considers also that it is women in nearly all cultures who are responsible for most cooking and for most care of small children, it follows that two of the most vulnerable population groups are most heavily affected. Given that it is the urban and rural poor who are most dependent on solid fuels and simple stoves, it also most heavily exposes populations with among the lowest socioeconomic (and political) status globally (Wilkinson et al, 2007).

NEED FOR EPIDEMIOLOGIC EVIDENCE

In a developed country, toxicological and exposure evidence of the sort noted above would be more than enough in themselves to warrant major and urgent protective action, even without detailed direct evidence on human health effects. Hundreds if not thousands of epidemiologic studies in Europe, Asia, North America and Latin America detail the health impact of the same pollutants in urban outdoor settings. These studies are remarkably consistent around the world (Cohen et al, 2004; ISOC, 2004; PAHO, 2005) and show major health effects even at particle levels as low as 10μg/m^3. A recent review by the WHO, for example, found that a long-term drop from 25 to 15μg/m^3 produces a 6 per cent drop in non-accident-related mortality rates (WHO, 2005).

Woodsmoke particles are somewhat different from the fossil fuel particles that pollute the atmosphere in developed-country cities, where these studies have mostly been done, and thus might be thought to have less effect. Although studies of pure woodsmoke exposures in outdoor settings are difficult to do, current toxicological, controlled human exposure, occupational and epidemiologic evidence is not supportive of treating woodsmoke particles as significantly different (Samet and Brauer, 2006; Naeher et al, 2007).

Seeing the high exposures in developing-country households, air pollution specialists may wonder why more research would be needed. After all, we know that particle levels 10 or 100 times lower cause significant effects in healthy populations (WHO, 2006b). Rather than waste time pinning down the effects precisely, it could be argued, the needed interventions could be deployed now. Indeed, it has even been argued that it would be unethical to do intervention trials because we already know that there must be important impacts (Last, 1992).

Perhaps paradoxically, however, in developing countries where health and environmental conditions are worst, the need for strong evidence is greatest. This is because of the extreme scarcity of resources for addressing health problems. In 2003, for example, India spent just US$7 per capita on health (WHO, 2006a), and thus those who decide on its allocation must be extremely pragmatic and highly critical of the evidence to ensure that available funds will be spent well. To argue within the health community that funds should be reallocated to improved fuel or stoves and, consequently, away from antibiotics, vaccines and clean water/sanitation, for example, requires much more than just evidence from laboratory studies and extrapolation of human health studies in developed countries. It requires the highest-quality biomedical research done directly with the populations of concern: developing-country households that depend on biomass.

CURRENT EPIDEMIOLOGIC EVIDENCE

Current epidemiologic evidence can be divided into three categories:

1 Well-accepted, although not completely quantified: supporting evidence from studies of active smoking, passive smoking and outdoor air pollution; human exposure and animal studies; and more than a dozen good epidemiologic studies in developing-country biomass-using households.
2 Highly suggestive: some supporting evidence from other sources and at least three epidemiological studies in biomass-using households.
3 Speculative: strong evidence from tobacco and outdoor air pollution studies, but no relevant studies yet in biomass-using households.

The major diseases associated with biomass smoke and a summary of the evidence are shown in Table 5.1.

Table 5.1 Health effects of the use of solid household fuels in developing countries

Disease	*Population affected*	*Relative risk (95% confidence interval)*	*Strength of evidence*
Chronic obstructive pulmonary disease	Females >15 years	3.2 (2.3, 4.8)*	Strong
	Males >15 years	1.8 (1.0, 3.2)*	Intermediate
Acute lower respiratory infections	Children <5 years	2.3 (1.9, 2.7)*	Strong
Lung cancer (coal smoke only)	Women >15 years	1.9 (1.1, 3.5)*	Strong
	Men >15 years	1.5 (1.0, 2.5)*	Intermediate
Blindness (cataracts)	Females >15 years	1.3–1.6**	Intermediate
Tuberculosis	Females >15 years	1.5–3.0**	Intermediate

Notes: For illustration, a relative risk of 1.5 indicates that a population living in solid fuel-burning households have a rate of the disease in question 1.5 times that of people living in clean fuel burning households.

* Based on formal meta-analyses.

** Range of results in published studies.

Source: Based on review and meta-analysis of published epidemiologic studies; Smith et al, 2004

Well-accepted effects

Chronic obstructive pulmonary disease (COPD)

Apparently, the first people to identify woodsmoke exposures as a hazard for women cooks were local physicians concerned with a particular kind of heart disease commonly thought to be nearly entirely due to smoking. *Cor pulmonale*, which is often fatal, is an unbalanced and enlarged heart, usually secondary to chronic lung disease. Independently in the 1970s, a general practitioner in India and a cardiologist in Nepal noted high rates of this condition in relatively young non-smoking women and traced it to correspondingly high rates of COPD, such as chronic bronchitis, in rural non-smoking women cooking with biomass fuels. Subsequently, there have been several dozen published studies pinning down this relationship, not only in terms of diagnosed COPD, but also in lung function and other changes that precede the development of this disease. COPD is a highly disabling disease with no known cure and inadequate palliative treatments available for low-income populations. It is also one of the most important causes of premature death in the world, killing perhaps 2.7 million annually (WHO, 2004). Although much of the overall burden is due to smoking, it is now thought also to be an outcome of long-term exposure to biomass smoke in developing-country households, particularly for women cooks. Of the premature mortality identified in the WHO-managed comparative risk assessment studies, as summarized in the lost life-years shown in Figure 5.1, about one-third is due to COPD.

Acute lower respiratory infections (ALRI)

The largest single impact of biomass smoke exposures is thought to be ALRI in young children, who receive most of their exposure while being cared for by their cooking mothers. As pneumonia, ALRI is the chief cause of death among the world's children, killing nearly 2 million per year. All children around the world experience the usually self-limiting and rarely fatal acute *upper* respiratory infections at similar rates, but the rate of child ALRI and pneumonia in developing countries is hundreds of times greater than in developed countries. Because it affects the youngest members of the population, globally it causes more lost life-years than any other disease (WHO, 2004).

For such an important disease, unfortunately, neither its microbiology nor its risk factors are well understood. In general, most fatal ALRI infections are thought to be due to organisms that are present all the time in the environments where poor children live, rather than being the result of epidemics. The main proximate causes, therefore, are those that affect the ability of children's immune systems to withstand infection and to limit the severity of infections when they occur. General malnutrition, along with specific micronutrient deficiencies, are thought to be the main problems. It is also known that children weakened by other diseases, including measles, diarrhoea and HIV, are more likely to contract ALRI. Insufficient parental education and access to emergency medical care are clear factors in ALRI mortality, which can occur within a day or two of a malnourished child becoming ill. Even including such suggested factors as crowding and chilling, however, established risk factors do not completely account for the worldwide incidence and mortality of this dangerous disease.

Starting with early (although negative) studies in Papua New Guinea in the 1970s, continuing with a few studies in the 1980s, and occurring at an increasing rate in the 1990s and this decade, several dozen studies have been published in the international biomedical literature linking household biomass smoke exposures with signs and symptoms of child respiratory infections and pneumonia. A review of such studies as part of the WHO comparative risk assessment found that children living in households that burned solid fuel seemed to have about twice the disease rate of those living in households with cleaner fuels (Smith et al, 2004). If studies of those carried on the mother's back during cooking are included, children in smoky conditions had about 2.3 times more risk. These calculations are the basis of the results in Figure 5.1, which estimated about 1 million premature ALRI deaths annually from biomass smoke exposures in the world, with an uncertainty of ±40 per cent.

All studies to date have been 'observational' in that they carefully observe existing populations with no attempt to conduct experiments under controlled conditions. Although easier and less expensive, observational studies are not fully able to distinguish associations from causal relationships. For the most hard-nosed funders, randomized control trials, the gold standard of epidemiology, are required to make the strongest case, as is expected for vaccine and nutrition supplement trials. Randomized trials are not possible with long-term effects, such as COPD, but can be done with relatively common acute effects, such as ALRI. They are expensive and lengthy, however, and must be conducted under strict ethical criteria.

The first randomized air pollution trial ever done with any normal population has just finished in Guatemala, and its findings for ALRI and woodsmoke are just emerging from analyses of the data (Smith et al, 2006). We found that the introduction of a well-operating chimney woodstove caused the greatest reduction in the most serious kinds of ALRI – those of bacterial origin – which are most likely to lead to death and thus lost life-years. Reduction in serious pneumonia was about 40 per cent, with a reduction of about 50 per cent in the smoke exposures of babies. The earlier observational studies probably have overstated the total effect on ALRI, perhaps because of the difficulty of distinguishing upper and lower acute respiratory infections in resource-constrained field studies, and residual confounding (confusing the effect of poverty with that of the smoke). With an intervention of improved fuels or low-emissions combustion of wood, however, the benefit would perhaps be larger.

Suggestive evidence

Evidence is mounting of several other health effects from biomass smoke exposures (Bruce et al, 2000; Smith, 2000; Smith et al, 2004). Here there is only space to briefly summarize what is known.

Tuberculosis (TB)

One type of biomass smoke exposure, that from tobacco in active smokers, is a clear risk factor for TB around the world, probably because it suppresses the body's immune systems, thereby increasing the chances that one will acquire the infection from others and that current latent infections will blossom into active TB (Lin et al, 2007; Bates et al, 2007). Woodsmoke exposures in households are much lower, of course, but evidence is growing of a similar if smaller effect. Two older studies, one in India and one in Mexico (Perez-Padilla et al, 2001), have found such an effect and more are underway. Although not likely to be a major cause of TB in itself, pinning down the effect of biomass smoke on TB would be useful because additional approaches to this ancient scourge are urgently needed. Current control measures, which are based solely on drug therapy, are inadequate in many poor countries, and TB is one of the few infectious diseases in the world that is on the increase because of its strong link with HIV and the emergence of drug-resistant strains.

Eye disease

Tobacco smoke is a risk factor for both cataracts and macular degeneration, debilitating and progressive conditions that can lead to blindness. Cataracts in particular are a problem in developing regions such as South Asia, where one-third of all cataracts occur. Known risk factors (age, sunlight, smoking) do not fully account for its prevalence. Several studies in South Asia have found a strong relationship with biomass smoke exposures, and cataracts have been triggered in rabbits by woodsmoke in laboratory studies (e.g., Pokhrel et al, 2005). Additional studies of biomass smoke and both cataracts and macular degeneration are underway in South Asia.

Cancer

Although a number of chemicals in woodsmoke are known to cause cancer in humans, there has not been definitive epidemiologic evidence yet of the risk of woodsmoke itself. It has thus recently been classified as a *probable* human carcinogen by the International

Agency for Research on Cancer (Straif et al, 2006). This is distinct from household coal smoke, which is well established as a cause of lung and other cancers from dozens of studies in China (Zhang and Smith, 2007) and classified as a *known* human carcinogen by this prestigious international agency. Tobacco smoke as well has been established as a carcinogen, not only in active smokers but also in passive smokers, who are exposed to smoke levels much lower than those in biomass-burning households. The 'probable' classification can be interpreted as indicating that woodsmoke is a weak carcinogen, and thus its effects are difficult to detect in the small epidemiologic studies that have been done to date. Given the many carcinogenic compounds in woodsmoke, as shown in Box 5.1, however, it seems likely that better and larger human studies will pin this effect down more firmly.

Adverse pregnancy outcomes

A small number of studies have shown an effect on the birth outcomes of women exposed to biomass smoke during pregnancy. Increased rates of stillbirth and low birth-weight have both been reported (WHO, 2007). Low birth-weight is a particularly prevalent and important problem in developing countries because it not only increases the chance of infant and child disease and mortality, but also seems to have a lifetime negative impact on cognitive development and health. Although several chemicals in woodsmoke are candidates for such an effect, carbon monoxide is likely to be the chief culprit because it interferes with the oxygen supply to the foetus. Given the importance of this condition, more studies are needed to pin down the effect of biomass smoke, although such studies are not easy because of confounding with nutrition, which is the major cause of low birth-weight. (Because they are poorer, biomass-using women tend to be more malnourished than those using other fuels.)

Speculative evidence

Cardiovascular disease

The chief health effect usually identified in active and passive tobacco use, as well as in outdoor air pollution studies, is cardiovascular disease. The increase from exposure may not be as great as for some other diseases, but because the background rate of heart disease is so high, the resulting public health impact is also high. No studies of cardiovascular disease seem yet to have been done in developing-country biomass-using households. Recently, however, a well-designed study in Guatemala found a definite effect on blood pressure among women cooks (McCracken et al, 2007). Although not a disease itself, of course, high blood pressure is an indicator of increased heart disease risk in all populations where it has been studied. Direct studies of woodsmoke and cardiovascular health would seem to be well warranted.

Asthma

Evidence is growing that asthma, particularly in children, is associated with outdoor air pollution in developed countries. A few studies have been done in developing-country biomass-using households, but they have not had sophisticated designs and the results are hard to interpret. Although asthma rates have increased in recent years to alarming levels in some developed countries, perhaps counter-intuitively, they are not thought to be high in most developing countries, despite some evidence of increases. The evidence overall is confusing and inconsistent, but many observers believe that asthma may actually be precipitated by clean environments (the 'hygiene hypothesis'), in that the immune systems of children in developed countries may now not be sufficiently challenged in early life to properly develop. Research in developing-country households could help throw light on this perplexing disease of worldwide interest.

BENEFITS OF WOODSMOKE

Paradoxically, household woodsmoke may provide benefits from pest control in some parts of the world. Most commonly mentioned are the continual fumigation of thatched roofing materials and mosquito repellence from smoke indoors. No studies of the former benefits seem to have been done, although of course such fumigation could be arranged at times when the house is empty and achieve the same benefit without high health-related exposures. The few studies of household biomass smoke's impact on mosquitoes indicate a drop in biting frequency, depending on biomass type burned (Paru et al, 1995). A recent detailed review supports the early finding (Snow et al, 1987) that there is no measurable impact on malaria prevalence of indoor smoke and that there is no net health benefit to maintaining smoke in malarial areas (Biran et al, 2007). Increased comfort from less biting would have to be weighed against the health impacts of the smoke and the need to find other, more effective ways of mosquito control (see Chapter 9, this volume).

WHAT CAN BE DONE?

Alleviating the health impacts of woodsmoke is not impossible. Although the ultimate cause of ill-health from woodsmoke is poverty, which prevents people from obtaining clean fuels and purchasing safe stoves, it does not necessarily follow that the best solution is poverty alleviation. The art and science of public health is finding ways of making people healthy before they are wealthy, through such 'magic bullets' as vaccines, targeted technologies such as clean water and sanitation, or women's education. Simple improvements in income, while eventually improving health, are usually extremely slow by comparison and much more expensive (inefficient) in achieving health goals. In addition, of course, economic goals will be more easily achieved with healthy and educated populations.

Four technological fixes are feasible:

1 improved ventilation of households and direct venting of stoves (chimneys);
2 improved stoves that reduce the production of pollution through better combustion of wood;

3 acceleration of the natural transition to clean fossil fuels, particularly liquefied petroleum gas (LPG); and
4 development of alternative gaseous and liquid fuels from biomass or coal that can be burned cleanly.

Efforts have been made in all four areas, but only the first and third in the list above have had significant sustained support by implementing agencies, and with only mixed success.

Improving general household ventilation by opening the eaves and related methods can somewhat reduce indoor pollution levels, but not nearly as much as the approaches above. There are also often limitations on the degree of change possible because of considerations of security, privacy and construction materials. Although it is generally true that the indoor air pollution problem is inversely related to income, however, the poorest populations often live in poor and open housing (and cook less), and thus may have somewhat lower exposures to the smoke than populations living in more substantial houses.

Perhaps the most successful effort was the National Improved Stove Programme in China, which introduced some 180 million improved stoves with chimneys over 15 years – probably one of the largest household-level development efforts in history (Sinton et al, 2004). The programme enjoyed good technical input, long-term (16 years) government support and a significant allocation of resources (although still less than a 15 per cent subsidy on average). Such characteristics have not been found in any other programme in the world.

Although focused on fuel efficiency, the Chinese programme nevertheless seems to have achieved a consistent reduction in smoke exposure in many households. Nevertheless, as has been found in smaller stove programmes, the reduction was not sufficient to bring exposure levels down to what would be considered healthy by WHO and national health-based standards (Edwards et al, 2007). The primary reason is that the stoves do not actually reduce emissions, but merely vent most of them immediately outdoors, where they can still expose people in the household environment.

Subsidizing LPG and kerosene for the poor has been a policy in many developing countries, although largely for political and equity reasons as well as a desire to reduce the impacts on natural ecosystems of fuel harvesting. Such direct subsidies, however, have generally been quite inefficient in achieving their stated goals, often operating in such a way that they do not adequately reach the poor, and at huge expense in some countries. India and Indonesia, for example, have spent more annually on such subsidies than they spend on all primary education (Smith et al, 2005). The programmes have thus become quite unpopular among development agencies and the international financial institutions.

Nevertheless, there would be significant benefits in finding smarter ways to promote these fuels among populations that are close to affording them on their own, particularly LPG, which is so clean, easy to use and efficient. The small degree to which cooking by the poor would add to the demand for petroleum by the rich, and thus exacerbate climate change and international energy supply problems, is minuscule compared with the potential benefits (Smith, 2002).

The global scale of the health impacts of woodsmoke is becoming better understood and accepted by scientists, governments and donor agencies, yet substantial scepticism about the cost-effectiveness of the available interventions remains, partly because of the mixed success of programmes to date. Several points can be made in response, however.

First, because it was done without foreign funding, technical input or coordination, the success of the Chinese stove programme is little known. China's experience can provide important lessons that could be adapted and applied today in many countries (Sinton et al, 2004).

Second, to date, the major programmes, either for improved stoves or for clean fuel promotion, have had not health as an objective, but rather control of deforestation, improved energy efficiency or fuel subsidy for the poor. Thus, it is probably not surprising that it has been hard to show health benefits for any of them.

Third, the total amount of money spent on designing, testing and implementing improved technology to reduce household smoke exposures is tiny by international comparisons: far less than the incremental cost of air pollution control on just one of the 100 coal-fired power plants being built in 2007, for example. It is not at all commensurate with the scale of the problem, which is probably greater in health terms than from all coal-fired plants put together. In addition, only sporadic and limited programmes have ever been attempted in the second and fourth technological areas noted above. Thus, perhaps it is not surprising that neither the technology nor the dissemination methods are well developed.

Finally, in common with other important environmental health interventions – for example, clean water and sanitation – the cost-effectiveness of clean fuels and stove technology is marginal for achieving health benefits alone, thus limiting the degree to which the health sector, with its many resource constraints and severe problems to address, is able to respond. Unlike vaccines and other highly targeted health interventions, however, in many populations improved household fuel technology has non-health benefits as well, including: time savings in fuel harvesting; lower pressure on natural ecosystems; improved kitchen safety, hygiene and ergonomics; the enhanced status of women; and even reduced greenhouse emissions (Smith, 1995; 2006).

The last point indicates that we need a way to evaluate and combine multiple benefits into a cogent overall package and find an agency or group of agencies willing to undertake the research and development for, and implementation of, needed interventions, even if they do not achieve any one objective in a highly effective manner but do quite well for the suite of benefits.

In the modern climate of evidence-based health and development programmes, we must verify the benefits of a stove programme or other widespread intervention designed to lower air pollution exposures and improve health. This is not an easy task for a programme that should involve millions of households spread across vast rural areas with poor infrastructure and communications. Progress is being made, however, in developing standard monitoring and evaluation methods that can be reliably applied in these settings (Smith et al, 2007).

CONCLUSION

With oil prices rising and fuel subsidies dropping in developing countries, it is likely that the health problems due to household smoke have actually increased recently; more households have been forced back down the 'energy ladder' to biomass fuels, such as wood. Even without these recent trends, analysis shows that the number of biomass households is expected to remain about constant, although their proportion of the world's population will slowly fall as population grows and simple economic growth moves more people up the energy ladder (Smith et al, 2004). Thus the scale of exposure will remain large.

Current estimates for child pneumonia, thought to be the largest health impact of household fuel use, will probably be revised downwards because of the latest research results, but new studies are being published monthly showing a range of other kinds of effects, the quantification of which will add to the health burden. Although by no means completely understood, the health impacts of such exposures are substantial.

Finding the most effective way to deal with this problem will require new technology, new social understanding and new organizational approaches, as well as clearer understanding of the multiple benefits that are possible. What is less clear is whether and how the international and local health and development communities will respond and apply the resources needed to address this ancient and significant source of human ill-health.

REFERENCES

Bates, M., Khalakdina, A., Pai, M., Chang, L., Lessa, F. and Smith, K.R. (2007) 'The risk of tuberculosis from exposure to tobacco smoke: A systematic review and meta-analysis', *Archives of Internal Medicine* 167: 335–342

Binswanger, H.C. and Smith, K.R. (2000) 'Paracelsus and Goethe: Founding fathers of environmental health', *Bulletin of the World Health Organization* 78(9): 1162–1164

Biran, A., Smith, L., Lines, J., Ensink, J. and Cameron, M. (2007) 'Smoke and malaria: Are interventions to reduce exposure to indoor air pollution likely to increase exposure to mosquitoes?', *Transactions of the Royal Society of Tropical Medicine and Hygiene* 101: 1065–1071

Bruce, N., Perez-Padilla, R. and Albalak, R. (2000) 'Indoor air pollution in developing countries: A major environmental and public health challenge', *Bulletin of the World Health Organization* 78(9): 1078–1092

Cohen, A.J., Anderson, H.R., Ostro, B., Pandey, K.D., Krzyzanowski, M., Kuenzli, N., Gutschmidt, K., Pope, C.A., Romieu, I., Samet, J.M., Smith, K.R. (2004) 'Mortality impacts of urban air pollution', in M. Ezzati, A.D. Rodgers, A.D. Lopez and C.J.L. Murray (eds) *Comparative Quantification of Health Risks: Global and Regional Burden of Disease due to Selected Major Risk Factors*, World Health Organization, Geneva, pp1353–1433

Edwards, R., Liu, Y., He, G., Yin, Z., Sinton, J., Peabody, J. and Smith, K.R. (2007) 'Household CO and PM measured as part of the review of China's National Improved Stove Program', *Indoor Air* 17(3): 189–203

Ezzati, M., Lopez, A.D., Rodgers, A., Vander Hoorn, S. and Murray, C.J.L. (2002) 'Selected major risk factors and global and regional burden of disease', *Lancet* 360(9343): 1347–1360

ISOC (International Scientific Oversight Committee) (2004) *Health Effects of Outdoor Air Pollution in Developing Countries of Asia: A Literature Review*, Special Report 15, Health Effects Institute, Boston, MA

Last, J. (1992) 'Epidemiological considerations', in K.R. Smith (ed) *Epidemiological, Social, and Technical Aspects of Indoor Air Pollution from Biomass Fuel*, World Health Organization, Geneva, Switzerland

Lin, H., Ezzati, M. and Murray, M. (2007) 'Tobacco smoke, indoor air pollution and tuberculosis: A systematic review and meta-analysis', *Plos Medicine* 7(4): 1–17

McCracken, J.P., Smith, K.R., Mittleman, M., Diaz, A. and Schwartz, J. (2007) 'Chimney stove intervention to reduce long-term air pollution exposure lowers blood pressure among Guatemalan women', *Environmental Health Perspectives* 115(7): 996–1001

Naeher, L.P., Brauer, M., Lipsett, M., Zelikoff, J.T., Simpson, C., Koenig, J.Q. and Smith, K.R. (2007) 'Woodsmoke health effects: A review', *Journal of Inhalation Toxicology* 19(1): 1–47

PAHO (2005) *An Assessment of Health Effects of Ambient Air Pollution in Latin America and the Caribbean*, Pan American Health Organization, Washington, DC

Paru, R., Hill, J., Lewis, D. and Alpers, M.P. (1995) 'Relative repellency of woodsmoke and topical applications of plant products against mosquitoes', *Papua New Guinea Medical Journal* 38(3): 215–221

Perez-Padilla, R., Perez-Guzman, C., Baez-Saldana, R. and Torres-Cruz, A. (2001) 'Cooking with biomass stoves and tuberculosis: A case control study', *International Journal of Tuberculosis and Lung Disease* 5(5): 441–447

Pokhrel, A.K., Smith, K.R., Khalakdina, A., Deuja, A. and Bates, M.N. (2005) 'Case-control study of indoor cooking smoke exposure and cataract in Nepal and India', *International Journal of Epidemiology* 34: 702–708

Samet, J.M. and Brauer, M. (2006) 'Particulate matter', in *WHO Air Quality Guidelines for Particulate Matter, Ozone, Nitrogren Dioxide, and Sulfur Dioxide: Background Materials for the Global Update 2005,* A.S. Committee, Regional Office for Europe of the World Health Organization, Copenhagen, Denmark, pp217–306

Sinton, J.E., Smith, K.R., Peabody, J.W., Liu, Y., Zhang, X., Edwards, R. and Gan, Q. (2004) 'An assessment of programs to promote improved household stoves in China', *Energy for Sustainable Development* 8(3): 33–52

Smith, K.R. (1987) *Biofuels, Air Pollution, and Health: A Global Review*, Plenum, New York, NY

Smith, K.R. (1995) 'Health, energy, and greenhouse-gas impacts of biomass combustion', *Energy for Sustainable Development* 1(4): 23–29

Smith, K.R. (2000) 'National burden of disease in India from indoor air pollution', *Proceedings of the National Academy of Sciences of the United States of America* 97(24): 13286–13293

Smith, K.R. (2002) 'In praise of petroleum?', *Science* 298(5600): 1847

Smith, K.R. (2006) 'Women's work: The kitchen kills more than the sword', in J.S. Jaquette and G. Summerfield (eds) *Women and Gender Equity in Development Theory and Practice,* Duke University Press, Durham, NC, pp202–215

Smith, K.R., Bruce, N. and Arana, B. (2006) 'RESPIRE: The Guatemala Randomized Trial', *Epidemiology* 17(6): S44–46

Smith, K.R., Dutta, K., Chengappa, C., Gusain, P.P.S., Masera, O., Berrueta, V., Edwards, E., Shields, K.N. and Bailis, R. (2007) 'Monitoring and evaluation of improved biomass cookstove programs for indoor air quality and stove performance: Conclusions from the Household Energy and Health Project', *Energy for Sustainable Development,* 15(2): 5–18

Smith, K.R., Mehta, S. and Maeusezahl-Feuz, M. (2004) 'Indoor smoke from household solid fuels', in M. Ezzati, A.D. Rodgers, A.D. Lopez and C.J.L. Murray (eds) *Comparative*

Quantification of Health Risks: Global and Regional Burden of Disease due to Selected Major Risk Factors, World Health Organization, Geneva, Switzerland, pp1435–1493

Smith, K.R., Rogers, J. and Cowlin, S.C. (2005) *Household Fuels and Ill-Health in Developing Countries: What Improvements can Be Brought by LP Gas?*, World LP Gas Association and Intermediate Technology, Paris, France

Smith, K.R., Zhang, J., Uma, R., Kishore, V.V.N., Joshi, V. and Khalil, M.A.K. (2000) 'Greenhouse implications of household fuels: An analysis for India', *Annual Review of Energy and Environment* 25: 741–763

Snow, R.W., Bradley, A.K., Hayes, R., Byass, P. and Greenwood, B.M. (1987) 'Does woodsmoke protect against malaria?', *Annals of Tropical Medicine and Parasitology* 81: 449–451

Straif, K., Baan, R., Grosse, Y., Secretan, B., El Ghissassi, F. and Cogliano, V. (2006) 'Carcinogenicity of household solid-fuel use and high-temperature frying', *Lancet-Oncology* 7: 977–978

WHO (2002) *World Health Report 2002: Reducing Risks, Promoting Healthy Life*, World Health Organization, Geneva, Switzerland

WHO (2004) *World Health Report 2004: Changing History*, World Health Organization, Geneva, Switzerland

WHO (2005) *WHO Air Quality Guidelines: Global Update for 2005*, Regional Office for Europe of the World Health Organization, Copenhagen, Denmark

WHO (2006a) *World Health Report 2006: Working Together for Health*, World Health Organization, Geneva, Switzerland

WHO (2006b) *WHO Air Quality Guidelines for Particulate Matter, Ozone, Nitrogen Dioxide, and Sulfur Dioxide: Background Materials for the Global Update 2005*, Regional Office for Europe of the World Health Organization, Copenhagen, Denmark

WHO (2007) *Indoor Air Pollution from Solid Fuels and Risk of Low Birth Weight and Stillbirth*, World Health Organization, Geneva, Switzerland

Wilkinson, P., Smith, K.R., Joffe, M. and Haines, A. (2007) 'A global perspective on energy: Health effects and injustices', Series on Energy and Health no. 1, *Lancet* 370: 5–18

Zhang, J. and Smith, K.R., (2007) 'Household air pollution from coal and biomass fuels in China: Measurements, health impacts, and interventions', *Environmental Health Perspectives*, 115(6): 848–855

Zhang, J., Smith, K.R., Ma, Y., Jiang, F., Qi, W., Liu, P., Khalil, M.A.K., Rasmussen, R.A. and Thornelow, S.A. (2000) 'Greenhouse gases and other airborne pollutants from household stoves in China: A database for emission factors', *Atmospheric Environment* 34(26): 4537–4549

6

Forest Women, Health and Childbearing[1]

Carol J. Pierce Colfer, Richard G. Dudley and Robert Gardner

In recent years, there has been a certain reluctance in some development and conservation circles to acknowledge the significance of population issues. Westerners are cognizant of their own roles in consuming the world's resources and understandably consider it inappropriate to warn others about population expansion. Additionally, there is growing recognition that, for instance, in forest management, public participation is often used for the purposes of the managers, donors and project leaders, rather than for local people's purposes (for a thorough examination of these processes as they apply to women's reproductive rights, see Braidotti et al, 1994; Reardon, 1995; Rocheleau and Slocum, 1995; Turshen, 1995). And finally, forest managers – usually men, and usually outsiders – typically feel uncomfortable dealing with the women who live in forests. Childbearing behaviour, they believe, is too personal, and outside the realm of forestry or ecological expertise. For all these reasons, many people are fearful of addressing broader population issues in remote forested areas and may even consider it unethical.

In this chapter, we begin by making the case, using arguments from Smail (2002a; 2002b), that the population situation is critical. Although population densities in forests are relatively low, the importance of keeping them low – for maintaining forest-based ways of life and biodiversity, as well as for aesthetic and moral reasons – is obvious to forest managers.

We next argue that, in fact, lower fertility among women who live in forests will effectively address many of these women's own concerns and enhance their own wellbeing (both 'real' and perceived), and we endeavour to show some of the links among population, health issues and women who live in forests. These links pertain to women outside forests as well. We hope that such clarification will stimulate others to look for creative ways to work more effectively with local forest women to stabilize population and improve local health in ways that will help both the people and the environment.

The 'Population Problem'

Smail (2002a; 2002b), following many others, has made a convincing argument that we ignore the current rate of population increase at our own (and our environment's) peril. Indeed, he argues that we need to be reducing our numbers, not just slowing the increase. He bases his thinking on five demographic observations and five observations pertaining to the Earth's carrying capacity, which we summarize here.

With regard to demography, Smail makes the following estimates:

1 The Earth's population will have grown from the current 6.2 billion to 9 billion by mid-century.
2 Despite reductions in the rate of population growth, the current total fertility rate in the developing world (3.7 children per woman)[2] is almost double that needed for eventual zero population growth.
3 Populations will be growing older, with perhaps as many as 20 to 25 per cent in the over-60 category by mid-century. Smail attributes this to falling death rates. However, although falling mortality is an element in this process, demographers have shown that falling fertility is a more important cause. Regardless, populations are ageing.
4 The quantitative scale, geographic scope, escalating pace and functional interconnectedness of these changes in population are unparalleled in human history, providing us with no precedents to guide us (Smail likens human population growth to a cancer).
5 We have a narrow window – between now and 2050 – to stabilize our population in a conscious and, we may hope, benign fashion, avoiding a Malthusian scenario.

From the environmental standpoint, Smail makes these arguments:

1 The Earth's resources are finite.
2 The Earth's true (optimal) carrying capacity – with people in long-term, adaptive balance with their environment, resource base and each other – may have already been exceeded by a factor of two.
3 About 20 per cent of the world's population has a 'generally adequate' standard of living, and the remaining 80 per cent, representing the fastest-growing populations, are also striving for higher standards of living (with the accompanying projected increase in consumption).
4 Using the equation 'impact = population × consumption × technology' (obtained from Hardin, 1999, and others), our total impact on the Earth's already strained ecosystems could easily quadruple by 2050. To quote from Smail:

> *The total impact of human numbers on the global environment is often described as the product of three basic multipliers: (1) population size; (2) per capita energy and resource consumption (level of affluence); and (3) technological efficiency in the production, utilization and conservation of such energy and resources.* (Smail, 2002b, p28)

5 There is significant potential for irreversible damage, including loss of wilderness and biodiversity, which is important on pragmatic, aesthetic and moral grounds.

Although one may quibble with one or another specific estimate in this scenario (and many do, among them the late economist Julian Simon, who won a wager with Paul Ehrlich, author of *The Population Bomb*, over whether resource costs would rise within a certain number of years), we believe there is a great deal of truth to it; sufficient to suggest that we should marshal more of our resources and energies to address this problem in a humane and effective way.

Demographers and others who specialize in population issues, like those specializing in health more generally, tend to ignore forested areas because of the comparatively minimal global, demographic effect of these sparsely populated areas. The six countries in generally well-forested Central Africa, for instance, have population densities ranging from 1.6 to 3.1 per cent (Population Reference Bureau, 2007). From the standpoint of individual forests, changes can be dramatic. In the Long Segar area of East Kalimantan, population density was estimated at two to three people per km^2 in the 1980s. By the 1990s, density had increased to around 60 per km^2 (Colfer, 1995), primarily because of government-sponsored transmigration programmes. Similarly, in the forested central and southern provinces of Cameroon, a 1987 rural population density of 17.4 persons per km^2 had increased to 22.4 persons per km^2 by 1997 (Kemajou and Sunderlin, 1999, p9), partly because of dislocations caused by the economic crisis of the early 1990s. For this large, forested area, the annual rate of population increase was 0.72 per cent between 1976 and 1987, changing to 4.10 per cent for the period 1987–1997. Leach (1994) reports rates of increase between 1963 and 1985 in the Gola forest region in Sierra Leone as ranging between 2.2 and 2.9 per cent. One study of residents in the Colombian rainforest surveyed 93 women: 20 had had more than 10 pregnancies, 6 had more than 10 living children, and 27 were pregnant or lactating (Townsend and de Acosta, 1987, p253).

The 'population problem' is vitally important for forest maintenance, and both forest managers and local women have excellent reasons for limiting population. The current trend toward more participatory forest management presents an entry point for beginning to stabilize population levels if partnerships can be developed between foresters and local women. The current public participation emphasis – on income generation alone – carries with it the possibility of greater population increases, particularly through in-migration, as improvements in local livelihoods may draw in others from afar.

Here, we attempt to portray our perceptions of the relationships among forest women, population and health using causal loop diagramming (a technique from the field of system dynamics). Both population studies and system dynamics have been lambasted dramatically by Braidotti et al (1994, pp143–147) who see system dynamics approaches to population studies as simplistic, dangerous and prone to use in justifying draconian measures to ensure fertility reduction by individual women. We acknowledge the simplistic aspect of all models, but argue that such simplification is useful, probably even necessary, in clarifying complex interrelationships. We explicitly renounce any attempts to apply force to individual women in their reproductive decisions. Instead we are seeking to understand the dynamic links among factors that can – by means of individual choices – benefit both women and forests.

Our purpose is to counter a common perception that may over-simplify the relationships among forests, population and health and foster a passive approach among forest managers. This common (Malthusian) view effectively paralyses foresters, biologists and ecologists when they think of human population issues. We then provide our view of the interactions, built on an extensive literature review and long-term field experience. Our interpretation provides a constructive approach and optimistic view for dealing with population reduction, environmental improvement and human wellbeing.

RETHINKING MALTHUS

In 1994, Emery Roe (1994) introduced the idea of a policy narrative. A policy narrative is a kind of simplified 'story' about how the world works. Because reality is too complex for policy-makers (or anyone else) to understand in its entirety, human beings make do with simpler stories that can help guide them in decision-making. That reality is too complex for us is an idea that resonates within the field of system dynamics as well (Forrester, 1971; Sterman, 2000). System dynamics allows our mental models of the world around us to be explicitly, and where possible quantitatively, mapped out for constructive discussion and analysis (cf. Sterman, 2000). All of us (including scientists) make use of simplifying stories, or mental models, and one example is a very pervasive but inaccurate story about population.

In the Malthusian view of population dynamics, population will increase exponentially while resources will not, resulting inevitably in disastrous consequences for humanity. From an environmentalist's perspective, there is an added poignancy to this story, with its air of inevitability. As population increases, environmental degradation proceeds apace and results in what system dynamicists call 'eroding goals'. As the decades pass and a continual process of environmental degradation is underway, people gradually become accustomed to lower standards. Higher levels of pollution and noise, reduced plant and animal populations and diversity, more urban sprawl and social conflict become 'normal'; people forget (or are never exposed to) the higher environmental standards that were once the norm (cf. Terborgh, 1999).

Despite the considerable uncertainty about the precise relationship between population and resources, we agree that continued population increase would eventually reach a point that the Earth's resources could not sustain, regardless of likely improvements in technology. We also think that reducing population would have beneficial effects (such as simplification of governance, improved natural habitats, reduced levels of violent conflict, a greater possibility of more equitable distribution of resources: generally, improved health for people and their environments).

But we disagree about the inevitability of Malthus's view. There are important elements in the population–environment interface that can turn this trend around. Many of those who are concerned with managing forests have characterized the 'population problem' as someone else's concern, albeit one with a huge effect on the forests they care about. But such compartmentalization ignores the interconnectedness of people's lives and the reality of human agency (the ability to act). Women (and men) make decisions about their fertility that cumulatively affect their environment, which in turn can affect their own wellbeing.

Over the past decade, there has been an increasing awareness of the interconnections between people living in and around forests and the conditions of the forests themselves. Between 1994 and 1998, the Center for International Forestry Research (CIFOR) investigated these relationships through interdisciplinary, international field teams in Austria, Brazil, Cameroon, Côte d'Ivoire, Gabon, Indonesia and the USA (Prabhu et al, 1996; 1998; Colfer and Byron, 2001). In that research, we began with the assumption that the maintenance or enhancement of environmental quality and human wellbeing was essential to sustainable forest management. Our research yielded a generic set of principles, criteria and indicators of sustainable forest management, as well as manuals for assessing the sustainability of particular forests and forest communities (CIFOR, 1999). In the human sphere, we concluded with three main necessities for sustainable forest management:

1 intergenerational access to resources is secure;
2 rights and responsibilities to manage equitably and cooperatively are clear; and
3 the health of people, cultures and forests is maintained.

This chapter focuses on the interactions among human wellbeing, population, and forest and human health, with special reference to women's real and potential roles. The actions and preferences of women, if supported over time, can contribute significantly to bringing about their own improved wellbeing and health, as well as stabilizing or even reducing the population. Our discussion is also based on a vision of the future that includes improved human wellbeing and the maintenance of significant amounts of healthy forest.

Recognition of the importance of collaborating with local people, rather than trying to plan for them, opens the door to management approaches that integrate the interests of people living in forests with the interests of outsiders concerned about the forests, as we tried to do in CIFOR's Adaptive Collaborative Management Program (see Hartanto et al, 2003; Kusumanto et al, 2005; Colfer, 2005a; 2005b; Fisher et al, 2007; Guijt, 2007). In the following pages, we share our perspective on some common elements of women's lives that have been shown to affect their childbearing behaviour and, in turn, population growth and forest health. The influence of Germain (1975) will be clear in the analysis that follows. Youssef (1978), Caldwell (1979; 1986), and Venkateswaran (1995) – as well as numerous gender-sensitive ethnographies – also document many of the same links described in this chapter, as does Colfer's extensive ethnographic experience. Our expectation is that awareness of these links can guide our resource allocation and attention. It is of course important to bear in mind that a given population size results from the previous population and the combined effects of births, deaths and migration (in and out). In some forest contexts, these other factors have more important effects on population size and growth than childbearing alone.

It is time to pay greater attention specifically to the factors that affect women's propensity to bear children. We argue that there is often an inverse relationship between women's health, level of education, amount and type of work, and public status on the one hand, and women's likelihood of bearing children on the other. Westoff and Bankole (1999), in talking about South Asian countries, say:

> *Theories are abundant to explain the spread of contraception and the decline of fertility in the developing world. The changes have been attributed to factors that include increases in income and in education, the improvement in the status of women, the decline of infant and child mortality, and the erosion of religious and traditional authority. All are plausible explanations, and all no doubt have some validity.* (p32)

We deal with most of these issues here, and we link them to the concept of human wellbeing that has been widely linked to sustainable forest management in general. We argue, as others have done before, that truly sustainable forest management is dependent on some level of human wellbeing. We further argue that reduced fertility can improve the health of women and their families, as called for in the Millennium Development Goals. Improved access to education and employment and increased public status are also typically seen as contributing to human wellbeing. Our fundamental argument is that benefiting forest women can also benefit the forests they inhabit. Given women's reproductive specialization, this is true to a greater degree than would be the case for men.

WOMEN, POPULATION AND CAUSAL LOOPS

Human population increase derives from natural increase (the balance of births over deaths in an area) and net in-migration. In this chapter, we focus on natural increase, the macro-level population component most directly linked to women's micro-level interests and decision-making.

In the following discussion, we use causal loop diagrams as a mechanism to portray our understanding of the interactions between women's lives and population. Understanding these interactions is important for those who strive to protect forested environments while working with local communities. Such understanding is even more important for those concerned about population growth, equity and human health in forested landscapes. Causal loop diagrams were invented to help system dynamicists develop and portray the conceptual foundations for their mathematical models of systems. Our own diagrams seek to illustrate the points of our argument; a quantified model is beyond the scope of this chapter.

Causal loop diagrams are simplifications of reality – as will become clear as we examine the diagrams – but they are useful to point out important causal feedback relationships affecting a problem or issue. They allow us to indicate our views of the direction of causality and the relative strength of the causal connection (by the width of the connecting arrow). Clearly, since this discussion is at a global scale, we can expect important variation in local conditions, but we have tried to represent fairly general processes seen and documented in many parts of the world. We want to stress that all models are partial and subject to revision.

A critical feature of causal loop diagrams is their capacity to help us look at feedback (see Richardson, 1991 for a thorough exposition on the background and importance of feedback thinking). Feedback loops can be either reinforcing (as in a vicious circle) or balancing (as in thermostats). Such loops are important when we are thinking about how and why populations grow, decline or stabilize.

In the diagrams to follow, there are two kinds of links between diagram components: positive and negative. In Sterman's words (2000, p39),

> *A positive link means that if the cause* increases, *the effect* increases *above what it would otherwise have been, and if the cause* decreases, *the effect* decreases *below what it would otherwise have been. … A negative link means that if the cause* increases, *the effect* decreases *below what it would otherwise have been,*

> *and if the cause* decreases, *the effect* increases *above what it would otherwise have been.* [Emphasis in the original]

A positive link is indicated by a plus sign on the arrow, and a negative link by a minus sign on the arrow.

Figure 6.1 (cf. Sterman, 2000) illustrates the way biologists tend to view population. Although many biologists would select other ways to portray their views, most would understand it and could apply it to the animal world. We will argue here that human population includes many other significant factors that affect growth or decline. The argument is not that biologists have a simplistic view of population, but rather that they focus on the equally complex biological aspects of population growth and decline and do not tend to think about those aspects that are peculiar to human beings (as discussed below).

Biologists tend to see the population problem in terms of a disruption of natural factors controlling population size. In Figure 6.1, biologists would emphasize the reinforcing effect of the birth-rate loop on population growth (R) and the balancing effect of the death-rate loop (B1). They would also focus on the many feedback effects of population size that can limit births and reduce average lifetime (balancing loops B2 and B3). They might also focus on external factors, such as climatic conditions. Unlike other kinds of animals, however, humans can actively promote balancing effects on births (loop B2) under the right conditions, and many of these conditions are intimately tied to local conditions. Demographers and anthropologists focus much more attention on the bundle of factors subsumed by biologists under 'other external factors', as we discuss below.

As with many interdisciplinary endeavours, differences in terminology between biologists and demographers can lead to serious misunderstandings. Where demographers consider a birth rate to be births per population, assumed (unless otherwise stated) to be an annual measure, biologists call this a fractional birth rate or a birth fraction, reserving 'birth' rate for the number of births per year. In Figure 6.1, since we are attempting to show

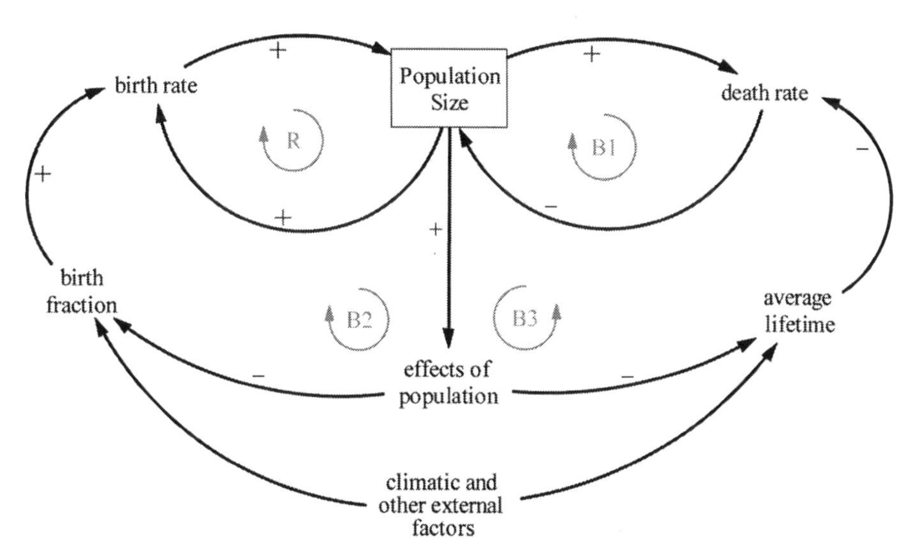

Figure 6.1 *Simplified biological view of population dynamics*

the biologists' point of view, we use biological terminology. But in subsequent mention of birth rates, we will be using the demographers' definition, because most of the literature about human population uses that definition.

The R encircled by an arrow refers to a reinforcing loop, and the B encircled by an arrow represents a balancing loop. Again, an example of a reinforcing loop is a vicious circle; a balancing loop is a self-regulating, homeostatic device.

A final point on causal loop diagrams: the links between these various components in the diagrams may involve time lags. As one component changes, time may pass before the effect is felt or seen in the related components. The inherent complexity of the interactions shown in the diagrams is exacerbated, from the standpoint of human understanding and action, by these time lags, which preclude easy identification of cause and effect.

Births, and by extension, human population, derive directly from the cumulative childbearing activity of women (plus mortality and migration). We have identified five issues that we feel are closely related to decisions about childbearing:

1 availability of birth control;
2 use of birth control;
3 desire to bear children;
4 typical number of children born per woman; and
5 time used for reproductive activity.

Because we are focusing here on factors affecting typical individual forest women's perceptions and decisions – with the idea that these are subject to change – we have not used two other common and closely related demographic concepts: total fertility rate and desired family size. These tend to be cross-sectional and aggregate, whereas we are emphasizing the longitudinal and individual aspects. By reproductive activity, we refer to domestic work as described by Momsen:

> *Goods and services must be produced for human use; human life and society must be* reproduced *to continue in existence. ... Children must be cared for and taught. ... But there is more to* social *reproduction than this: food, water, warmth, clothing, shelter, hygiene and care for the sick must be provided as well as personal support and comfort.* (Momsen, 1987, p39) [Emphasis in original]

An equally legitimate and more demographic definition of reproductive activity would be 'all activities that women/people with (or planning to have) children do that those without children do not do'. In this chapter, though, we are using Momsen's broader, more anthropological definition.

Demographers focus on still more proximate causes of population growth, like age at marriage, onset of sexual activity, length of breastfeeding, rates of abortion and contraceptive use (cf. Bongaarts, 1978, and later works). These issues are obviously also important, but they are issues in which women can potentially have active, decision-making roles. This is not to underestimate the significance of the various societal and contextual factors that influence women's decision-making (e.g., availability of contraception, societal norms, legal strictures). To bring about the potential positive changes in forests and human wellbeing implied by this analysis (to tap into the potential positive feedback loops), forest women will need support and encouragement from natural resource managers and health professionals to enhance maternal and child health; educational, employment and political opportunities; and women's status.

The five issues listed above appear in each of the diagrams below. Additionally, each individual woman has a finite amount of time and energy to use each day, and today most women have some role in decisions about how they allocate that time and energy (cf. Sanday, 1974, p189). The other topics that we discuss (health, education, work and status) all pertain to these women's decisions (or potential decisions) about how to allocate their time and energy.

In the subsequent discussion we focus on these four broad topics that we see as closely related to the population issues outlined above, one by one. The diagrams characterize ways in which these four topics can interact with the childbearing behaviour of women. Our intent is to portray dynamically or longitudinally some of the interactions among factors that individual women take into account as they think about their own reproductive behaviour. Space does not permit discussion of every loop in the diagrams. We hope that readers will consider these hypothesized relationships and communicate critiques and additional insights to us.[3]

Health

Links between health and childbearing are illustrated in one straightforward balancing or negative feedback loop, 'Fertility affects women's health', which links women's health and the typical number of children born per woman (Figure 6.2). Repeated pregnancies, particularly if closely spaced, have adverse effects on women's own health and, by extension, on the health of those they care for. Continued ill-health can affect women's childbearing capacity as well.

A reinforcing loop, 'Child survival affects reproduction', links a typical number of children born per woman, the desire to bear children and children's health. It is important to remember that we are talking in dynamic, longitudinal terms. As the typical number of children born per woman goes up, so does the time used for reproductive activity; as that time goes up, so does the total time spent on care of children, though not the time spent *per child* (see also the other factors affecting time used for reproductive activity, in Figures 6.3 to 6.6). The negative link between typical number of children born per woman and children's health captures the undesired impacts of large numbers of children on child health. A woman who is exhausted from repeated pregnancies and the care of many children cannot provide as much care to her family as can a woman with fewer children, potentially leading in a vicious cycle to increased child mortality and general morbidity within the family. When the survival rate of existing children is low, the desire to bear more children goes up, increasing women's childbearing (Pritchard and Sanderson, 2002). Children serve useful functions

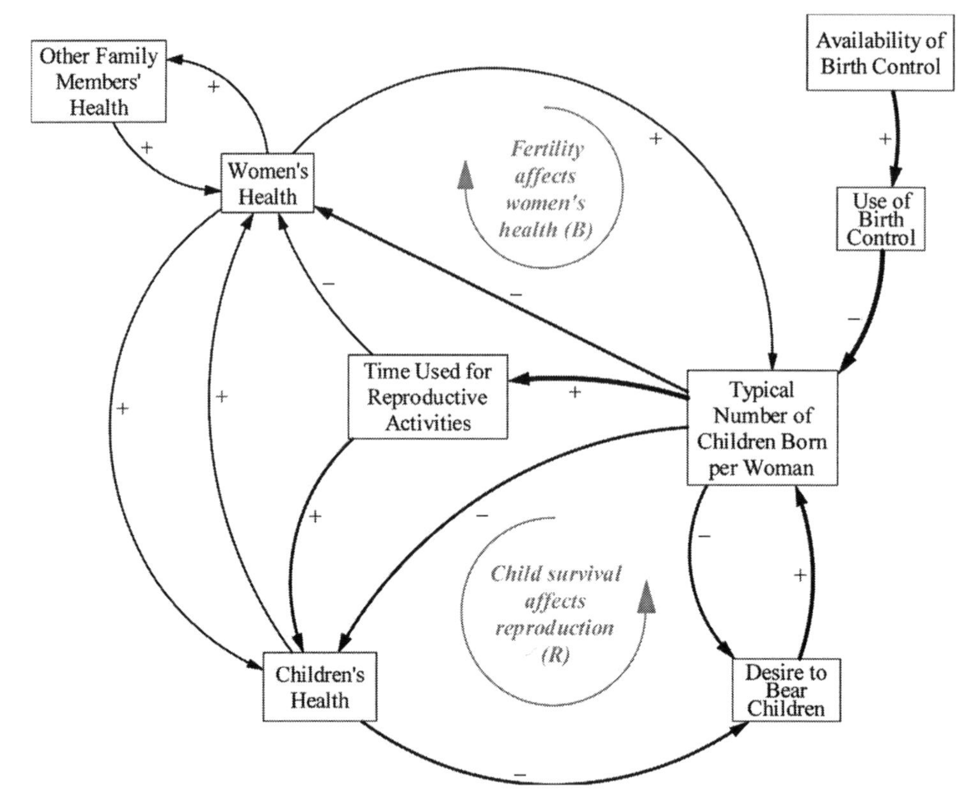

Figure 6.2 *Health and childbearing*

in many parts of the world: as insurance and companionship in old age, as workers on the farm, as helpers around the house and with other children (Barkat-e-Khuda and Hossain, 1996). On the other hand, when children's health status improves, the same links mentioned above can result in a lowering of the typical number of children born per woman.

The links between people's health and the birth rate have been long recognized (Myers, 1985). In most cultures, women play important roles in family health through caregiving and the provision of nutritious (or non-nutritious) meals (e.g., Repetto, 1985; Pearson, 1987; Shiva, 1989; Venkateswaran, 1995; and multitudinous ethnographies). Strengthening women's abilities to enhance their own and their families' health can also be instrumental in lowering birth rates.

Education

With the Middle East a notable exception (Youssef, 1978), increased education for women has been shown to result in lower fertility levels (Caldwell, 1979; 1986; Myers, 1985). We show four interesting feedback loops pertaining to this issue in Figure 6.3.

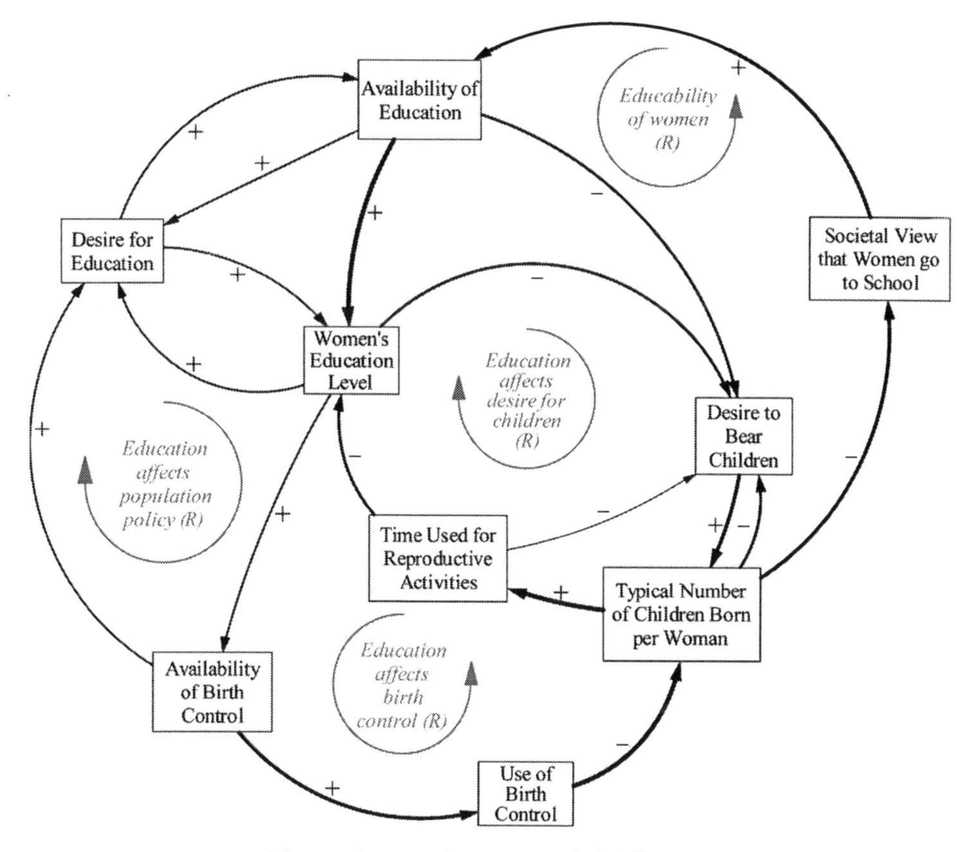

Figure 6.3 *Education and childbearing*

Education affects birth control

As women's educational levels go up, they are likely to have more access to birth control (Castro-Martín, 1995; United Nations Population Division, 1995; Dissanayake, 1996). This may happen for several reasons. The educational process makes more information available to women and may expose them to increased understanding about population-related issues, like their own and their children's health, further educational opportunities, and global issues like over-population. It may also make increased income available, making contraception more affordable. More highly educated women may marry more highly educated men, who may be more open to contraceptive use. Educated men may also be more open to smaller family sizes, which may affect desired family size. Women's educational level can directly affect the use of birth control, and use of birth control is strongly affected by the desire to bear children. Many similar links not shown here will be obvious in any given locale. With contraception, fertility can be postponed or controlled, and schooling becomes a viable possibility. This increases women's desire for education, which in turn increases the likely availability of education, thereby raising women's educational level.

Education affects desire for children

As women's educational levels rise, their desire for children typically decreases. The result is that the typical number of children born per woman also goes down (Barkat-e-Khuda and Hossain, 1996). Zlidar et al (2003) recently conducted a comprehensive survey and found that '[i]n nearly every surveyed country, the more years of school that women have completed, the lower their fertility'. This global trend derives from such factors as the postponement of childbearing to attend school, the greater likelihood that educated women know about and approve of birth control, and the opportunities for employment and involvement in public affairs available to educated women. As the number of children goes down, access to education becomes easier (both through societal recognition that women go to school, and because women have more time to do so). This availability of education, in turn, increases women's education levels still further.

Educability of women

If the typical number of children born per woman goes down, society's perceptions about the appropriateness of education for women goes up, leading to the increased availability of education, decreasing women's desire to bear children, and reducing the typical number of children born. If, on the other hand, the typical number of children born per woman goes up, society's view that education is appropriate for women goes down, bringing the availability of education to women down with it. This increases women's desire to bear children (since other options for their time and energy are unavailable to them), which in turn results in an increase in the typical number of children born per woman.

Education affects population policy

Finally, women's educational levels strengthen both their desire for birth control and their ability to demand the availability of birth control. The availability of birth control, in turn, widens women's perceptions of what their opportunities include, thereby strengthening their desire for education.

These relationships between education and fertility are affected by the postponement of marriage to pursue education, increased knowledge about family planning, increased status due to education, and/or increased interest in and qualifications for employment outside the home. An educated woman with fewer children to care for (and thus fewer demands on her own children for help in childcare) may also allow her daughters to gain an education, which further reinforces the cycle of lower fertility and increased educational levels for women.

Work

Women's involvement in productive work, whether paid or subsistence labour, affects their involvement in the reproductive sphere (cf. Sanday, 1974). Muhuri et al (1994), for instance, found in a study of 33 countries that most women who worked for cash for a non-family enterprise had lower fertility.

Because human beings have finite amounts of time and energy, the effort women expend in the reproductive sphere is not available for them to expend in the productive sphere (and vice versa). Here we divide productive work into two types: subsistence work (Figure 6.4) and paid work (Figure 6.5). We have separated subsistence and paid labour because of some additional complexity introduced by the latter in the relationships between production and reproduction.

In Figure 6.4, there are two interesting loops, both reinforcing. The first loop, 'Production affects reproduction', simply reflects the fact that involvement in reproductive activity limits the time used for subsistence production, and vice versa.

The second loop, 'Desire to work affects childbearing', shows how the desire to work negatively influences the desire to have children, which in turn would reduce the typical number of children born per woman. That in turn would reinforce women's desire to work, since they have the time and energy to do so. Conversely, if the desire to work goes down, the desire to have children typically goes up, which increases the number of children born per woman and decreases further the woman's desire to work, as she has no time.

There are three important reinforcing loops in Figure 6.5. First, we see again the reinforcing loops, 'Production affects reproduction' and 'Desire to work affects childbearing'. Although many connections remain the same, their strength is increased in cases where wage labour is an issue (cf. Barkat-e-Khuda and Hossain, 1996). Where subsistence labour is the norm, for instance, modern birth control is less likely to be available; we have highlighted this difference by thickening the line between wage labour

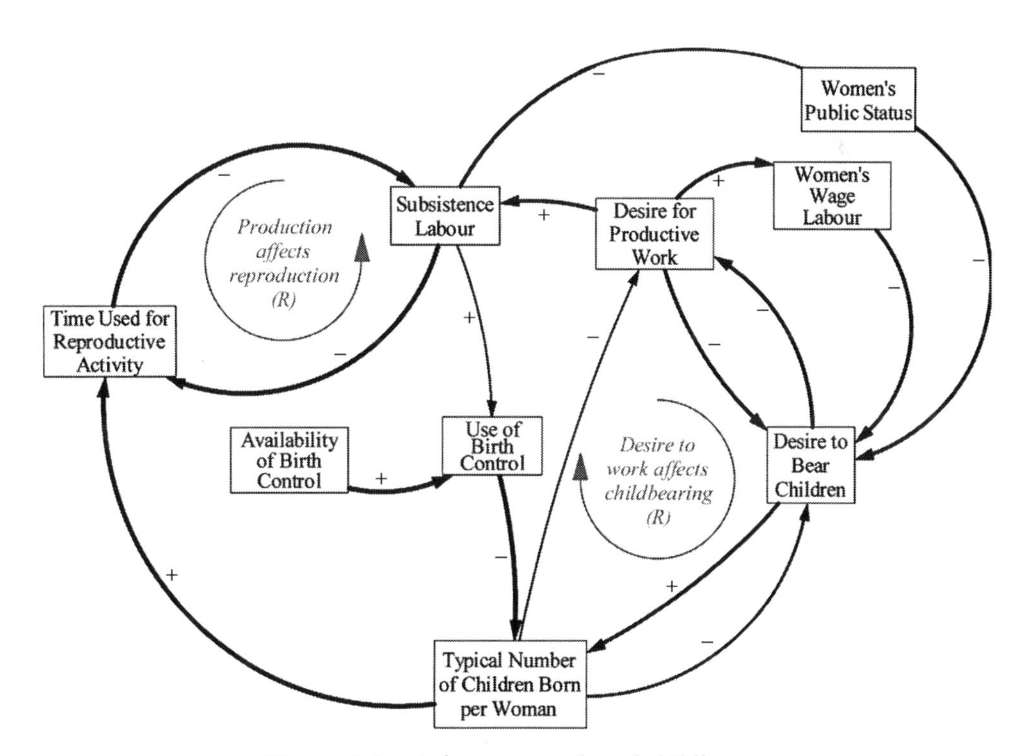

Figure 6.4 *Subsistence work and childbearing*

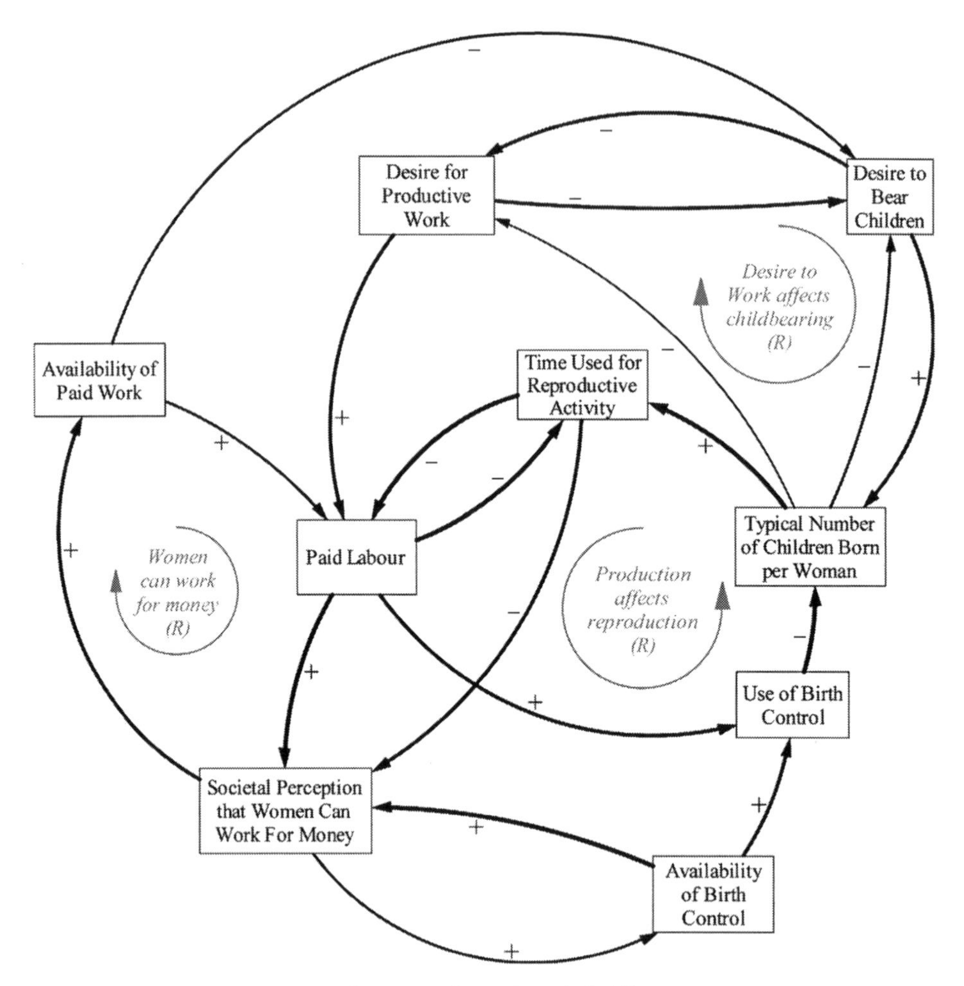

Figure 6.5 *Paid work and childbearing*

and use of birth control (Figure 6.5), compared with that linking subsistence labour and use of birth control (Figure 6.4). The availability of paid work adds a significant opportunity cost to childbearing, further reducing women's interest in childbearing compared with subsistence labour, which is usually more compatible with childcare.

A third important loop, 'Women can work for money', is new in Figure 6.5. Here, women's involvement in the labour market increases the societal perception that women should be able to work for wages. This perception in turn increases the availability of work to women, which in turn increases their involvement in wage labour. But there is an added dimension of relevance for fertility: if the time used for reproductive activity goes up because fertility goes up, the societal perception that women can work for money goes down, causing a reduction in paid labour, which feeds back to an increase in the time used or available for reproductive activity.

In general, direct contributions to family welfare – through production or earnings – increase a woman's value to the family, and in many cases give her a greater voice in decision-making about reproduction, along with a greater motivation to reduce her number of pregnancies. Such productive involvement also strengthens women's autonomy and provides insurance in case of abandonment or the ill-health of other productive family members.

Public and private status and autonomy

Status can include everything from the chivalrous idea of placing women 'on a pedestal', to pragmatic power, to formal authority (cf. Dubisch, 1971; Rogers, 1978; Youssef, 1978; Colfer, 1985). There is evidence that in areas where women's status rises, fertility rates decline (Mason, 1987; 1997; cf. Riley, 2003, who sees this important issue as more complex than is often recognized).

In Figure 6.6, we address what Sanday (1974) referred to as female power and authority in the public domain (women's public status and autonomy), and also women's private status and autonomy. Sanday follows Smith (1960), who defines power as 'the ability to act effectively on

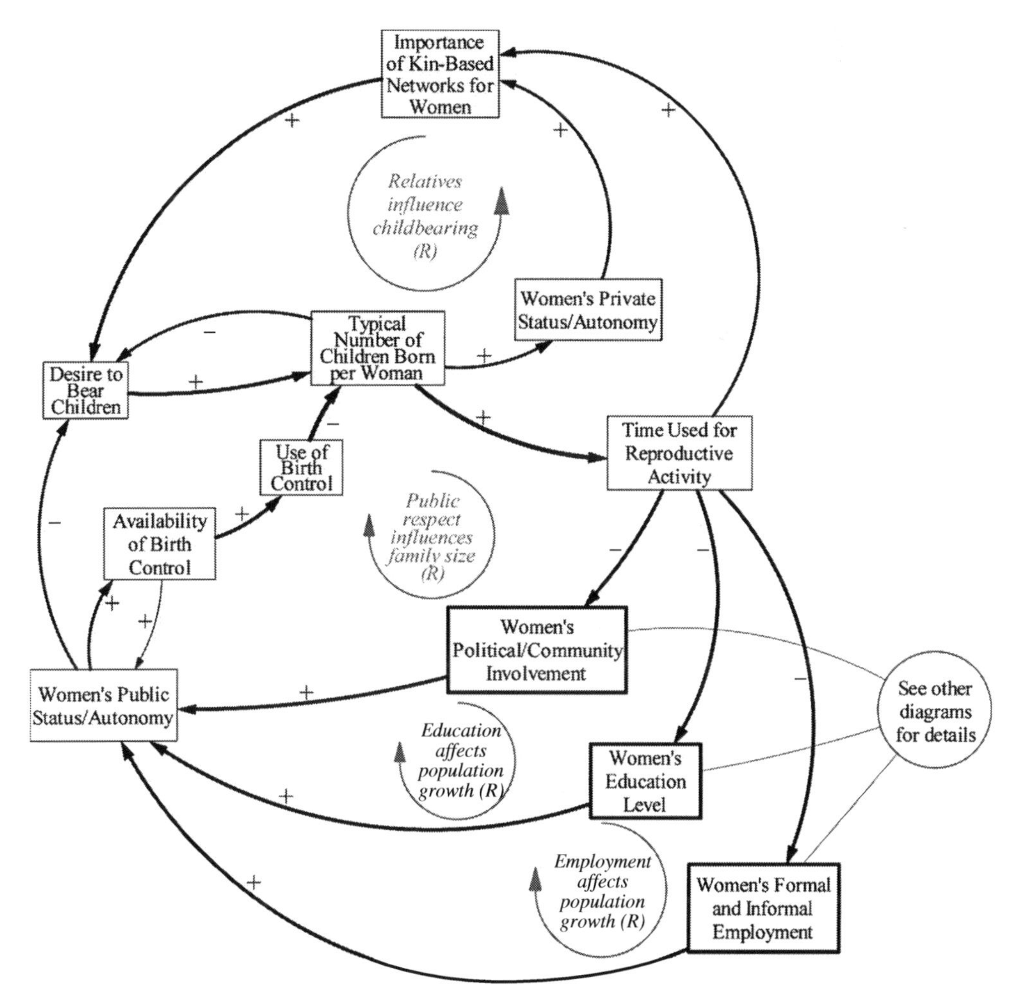

Figure 6.6 *Public and private status/autonomy and childbearing*

persons or things, to take or secure favourable decisions which are not of right allocated to the individuals or their roles', and authority as 'the right to make a particular decision and to command obedience'. There is an ongoing debate within anthropology about the appropriateness of dividing the world into public and domestic spheres, but there do seem to be some important implications for population regarding women's direct involvement in one or both of these spheres. As many have argued, in real life these two domains overlap and intersect, but in a model such as this, it is convenient to separate them, since the factors affecting these two kinds of status differ significantly in their implications for population growth.

In the first loop, 'Relatives influence childbearing', which pertains to women's status in the private sphere, the typical number of children born per woman increases the time used for reproductive activity. This increases the importance of kin-based networks for women. Youssef (1978) provides an excellent summary of how this has operated among Muslim women. Kin-based networks are the most direct way to obtain help with domestic duties. As the strength of kin-based networks increases, so does the typical number of children born per woman, through 'Desire to bear children'. The propensity in many cultures for mothers-in-law and husbands to encourage young brides to reproduce is well documented. In such situations, the typical number of children born per woman commonly increases, as ultimately women's main avenue to power and authority is through their children.

A decrease in the importance of kin-based networks for women will lessen the pressure to reproduce and lower fertility. This in turn reduces the time used for reproductive activity and leads to increased female activity in the political, community, educational and productive spheres. These increases result in increased status and autonomy for women in the public sphere ('Public respect influences family size'), which decreases their desire for more children and ultimately reduces the typical number of children born per woman.

Wipper (1995) reports a study by Charles Hammerslough (1991):

> *In one study of voluntary associations and the use of contraceptives in rural Kenya, group members were found to be 33 per cent more likely to be current contraceptive users (controlling for age, education and urban-rural residence), were more likely to have discussed family planning with their husbands, and to know more about methods and sources of supply than non-members. ... Non-members who live in areas with strong economically-oriented women's groups were more likely to use them than women in areas without these groups.*

In sum, women with higher public status are likely, almost by definition, to have a louder voice in family decision-making, including in reproductive decision-making. Lower personal fertility levels may grant them access to otherwise-unavailable educational and income-generating opportunities, which may in turn reinforce decisions to limit the number of children they bear.

BRINGING IT ALL TOGETHER

We have tried to portray some recurring patterns relating to cause and effect with regard to women, population and health. Our particular interest here has been forested contexts, where external stakeholders are concerned about population growth and health and where forest women themselves often have similar concerns relating to family size and wellbeing, from a more personal perspective. By including population elements (such as the desire to bear children, or typical number of children born per woman) in each of the diagrams, we hoped to demonstrate the kinds of links and feedback loops that affect both women and population size. One could combine these diagrams into a larger diagram to convey the interconnectedness of all the issues (health, education, work and status), but the complexity of the resulting diagram would make it comprehensible only to devoted advocates of causal loop diagramming and population concerns.

We have also hoped to convey the dynamic and active aspect – represented by human agency – that can counteract the depressing and hopeless scenarios of global population growth. Women, their families and their societies all represent a source of human creativity and potential cooperation in efforts to improve people's lives and protect their own habitats. Women's potential roles, particularly, have been under-recognized and underused, to their own and their environment's loss.

SUMMARY AND CONCLUSIONS

In this chapter, we have portrayed some of the interconnections between women and their childbearing behaviour, admittedly on a broad scale and ignoring many issues of importance in particular contexts. We began with the view that the Earth's population trajectory is a worrying concern, even in relatively sparsely populated forested areas. We then argued that women's childbearing behaviour, taken cumulatively, is important in determining the direction of future population growth or reduction; and that population growth is a legitimate concern for environmental scientists (as well as those more directly concerned about human populations themselves).

These same population, environment and human issues are legitimate and real concerns for individual rural women as they work towards improving their lives and those of their children. In our work with forest communities in CIFOR's Adaptive Collaborative Management Program, we have found consistent emphasis among rural peoples, including women, on issues pertaining to economic wellbeing and health. Based on our experience to date, we have developed a hypothesis for planned future work: *improving the capabilities of women to manage their environment, health and fertility – through improving their social capital (links to each other and other stakeholders) and their ability to adapt more quickly to external and internal changes – will result in improvements to rural health, livelihoods and the environment.* Essentially, this approach requires a continuing focus on the women themselves, beginning with their important roles in health and reproductive matters. In this way, we expect to build their confidence and trust, and strengthen their links both within communities and with outsiders. We anticipate that then building on their existing uses and knowledge of the forest will be easier. We expect to retain our dual focus, which values human and environmental wellbeing equally. From the individual woman's point of view, improved health, education and employment are already desired; they are recognized as part of improved human wellbeing. Strengthening women's access to these benefits will be widely appreciated, as will be (with varying lag times) the ability to control their fertility.

These interconnections, taken in conjunction with the growing awareness of the necessary involvement of local communities in forest and other natural resource management, suggest that we need not passively await a Malthusian catastrophe or a macro-economically determined 'demographic transition'. Instead, by working with forest women to strengthen their access to education and employment, improve their health and that of their children, and raise their general status within their communities, we can have a positive effect on the women, their families and the forests that provide their sustenance and habitat. We conclude by stressing the importance of fine-tuning these ideas in particular contexts, in cooperation with local people. Although the patterns and relationships identified in this chapter are common, the strength of connections between issues varies widely, as does the acceptability of particular strategies. Fundamentally, though, outside help with health, education, employment and status can serve as a powerful motivating force for forest women to cooperate in efforts to protect the environment and stabilize population growth, for themselves, for their children and for the world at large.

NOTES

1 The ideas presented in this paper originated in Colfer (2001), and an earlier version of the chapter was presented at the Second Worldwide Symposium on Gender and Forestry, August 2004, Mweka, Tanzania.
2 Robert Gardner considers this a slight overestimate.
3 Write to Carol Colfer at c.colfer@cgiar.org.

REFERENCES

Barkat-e-Khuda and Hossain, M. (1996) 'Fertility decline in Bangladesh: An investigation of the major factors', Extension Project (Rural) Working Paper No. 111, International Centre for Diarrhoeal Disease Research, Bangladesh

Bongaarts, J. (1978) 'A framework for analyzing the proximate determinants of fertility', *Population and Development Review* 4(1): 105–132

Braidotti, R., Charkiewicz, E., Hausler, S. and Wieringa, S. (1994) *Women, the Environment, and Sustainable Development: Towards a Theoretical Synthesis*, Zed Books/INSTRAW, London

Caldwell, J.C. (1979) 'Education as a factor in mortality decline: An examination of Nigerian data', *Population Studies* 33(3): 395–412

Caldwell, J.C. (1986) 'Routes to low mortality in poor countries', *Population and Development Review* 12(2): 171–220

Castro-Martín, T. (1995) 'Women's education and fertility: Results from 26 demographic and health surveys', *Studies in Family Planning* 26(4): 187–202

CIFOR (Center for International Forestry Research) (1999) *CIFOR C&I Toolbox*, R. Prabhu (ed), 9 volumes, CIFOR, Bogor, Indonesia

Colfer, C.J.P. (1985) 'Female status and action in two Dayak communities', in M. Goodman (ed) *Women in Asia and the Pacific: Toward an East–West Dialogue*, University of Hawaii Press, Honolulu, HI, pp183–211

Colfer, C.J.P. (1995) 'Social aspects of the C&I test in East Kalimantan, March 1995', CIFOR, Bogor, Indonesia

Colfer, C.J.P. (2001) 'Are women important in sustainable forest?', in G. Lidestay (ed) *Proceedings, Women and Forestry: How Can Gender Research Contribute to a More Sustainable Forest Management*, Swedish University of Agricultural Sciences, Department of Silviculture, Report No. 47, pp48–56

Colfer, C.J.P. (2005a) *The Complex Forest: Communities, Uncertainty, and Adaptive Collaborative Management*, Resources for the Future, Washington, DC/CIFOR, Bogor, Indonesia

Colfer, C.J.P. (ed) (2005b) *The Equitable Forest: Diversity, Community and Natural Resources*, Resources for the Future, Washington, DC/CIFOR, Bogor, Indonesia

Colfer, C.J.P. and Byron, Y. (2001) *People Managing Forests: The Links between Human Well-being and Sustainability*, Resources for the Future, Washington, DC/CIFOR, Bogor, Indonesia

Dissanayake, L. (1996) 'Relative impact of "starting, spacing and stopping fertility behaviour" in Sri Lanka', *The Journal of Family Welfare* 42(3): 1–7

Dubisch, J. (1971) 'Dowry and the domestic power of women in a Greek island village', paper presented at the 70th Annual Meeting of the American Anthropological Association, November, New York, NY

Fisher, R., Prabhu, R. and McDougall, C. (2007) *Adaptive Collaborative Management of Community Forests in Asia: Experiences from Nepal, Indonesia and the Philippines*, CIFOR, Bogor, Indonesia

Forrester, J.W. (1971) 'Counterintuitive behavior of social systems', *Technology Review* 73(3): 52–68

Germain, A. (1975) 'Status and roles of women as factors in fertility behaviour: A policy analysis', *Studies in Family Planning* 6(7): 192–200

Guijt, I. (2007) *Negotiated Learning: Collaborative Monitoring for Forest Resource Management*, Resources for the Future, Washington, DC/CIFOR, Bogor, Indonesia

Hammerslough, R. (1991) 'Women's groups and contraceptive use in rural Kenya', in T. Locoh and V. Hertrich (eds) *The Onset of the Fertility Transition in Subsaharan Africa*, Derouaux-Ordina, Liège, pp267–287

Hardin, G. (1999) *The Ostrich Factor: Our Population Myopia*, Oxford University Press, New York, NY

Hartanto, H., Lorenzo, C., Valmores, C., Arda-Minas, L. and Burton, E.M. (2003) *Adaptive Collaborative Management: Enhancing Community Forestry in the Philippines*, CIFOR, Bogor, Indonesia

Kemajou, J.P.W. and Sunderlin, W.D. (1999) 'L'impact de la crise économique sur les populations, les migrations et le couvert forestier du Sud-Cameroun', CIFOR Occasional Paper 25, CIFOR, Bogor, Indonesia, p29

Kusumanto, T., Yuliani, L., Macoun, P., Indriatmoko, Y. and Adnan, H. (2005) *Learning to Adapt: Managing Forests Together in Indonesia*, CIFOR, Bogor, Indonesia

Leach, M. (1994) *Rainforest Relations: Gender and Resource Use among the Mende of Gola, Sierra Leone*, Smithsonian Institution Press, Washington, DC

Mason, K.O. (1987) 'The impact of women's social position on fertility in developing countries', *Sociological Forum* 2(4): 718–745

Mason, K.O. (1997) 'Gender and demographic change: What do we know?', in G.W.E.A. Jones (ed) *The Continuing Demographic Transition*, Clarendon Press, Oxford, pp158–182

Momsen, J.H. (1987) 'Introduction', in J.H. Momsen and J. Townsend (eds) *Geography of Gender in the Third World*, State University of New York Press, Albany, NY, pp15–26

Muhuri, P.K., Blanc, A.K. and Rutstein, S.O. (1994) *Socioeconomic Differentials in Fertility*, Demographic and Health Surveys Comparative Studies, Macro International, Calverton, MD, p79

Myers, N. (1985) 'The global possible: What can be gained?', in R. Repetto (ed) *The Global Possible: Resources, Development and the New Century*, Yale University Press, New Haven, CT, pp477–490

Pearson, M. (1987) 'Old wives or young midwives? Women as caretakers of health: The case of Nepal', in J.H. Momsen and J. Townsend (eds) *Geography of Gender in the Third World*, State University of New York Press, Albany, pp116–130

Population Reference Bureau (2007) '2007 world population data sheet', available from www.prb.org/pdf07/07WPDS_Eng.pdf, accessed 30 December 2007

Prabhu, R., Colfer, C.J.P., Venkateswarlu, P., Tan, L.C., Soekmadi, R. and Wollenberg, E. (1996) *Testing Criteria and Indicators for the Sustainable Management of Forests: Phase 1*, CIFOR, Bogor, Indonesia

Prabhu, R., Maynard, W., Eba'a Atyi, R., Colfer, C.J.P., Shepherd, G., Venkateswarlu, P. and Tiayon, F. (1998) *Testing and Developing Criteria and Indicators for Sustainable Forest Management in Cameroon: The Kribi Test, Final Report,* CIFOR, Bogor, Indonesia

Pritchard, L. and Sanderson, S. (2002) 'The dynamics of political discourse in seeking sustainability', in L.H. Gunderson and C.S. Holling (eds) *Panarchy: Understanding Transformations in Human and Natural Systems,* Island Press, Washington, DC, pp147–172

Reardon, G. (1995) *Power and Process: A Report from the Women Linking for Change Conference, Thailand, 1994,* Oxfam Publications, Oxford, UK

Repetto, R. (1985) 'Population, resource pressures, and poverty', in R. Repetto (ed) *The Global Possible: Resources, Development and the New Century,* Yale University Press, New Haven, CT, pp131–169

Richardson, G.P. (1991) *Feedback Thought in Social Science and Systems Theory,* University of Pennsylvania Press, Philadelphia, PA

Riley, N.E. (2003) 'The demography of gender', in D. Poston and M. Micklin (eds) *Handbook of Population,* Kleuwer Academic/Plenum Publishers, Dordrecht, Netherlands, pp109–142

Rocheleau, D. and Slocum, R. (1995). 'Participation in context: Key questions', in R. Slocum, L. Wichhart, D. Rocheleau and B. Thomas-Slayter (eds) *Power, Process and Participation: Tools for Change,* Intermediate Technology Development Group Publishing, London, pp17–30

Roe, E. (1994) *Narrative Policy Analysis: Theory and Practice,* Duke University, Durham, NC

Rogers, S.C. (1978) 'Woman's place: A critical review of anthropological theory', *Comparative Studies in Society and History* 20(1): 123–162

Sanday, P. (1974) 'Female status in the public domain', in M.Z. Rosaldo and L. Lamphere (eds) *Woman, Culture and Society,* Stanford University Press, Stanford, CA, pp189–206

Shiva, V. (1989) *Staying Alive: Women, Ecology and Development,* Zed Books, London

Smail, J.K. (2002a) 'Confronting a surfeit of people: Reducing global human numbers to sustainable levels', *Environment, Development and Sustainability* 4: 21–50

Smail, J.K. (2002b) 'Remembering Malthus: A preliminary argument for a significant reduction in global human numbers', *American Journal of Physical Anthropology* 118: 292–297

Smith, M.G. (1960) *Government in Zazzau 1800–1950,* Oxford University Press for the International African Institute, London/New York/Toronto

Sterman, J.D. (2000) *Business Dynamics: Systems Thinking and Modeling for a Complex World,* McGraw-Hill, Boston, MA

Terborgh, J. (1999) *Requiem for Nature,* Island Press, Washington, DC

Townsend, J. and de Acosta, S.W. (1987) 'Gender roles in colonization of rainforest: A Colombian case study', in J.H. Momsen and J. Townsend (eds) *Geography of Gender,* State University of New York Press, Albany, NY, pp240–257

Turshen, M. (1995) 'African women and health issues', in M.J. Hay and S. Stichter (eds) *African Women South of the Sahara,* Longman Publishing, New York, NY, pp239–249

United Nations Population Division (1995) 'Women's education and fertility behaviour: Recent evidence from the demographic and health surveys', UNPD, New York, p113

Venkateswaran, S. (1995) *Environment, Development and the Gender Gap,* Sage Publications, New Delhi, India

Westoff, C.F. and Bankole, A. (1999) 'Mass media and reproductive behavior in Pakistan, India and Bangladesh', Demographic and Health Surveys Analytical Report, Macro International, Calverton, MD, p32

Wipper, A. (1995) 'Women's voluntary associations', in M.J. Hay and S. Stichter (eds) *African Women South of the Sahara*, Longman, New York, NY, pp164–186

Youssef, N. (1978) 'The status and fertility patterns of Muslim women', in L. Beck and N. Keddie (eds) *Women in the Muslim World*, Harvard University Press, Cambridge, MA

7

The Gender Agenda and Tropical Forest Diseases[1]

Pascale Allotey, Margaret Gyapong and Carol J. Pierce Colfer

The determinants of health can include social relations, the distribution of power and the ability to exercise agency and mobilize and utilize resources. The gender approach looks at the determinants of health and the social, cultural and economic reality in which men and women live and work, and explores whether that reality promotes health and prevents disease (Dias, 1996). In this chapter, we highlight those realities in tropical forested environments, particularly as they pertain to diseases common in such forests. Since the literature on disease often fails to specify the environment in which the study was carried out, we have used the survey done by Colfer et al (2006) to ascertain the diseases that represent significant dangers to humans in forested areas.

After this introductory section, we define our terms: forests, sex and gender. The next section looks at how men and women can experience illness and health differently; we then discuss the value of gender analysis for improving health. The following section looks at the sex and gender differences in specific diseases that are common in forested areas. We then turn to conditions that affect maternal and infant health, and finally, we examine the gender perspective on research and treatment. A brief discussion and conclusion point the way to future research and action.

Definitions

We use 'forests' in this context to describe a broad range of ecosystems: from the dense, humid, tropical forests of Central Africa, the Amazon Basin and Indonesia, to sparse, dry forests, such as the *miombo* woodlands of southeastern Africa and the dry forests in parts of India. We have also included areas undergoing rapid deforestation and the agroforestry systems that co-exist with many forests.

In our usage, 'sex' describes the biological attributes that differentiate males from females. 'Gender' captures the culturally and socially constructed aspects of being male or female, reflecting both relations between individuals at a personal level and the values and norms that permeate the broader social structure. The consequences of this differentiation for forest-dwellers' health are the subject of this review.

Biomedicine has focused on the male body as the norm, with predictable implications for women's care (Schiebinger, 1993). Gender discrimination has meant that men have commanded greater access to resources that promote health and prevent and cure disease (Raikes et al, 1992; Okojie, 1994; Vlassoff and Bonilla, 1994; Vlassoff, 1995; Vlassoff and Manderson, 1998; Allotey and Sundari Ravindran, 2002; Vlassoff and Moreno, 2002). The study of gender is a critical issue of equity, particularly for women.

GENDER ISSUES IN EPIDEMIOLOGY

We briefly highlight five recurrent differences in the ways men and women can experience illness and health.

Roles and exposure

Gender influences exposure to infection and the risk of disease (Manderson et al, 1996). Men are often responsible for heavy physical labour in forest areas and are thus exposed to industrial accidents and disease-transmitting vectors. For women, domestic activities such as cooking over wood fires can have a protective effect against vectors such as sand flies and mosquitoes because smoke acts as an insect repellent (Vlassoff and Bonilla, 1994; but see Chapter 5, this volume, on exposure to air pollution).

The effects of gender on exposure are also evident in local explanatory models of disease. 'Spiritual war' provides the local explanation for elephantiasis (see below, under 'filariasis') in Ghana (locally called *natintim* or *napimpim*: Allotey, 1995; Gyapong et al, 1996a) and Haiti (Eberhard et al, 1996; Coreil et al, 1998).

Female modesty in some Islamic countries means that for diseases that rely on direct contact of a vector with the skin, such as the sand fly in cutaneous leishmaniasis, women may be less affected clinically. However, disfiguring lesions on the exposed parts of the face in women present a different set of problems.

Perceptions of disease

All cultures identify illness and classify diseases based on their own cultural experiences (Nations, 1986). Local perceptions and interpretations of disease affect cultural reactions to disease (Helitzer-Allen et al, 1993). The different cultural knowledge and attitudes of men and women affect their response to illness and disease management (Weigel et al, 1994; Cobra et al, 1995; Bandyopadhyay, 1996; Velez et al, 1997; Gyapong et al, 2000).

Perceptions of disease also influence the levels of stigma associated with the condition. Stigma experienced by individuals is often in excess of the risk of contagion (Liefooghe et al, 1997; Velez et al, 1997). In a multi-centre study of stigma related to onchocercal skin disease, Vlassoff et al (2000) found stigma to be qualitatively different for men and women (Lu et al, 1988; Feldmeier et al, 1993; Vlassoff and Bonilla, 1994; Allotey, 1995; Rao et al, 1996; Vlassoff et al, 1996; Eberhard et al, 1996; Oliveira and de Pessini, 1997; Coreil et al, 1998; Cofie and Adjei, 1999; Vlassoff et al, 2000). When stigma brings with it discrimination, the sick can be deprived of material and social goods within their community, affecting access to health services, basic rights and ultimately, the severity of their conditions.

Responses to illness and disease

Gender-differentiated responses to illness include actions taken in recognizing illness, diagnosis of disease, health-seeking behaviour and accessibility to health services. There is a growing body of literature on the health effects of caring for relatives with chronic and infectious diseases (e.g., Twigg et al, 1990; Cancian and Oliker, 2000; Morrison, 2000). Caring responsibilities significantly restrict the types of employment available to individuals, the length of work hours, and opportunities for education and other forms of engagement with the broader community (Hibbard and Pope, 1983; Australian Bureau of Statistics, 2001). Lack of decision-making power to seek external health care has been recorded in several studies on gender and health in Asia, Africa and Latin America (Okojie, 1994; Kisekka, 1995).

Women's multiple roles and the resulting excessive burdens (Hibbard and Pope, 1983) cannot be denied, but roles are not static. Dramatic changes to family structures and social networks have occurred, as with the HIV epidemic and conflict (Isiugo-Abanihe, 1994). Significant gaps remain in our knowledge about how power is negotiated within households (Fatima, 1991; Ezeh, 1993; Velez et al, 1997). Urbanization and a greater emphasis on cash incomes (Potash, 1995) have increased women's reliance on men and gender inequality (Morrison, 2000) in some instances but have increased women's opportunities in others.

Women are more likely than men to spend available resources not only on curative care but also on prevention of disease and on health-promoting activities (Agyepong, 1992; Thomas, 1997; Lampietti et al, 1999; Rashed et al, 1999). Women also have a greater tendency to consult with traditional healers than to visit health centres for tropical and other diseases (Hudelson, 1996). In particular, stigmatized diseases such as leprosy, TB and gynaecological conditions may first be referred to traditional healers (Kisekka, 1995; see also Weigel et al, 1994; Ager et al, 1996; Kaur, 1997; Coreil et al, 1998).

Quality of care

Expectations of quality of care from health services and in the treatment of patients by health-care providers also differ (Kaur, 1997). Women in both developed and developing countries, including forested areas, have expressed concerns about the lack of explanation, investigative and treatment procedures, and quality of interaction with health-care professionals (e.g., Raikes et al, 1992; Okojie, 1994).

Diagnostic techniques may present barriers (e.g., where providing urine, blood and stool samples is taboo; Feldmeier et al, 1993). Female modesty may be an obstacle to physical examinations by unfamiliar males (Abiose, 1992), and men with a swelling in the testicle (hydrocele), which may result from filarial infection, may refuse to be examined by a woman (Gyapong et al, 2000). Sometimes, the mode of presentation of information may be inappropriate (Vlassoff, 1997). Women have complained that health workers demanded bribes to provide care, were rude and condescending, blamed women for their ill-health and humiliated them publicly by failing to provide an atmosphere for private consultations and examinations (Allotey, 1995; Fonn and Xaba, 1995; Haaland and Vlassoff, 2001).

Compliance with treatment

The compliance with treatment in the control of infectious diseases (whether in mass administration or after case detection) is critical not only to ensure the management of the condition but also to reduce the risk that infective agents will mutate into drug-resistant strains, and to ensure the effectiveness of control programmes that require the destruction of the reservoir of infection within the community (e.g., in the mass treatment programmes employed in onchocerciasis and filariasis control). A study in Colombia to assess the effects of gender on leishmaniasis showed men more likely than women to comply with treatment. Health workers were not well enough trained to recognize and treat symptoms, and women could not access services as easily (Velez et al, 1997). Men from many rural communities attended markets in nearby towns on a weekly basis to sell produce, making incidental trips to the health centre for the required tests and diagnosis – and therefore compliance with treatment – for cutaneous leishmaniasis easier than for women.

VALUE OF GENDER ANALYSIS

Gender analysis makes an essential contribution to improving health policies and planning in forested areas by highlighting the points of intervention (Vlassoff and Moreno, 2002). Identifying the types of data needed to explore differences in health and disease and explaining available epidemiological data are both important. Accurate health-related data that provide sex-disaggregated figures for various age groups are generally lacking, particularly in remote, forested areas.

In the absence of robust quantitative data, current knowledge of the impact of gender in health has drawn heavily on qualitative data sources (Vlassoff, 1997; Vlassoff and Manderson, 1998). Techniques used in anthropology, such as observation, in-depth interviews, focus group discussions and participatory research techniques have yielded critical information that explains some sex differences in infectious disease epidemiology. Qualitative data have also identified sources of bias in available data and, importantly, have demonstrated differences in the experience of diseases even where there are no differences in the epidemiological data. In the following section, we present information on gender and eight tropical diseases.

GENDER ANALYSIS OF EIGHT DISEASES COMMON IN FORESTS

Communicable diseases in general, and tropical diseases of the forest in particular, can largely be controlled by improving environmental conditions and providing early diagnosis and treatment. However, with persistent poverty and increasing wealth disparity, they continue to be a major cause of death and disability, with the poorest 20 per cent of the world's population experiencing 47 per cent of deaths from communicable and related diseases (Gwatkin and Guillot, 1998). Public health efforts to control communicable diseases continue to be biased towards biomedical models that discount the systematic long-term evaluation of social and environmental interventions in disease control (see Chapter 14, this volume). Assessments of the outcomes of interventions are based primarily on reductions in illness and death. Although the global burden of disease methodology has been instrumental in redirecting the focus of health planners towards non-fatal outcomes of diseases (Murray and Lopez, 1996), a major shortcoming remains the exclusion of sociocultural contexts, such as gender (Allotey and Reidpath, 2002; Reidpath et al, 2001).

Projects such as the Special Programme for Research and Training in Tropical Diseases have enhanced communicable disease interventions by promoting basic and applied social science research in tropical diseases (Vlassoff and Manderson, 1998). Important foci include schistosomiasis, onchocerciasis, filariasis, malaria, African trypanosomiasis, Chagas disease, leishmaniasis and dengue. These conditions are neglected despite the high associated disease burdens. Reasons for their persistence include environmental conditions, poverty (reflected in inadequate housing, low literacy and poor nutrition) and lack of access to the often-inadequate health services. These are particularly common problems in forested areas. Chagas, for instance, is directly associated with quality of housing, malnutrition and diarrhoea (Dias, 1996). Data on dengue show a significant association with slum housing, lack of screening and wooden house construction (McBride and Bielefeldt-Ohmann, 2000).

Further reasons for the continuing high burden of these communicable diseases and for the re-emergence of previously controlled infections in non-endemic areas include easier travel, conflict, humanitarian emergencies and population displacement, all resulting in greater global movements of people (also emphasized in Chapter 12, this volume).

Within the limitations of available data, substantial evidence demonstrates some sex differences in the occurrence of most of the tropical diseases mentioned above. The information below summarizes the interdependence of biological and gender explanations of differences where they occur and where data are available. We begin with two diseases about which comparatively little is known, and proceed to those for which more gender-disaggregated data are available.

African trypanosomiasis (sleeping sickness)

African sleeping sickness is transmitted by the bite of bloodsucking male and female infected tsetse flies (*palpalis* group) (Smith et al, 1998; see Chapter 8, this volume, for a discussion of vector-borne diseases emphasizing the role of bats). Patz et al (2000) describe two types. *Trypanasoma brucei gambiense*, the main vector in Western and Central Africa, requires substantial levels of humidity and prefers densely forested riverine habitats. *T.b. rhodesiense* thrives in the open savannahs of eastern Africa. Cattle and other wild mammals act as reservoir hosts of the parasites. This mode of transmission makes herding cattle and hunting – predominantly male activities – high-risk occupations in endemic countries. Disease may also occur in those sporadically exposed to vectors, such as poachers, honey-gatherers, firewood-collectors and tourists (Smith et al, 1998). Transmission is also possible through contamination with infected blood or through the placenta.

There is a general dearth of sex-disaggregated information and on the effects of gender in African trypanosomiasis, which may in part reflect the limited number of female researchers in this area. So although there appear to be differences in exposure, this is not reflected in the available prevalence data. In one study comparing positivity in males and females among an indigenous Colombian population, Corredor Arjona et al (1999) found no difference in any age group after the application of the Montenegro skin test, suggesting that the exposure risk is equal among women and men. The paucity of data in this area needs to be addressed.

Chagas disease (American trypanosomiasis)

Chagas, or American trypanosomiasis, is also vector-borne, transmitted by the triatomine bugs (*Triatoma infestans*). The disease is endemic to Central and South America and causes the third-largest disease burden in the region. The insects' preferred habitat is the crevices of walls and roofs in mud homes in rural areas and urban slums (Moncayo, 1999); deforestation processes are exacerbating the problem in the Amazon (Coura, 2002). Transfusion of infected blood has also been identified as a significant mode of transmission, and up to 53 per cent of the blood in some blood banks is infected (Moncayo, 1992). Current data suggest that transmission may also occur through ingestion, which would explain the growing number of family micro-epidemics (da Silva Valente et al, 1999), although this mechanism is still not well understood.

Chagas is directly related to poverty, and women and children who remain close to home have a higher peri-domestic exposure to the vector (Azogue, 1993). There are very few sex-disaggregated data. One study reports that seropositivity of Chagas is higher in women than men, with a rate of 56.2 per cent in women and 43.8 per cent men (Kaur, 1997). Other gender differences are unclear.

Dengue

The dengue mosquito breeds mainly in artificially created water receptacles in peri-urban domestic settings, though in recent years it has spread to forested areas (Gratz and Knudsen, 1997). It is responsible for three presentations of the condition: dengue fever, dengue haemorrhagic fever and dengue shock syndrome. A detailed observation study noted the potential for higher exposure for women, who are largely responsible for the home and household tasks like water storage, laundry, cleaning and house pots (Whiteford, 1997). However, in areas where responsibility for maintaining 55-gallon drums kept outside the home fell on the men, risk increased for them (Whiteford, 1997). Transient or sporadic exposure also occurs through tourism, accounting in part for occurrence in non-endemic areas. Approximately 84 per cent of the contingent of predominantly male soldiers returning from Somalia in 1992 reported symptoms of dengue fever (McBride and Bielefeldt-Ohmann, 2000).

Data on male–female prevalence rates for dengue are inconclusive. Women in Thailand are hospitalized for dengue haemorrhagic fever at twice the rate of men (Halstead, 1990). In Singapore, however, 1.5 times more male admissions were recorded. A study in Malaysia reported a higher incidence of dengue haemorrhagic fever in males, but a higher case fatality rate in females (Shekhar and Huat, 1992). No reasons were proposed for the differences. Other studies find a more even distribution between males and females (Hayes and Gubler, 1992; Cobra et al, 1995).

McBride and Bielefeldt (2000) have suggested that biology may be a possible factor in the differences, but there is little evidence. There is, nonetheless, an association between the development of dengue haemorrhagic fever and shock syndrome and secondary-type antibody response that is affected by age and sex, with a higher incidence in young children and females than in males (Cobra et al, 1995).

Filariasis (elephantiasis)

Filariasis is a painful and disfiguring disease that has a major social and economic impact in Asia, Africa, the western Pacific and parts of the Americas (Ottesen et al, 1997). It is one of the leading causes of permanent and long-term disability in the world (Raviglione et al, 1995). Filariasis is caused by long, thin filarial worms transmitted at night, when female *Culex* and *Anopheline* mosquitoes take blood meals from individuals infected with microfilariae. These microscopic larvae of the worm infiltrate the lymph system, which is responsible for fluid drainage in the body. The filarial infection causes acute, then chronic lymphatic disease and, if untreated, results in chronic swelling as a result of the accumulation of watery fluid (lymphoedema).

Males and females were found to be equally susceptible to filarial infection in a study in Orissa, India. But prevalence of chronic lymphatic filariasis disease was significantly higher in males (17 per cent versus 6.2 per cent in females) (Sahoo et al, 2000), consistent with studies elsewhere (Gyapong et al, 1996b; Kazura, 1997; Pani et al, 1997). Reports of chronic disease show a higher prevalence among males, possibly due to the incidence of hydroceles (Michael et al, 1996). Both infection and lymphoedema cases are more prevalent in men in all endemic areas of Brugian filariasis (Michael et al, 1996).

The incidence of Bancroftian filariasis is on the increase in Thailand, with a high prevalence in predominantly male migrant workers from Myanmar (Triteeraprapab and Songtrus, 1999). Walsh et al (1993) suggest that deforestation, though inadequately documented, probably played a role in decreasing the incidence of the disease in Malaysia and Indonesia. In southern Ghana, women who fish and trap shrimp in the mangrove swamps are exposed to the filariasis mosquito, which may be a reason for the higher prevalence of lymphoedema in women than in men in that region (Gyapong, 2000).

For reasons not currently understood, lymphoedema in Bancroftian filariasis affects women more frequently than men (Coreil et al, 1998; Sahoo et al, 2000). Global patterns show lymphoedema occurring 1.5 times more in women than in men (Michael et al, 1996), possibly because of immunological factors. Sex differences in susceptibility are apparent only during the reproductive years, and de Almeida and Freedman (1999) suggest a pregnancy-associated immunological response. Further evidence of the body's response to filarial infections reveals that IgE and Ig4 antibodies are evident in children as young as 18 months in endemic areas. The prevalence of these antibodies is higher in males than females in a study of children under six in Indonesia. It is not clear, however, whether this difference is attributable to gender-related exposure, sex hormones or genetics (Terhell et al, 2000).

Limbs are the most common sites of infection. However, lymphoedema often affects the breasts and vulva in women and the scrotum in men (Pani et al, 1991). According to a study in Haiti, women are ten times more likely to have lymphoedema of the leg (Coreil et al, 1998). However, the high number of urinary and genital symptoms in men is an important cause of disability. The development of hydroceles is relatively common in men with filariasis, with as many as seven out of eight men with elephantiasis also having hydroceles in some studies (Simonsen et al, 1995). Lymph scrotum, which is experienced in a notable minority of cases, results in the leaking of lymph through the scrotal skin (Dreyer et al, 1997). Data suggest that hydroceles can alter testicular function or cause decreased spermatogenesis (Dreyer et al, 1997). The prevalence of uro-genital symptoms in women may be underestimated because of the reluctance of both women and health providers to examine the genitals of female patients. In a survey of the clinical manifestations of filariasis in Tanzania, physical examinations of males included the genitals, arms and legs. Examinations for women, however, were restricted to the arms and legs (Simonsen et al, 1995). There is evidence that women have a lower prevalence of microfilaraemia and lower worm densities, but this does not necessarily provide any information on clinical disease (Vlassoff and Bonilla, 1994).

Leishmaniasis (*kala-azar*)

The leishmaniases are a complex group of parasitic diseases caused by various forms of the *Leishmania* parasite, which is transmitted directly (from one human host to another) or zoonotically (from household animals and rodents) through the bite of forest-floor sand flies. Transmission reflects people's behaviour, and sand fly and reservoir activity (Wijeyaratne et al, 1994). There are three main forms of the disease: cutaneous leishmaniasis, which causes largely self-healing but disfiguring lesions; mucocutaneous leishmaniasis, which attacks the mucous membranes of the nose, mouth and throat, also resulting in mutilating lesions that require reconstructive surgery; and visceral leishmaniasis, the most severe form, which affects the liver and spleen and is fatal if untreated, with up to 90 per cent mortality (Bora, 1999; Klaus et al, 1999). The visceral form has re-emerged from near-eradication largely because of extensive population movements from civil unrest and agricultural and development projects, which have pushed humans into the natural environment of the sand fly (Wijeyaratne et al, 1994).

Leishmaniasis was declared an occupational disease affecting men because of its association with the building of roads and railroads (Klaus et al, 1999), oil and gold extraction, deforestation, farming, hunting and military service (Wijeyaratne et al, 1994); the WHO considers it a disease of forest workers and reports the burden from this disease to be 860,000 disability-adjusted life years (DALYs) for men and 1.2 million DALYs for women (www.who.int/tdr/diseases/leish/direction.htm). Most of the cases identified in a study in Jordan were soldiers serving in the Jordan Valley (Khoury et al, 1996). In Ethiopia and Kenya, most of those affected were livestock herders and coffee plantation workers. A longitudinal study by Weigle et al (1994) found that the greatest risk factors for cutaneous leishmaniasis in Colombia were entering the forests after sunset, hunting, lumbering and farming – all predominantly male activities. The risk for farmers was nine times higher than for those in other occupations. Armijos et al (1997) found a threefold difference based on occupational roles.

The risk of infection of visceral leishmaniasis for those under the age of seven years appears to be independent of sex, but males report higher rates as they get older (Velez et al, 1997). Ratios of up to five times more in men are evident in hospital morbidity figures (Bora, 1999), which may reflect a higher likelihood of reporting and treatment for males. Bora (1999) found higher rates of all forms of leishmaniasis in men in India, with post-*kala-azar* dermal leishmaniasis (PKDL, which develops after infection with the visceral form) affecting the face, trunk, scrotum and penal shaft. Ratios of 2:1 were found in the five-to-nine age group (Bora, 1999). Similar sex ratios are reported for visceral leishmaniasis in Bihar (Thakur, 2000). Community surveys reveal much higher rates among women than is evident from hospital records, indicating that visceral leishmaniasis and PKDL remain under-reported in the community. Bora (1999) suggests that this is due to the prevailing sociocultural and economic circumstances that hinder women's presentation in hospital. The extent to which the differences between men and women are due to exposure or to under-reporting is unclear. There is some indication that part of the difference may be attributable to limited access to services (Wijeyaratne et al, 1994).

Gender was not found to be significant in a village-level survey in Brazil of cutaneous leishmaniasis (Brandao-Filho et al, 1999). Other studies have found that male adults have 2.8 times the risk of women of getting the cutaneous form (Armijos et al, 1997). However, with active case-finding, Velez et al (1997) found a high infection rate around the home without preference for men. A study in Colombia of the prevalence of Chagas and visceral leishmaniasis reported similar estimates in men and women, suggesting equal exposure (Corredor Arjona et al, 1999).

Armijos et al (1997) suggest the possibility that women have an enhanced immune response related to hormonal influences, which may in part account for their apparently lower risk of leishmaniasis. There is some evidence of greater immunity to the visceral form in women, with seroprevalence studies demonstrating that exposed females are less likely to develop clinical symptoms than exposed males (Brabin and Brabin, 1992). However, although men appear to have a higher risk than women, this study reported no sex differences with respect to lesion number, size or type of parasite species in those already infected. Weigel et al (1994) suggest that the differences in risk are more likely to be due to exposure and therefore gender-related, since there was no evidence of elevated risk in males between six months and 17 years. Neither case detection nor exposure factors provide a satisfactory explanation for the lower rate of visceral leishmaniasis in women. Hormonal influences, response to treatment and differences in immune response are also possibilities (Wijeyaratne et al, 1994).

Malaria

Malaria, an infectious disease caused by the plasmodium parasite, is very common in tropical forests (see Chapter 9, this volume). It is endemic in 90 countries, 50 per cent of which are in Africa, where the World Bank (1993) ranks the disease as the leading cause of DALYs. The prevalence and geographical range of malaria have increased, facilitated in part by tourism and other travel, particularly of refugees from conflict areas, as well as more general deforestation, migration and climate change patterns (e.g., Walsh et al, 1993; Nchinda, 1998; Patz et al, 2000), although the links to deforestation are not entirely straightforward.

In general, biological susceptibility to malaria is universal; the bite from a single infected mosquito is enough to infect the host. In fact, however, the risk of acquiring infection and disease varies. In highly endemic areas, some adults develop immunity due to constant reinfection. Those at highest risk biologically are infants from six months to five years, pregnant women, populations in politically unstable malarial areas and travellers from non-malarial areas (Tanner and Vlassoff, 1998). The total number of deaths from malaria is slightly higher for males than for females up to 14 years. From 15 years onwards, more females than males die from malaria. The same is true for malaria incidence and for DALYs lost from malaria (WHO, 2001b).

Like other vector-borne diseases, malaria incidence relies on exposure to mosquitoes, and exposure is usually dependent on gender roles (Kunstadter, 2003 gives a thorough assessment of the links between malaria and gender). Adult males generally have a greater occupational risk of malaria (Reuben, 1993). In many farming communities, for example, men are more likely to stay in the fields during peak times of sowing and harvesting, sleeping out in the open with little protection. Similarly, higher mortality from malaria is reported for gem miners in Thailand, predominantly men, than for women. However, malaria deaths in children below age 15 and adults over 65 who remained in the village showed no significant sex difference (Wernsdorfer and Wernsdorfer, 1988). Observations in a village in India showed that while women and children slept indoors, men and boys over ten years slept outdoors, even in cold weather (Reuben, 1993). In communities in Cameroon, men tend to engage in cash-crop farming while women were responsible for subsistence farms. As in India, men were more likely to stay out overnight. However, women harvested maize before daylight to get the crop to the market by dawn, and were thus exposed during the peak biting times of the mosquito (Vlassoff and Manderson, 1998). Nakazawa et al (1994) examined people in four villages in the Gidra (lowland Papua New Guinea). Although both sexes suffered the disease, men who hunted in mangrove swamps and collected coconuts in the bush, and men who dived in rivers were found to be most frequently bitten, and cumulative prevalence was higher in males than females in a coastal village *(Plasmodium falciparum)* and in a riverine village *(P. vivax).*

Onchocerciasis (river blindness)

Onchocerciasis, or river blindness, is caused by the parasitic filarial worm *Onchocerca volvulus,* for which human beings are the only known reservoir. The vector is *Simulium* black flies, which breed in fast-flowing rivers and thrive in humid tropical rainforests (Thomson et al, 2000); they bite during the day and deposit the infective larvae in the skin. The adult worms produce millions of microfilariae, usually found in subcutaneous nodules, and have an average longevity of 10 or 11 years. The microfilariae are the main cause of the clinical manifestations of the disease, which include dermatitis (resulting in severe itching, loss of skin pigment and thinning of the skin) and infection of the lymph nodes, which may lead to a hanging groin and elephantiasis of the genitals. The most severe manifestation of onchocerciasis are irreversible ocular lesions of both the anterior and posterior segment of the eye, resulting in impaired vision and finally total blindness (Molyneux and Morel, 1998). Of people living where the condition is not controlled,

6.4 million victims live in areas with severe blinding onchocerciasis, and 8.6 million victims are in areas with the parasite strains responsible for severe skin disease. Onchocerciasis remains an important public health problem throughout much of sub-Saharan Africa, and is estimated to affect more than 17 million people in 26 African countries (Boatin et al, 1997; Ogunrinade et al, 1999). Onchocerciasis is also endemic in Yemen, Guatemala, Mexico, Venezuela and Colombia (WHO, 1995; Vlassoff et al, 2000).

Differences in prevalence of onchocerciasis relate to exposure to the vector. In a study in Nigeria, Brieger et al (1997) observed that variations in onchocerciasis prevalence are related to vector ecology and densities, which in turn are related to village distance from a river and the geography of the river itself (Brieger et al, 1997). Thomson et al (2004) used remote sensing linked with clinical studies in Cameroon to determine that infection rates of the closely related loiasis (or *loa loa*), which occurs only in western and equatorial Africa, increase with increasing age, certain elevations and male gender. Exposure in utero has also been documented (Brinkman et al, 1976). However, sex differences are also not consistent between studies. No difference was found in the prevalence of *Onchocerca volvulus* in an epidemiological study in Sierra Leone (Gbakima and Sahr, 1996). A study in Malawi, however, found a higher prevalence and intensity of infection in men than in women, reflecting the nature of infections as well as exposure during migrant work. For residents who had never lived outside the study area, prevalence for males and females was similar (Courtright et al, 1995).

The experience of symptoms may also differ for males and females. Women with reactive lesions of onchocercal skin disease tend to report itching more commonly than men. They also found the itching socially embarrassing, affecting their sexual and marital lives (Kaur, 1997).

Schistosomiasis (bilharzia)

Schistosomiasis, also known as bilharzia, is caused by five species of the water-borne flatworm *Schistosoma*. The worm is transmitted through contact with infected snails living in freshwater habitats, and consequently, people who engage in agriculture, fishing and the use of water for household chores – all very common in forested regions – are most at risk. Ecotourism also partly accounts for the persistence and, in some instances, increasing prevalence of the disease (Chitsulo et al, 2000).

Men in fishing and farming communities – common both inside and outside forests – report higher rates. Estimates show higher levels of infection and illness from schistosomiasis in men than in women in China (Booth et al, 1996) and in Egypt (El-Khoby et al, 2000). However, in communities where women wash utensils and clothes in rivers and obtain household water from the waterways infested with the vector, they show equal or higher rates of schistosomiasis than men (Kaur, 1997). In a study in Kenya, both male and female children were found to spend time playing in water. As they got older, playing in water reduced, but young girls continued to spend more time in water than the boys their age, washing dishes and clothes (Fulford et al, 1996). The different patterns of exposure were reflected in the prevalence. Brouwer et al (2003), in a study of children in Chikwaka communal land in Zimbabwe – where 60 per cent of the children were infected with *Schistosoma haematobium* – found boys to be at significantly greater risk than girls. Hunter (2003) found the same predominance of infected boys in his 1997 study of children in Ghana, where 54 per cent were infected, on average.

There are clear differences in the manifestations of schistosomiasis between men and women. Recent research has highlighted the implications of the pathological process of schistosomiasis as it affects women. Attention has recently been drawn to the localized effects of schistosomiasis around the genitals, rectum and uterus, and particularly the sexually transmitted disease implications of damage to the lining of the vaginal wall (Feldmeier et al, 1995; Polderman, 1995). Feldmeier et al stress the importance of the recognition of these symptoms, described collectively as female genital schistosomiasis, and the possible predisposition to developing malignant tumours. Current evidence suggests that genital lesions may either resolve spontaneously or may become chronic and result in ulceration that destroys the hymen and clitoris, with the most debilitating sequelae being incontinence and trauma to the bladder and vagina (Feldmeier et al, 1995). Other consequences, such as infertility, have an important social if not pathological impact (Anyangwe et al, 1992). The effectiveness of standard treatments, such as prizaquantel, on genital lesions is currently unclear because of lack of research in this area. It is also likely that the condition may be underdiagnosed because women would be more likely to present to a gynaecologist than an infectious disease specialist, and female genital schistosomiasis is not well recognized. Female genital schistosomiasis also increases the risk of HIV infection because the lesion provides easier access to deeper vaginal cell layers during intercourse with an infected partner. However, there are still major gaps in our understanding of the genital manifestations of schistosomiasis, including a lack of data on the possible psychological effects of painful intercourse and post-coital bleeding in women with genital lesions (Feldmeier et al, 1993).

MATERNAL AND INFANT HEALTH

The gender and sex interdependence is critical in maternal and child health because of the effect of pregnancy on the natural course of many conditions and the influence of diseases on pregnancy and its outcome. Pregnancy is associated with reduced immunity to infection, and in general, parasitic infections in pregnancy may lead to maternal disease and malnutrition, premature delivery, intra-uterine growth retardation and placental and peri-natal infection (Brabin and Brabin, 1992). In addition, pre-pregnancy and maternal health is related to nutritional status, age at pregnancy and parity, all of which are heavily influenced by roles within the community and are important factors in exposure and response to infectious diseases (see Chapter 6, this volume).

Conditions such as malaria, schistosomiasis and visceral leishmaniasis are associated with maternal anaemia (e.g., Wijeyaratne et al, 1994; Kunstadter, 2003). The alteration in immunity status during pregnancy is particularly marked in women giving birth for the first time, and increases their susceptibility to severe forms of *Plasmodium falciparum* malaria (Steketee et al, 1996; Steketee and Mutabingwa, 1999). Pregnancy is thus associated with a higher risk of infection, increased frequency of episodes, increased severity and complications, and increased fatality rates. Mortality due to cerebral malaria is 40 per cent in pregnant women, twice the mortality in non-pregnant women with the condition. Pregnant women with malaria also have higher parasite loads and are more likely to develop hypoglycaemia and pulmonary oedema (Reuben, 1993).

Schistosomiasis results in possible disruption of pregnancy, maternal and infant mortality and poor development in the foetus (Michelson, 1993). Haematuria – blood in the urine, a common symptom of schistosomiasis – can also result in severe anaemia during pregnancy. Similar results have been found for visceral leishmaniasis.

The density of infection in lymphatic filariasis and clinical disease is lower in women of reproductive age (Brabin, 1990). However, it is unclear whether this is due to pregnancy-related changes or gender. For many of these conditions, data on non-pregnant, reproductive-age women are lacking.

The assessment of gender and health in pregnancy also has important implications for the unborn child because of the strong links between the health of the mother and the outcome of the pregnancy. Some infectious diseases can be transmitted congenitally or through the placenta, and Chagas might be transmitted *in utero.* Cases have been identified in children under age 13 in urban slums in Asunción, Paraguay. Their mothers, most of whom had migrated from endemic areas, were serologically positive; none of the children had received blood transfusions or travelled to endemic areas, providing strong evidence for congenital transmission (WHO, 1997). Similar findings were reported from a study in Argentina. Azogue (1993) estimates congenital transmission in approximately 11 per cent of cases in children. Chagas programmes have now been integrated with maternal and child-health services to address the issue (WHO, 1997).

There is epidemiological evidence to support the concept of increased susceptibility to filarial infection by in utero exposure to the parts of the filarial worm that stimulate the immune response. A study in Haiti showed that children born to mothers with *W. Bancroft microfilaria* are almost three times as likely to be microfilaremic as children whose mothers did not have the antigens. There was no such relationship to paternal status (Hightower et al, 1993). There were no sex differences in the prevalence of microfilariae in children, although there was a difference in adults. The immunological mechanism of the increased risk is not clear.

The risk of African trypanosomiasis also increased for children if the mother had the disease, although it is not clear whether this familial clustering is due to exposure rather than genetic susceptibility (Khonde et al, 1997). Children are also more likely to be infected if their mothers have tuberculosis, although systematic research is needed (Hudelson, 1996). Studies of dengue in infants suggest that maternal dengue antibodies have a role in the development of dengue haemorrhagic fever in infants, and it has been hypothesized that maternal antibodies are protective before seven months and then are enhancing (Cobra et al, 1995).

Although pregnancy presents important gender-related issues in the control of infectious diseases, pregnant women are generally excluded from research programmes related both to etiology and natural history of disease and the evaluation of treatment protocols. This is largely to protect against possible risks to pregnancy outcomes for the women themselves, and abnormalities in the foetus. However, the exclusion means that little is understood about adequate control in women of reproductive age. In the last decade, the US National Institutes of Health and the US Food and Drug Administration have recommended the inclusion of women of childbearing age in trials. In 1998, drug and device companies were required to report on age, race and gender of trial participants, and in 2000, companies undertaking trials for life-threatening conditions were required to include women in research (Ferguson, 2000; McGowan and Pottern, 2000; Meinert et al, 2000; Vivader et al, 2000).

During a campaign to test Ivermectin in the mass treatment of onchocerciasis, some women were inadvertently treated during the first trimester of pregnancy (Yumkella, 1996). However, because of the general exclusion of women, the effects of Ivermectin on pregnancy and lactation are not well understood and need to be addressed for mass treatment campaigns aimed at destroying the reservoir of infection. Kunstadter (2003) reports research results showing more serious side-effects for women than for men from Mefloquine, an anti-malarial. Furthermore, there is some indication that onchocerciasis may reduce the response to the tetanus toxoid vaccination given routinely during pregnancy (Abiose, 1992), highlighting the importance of understanding pathological processes of tropical diseases in pregnancy.

GENDER PERSPECTIVES ON RESEARCH AND TREATMENT PROGRAMMES

HIV/AIDS (see Chapter 10, this volume) adds a level of complexity to health conditions in tropical forests. People who are HIV-positive have an increased risk of co-morbidities with most infectious diseases because their immune systems are compromised. Murray (1991) reports that risk of developing tuberculosis is 100-fold compared with those who are HIV-negative. HIV–leishmaniasis co-infections have been increasingly reported in Sicily (Cascio et al, 1997), a trend also observed in Spain and France (Alvar, 1994; Marty et al, 1994; WHO, 2000).

We could expect that co-infection of tropical diseases with HIV is highly gendered, and research should integrate such risk factors as commercial sex work, sexuality, negotiation of safe-sex relationships, injecting drug use and other high-risk behaviours with what is currently known about gender, forests and tropical diseases. The HIV pandemic continues to create rapid social and demographic change, and the high rates of illness and death present a significant challenge to public health control programmes. More than anything else, this challenge justifies incorporating gender considerations into all aspects of health in forests as a matter of course, rather than selective 'add-ons' to projects with a single disease focus.

The control of communicable diseases in forests requires that public health and other health professionals engage with community members. Scientific knowledge about the agent, the host or the environment cannot be applied without the participation of communities because the transmission and maintenance of these diseases rely on human behaviours, environmental conditions and sociocultural factors (see Part III, this volume). Even conditions that can be eradicated solely with biomedical interventions need mass treatment and therefore full community participation. Filariasis control, for instance, requires mass treatment of at least 80 per cent of an endemic population to be successful (de Almeida and Freedman, 1999). The need for ongoing engagement with communities provides an obvious rationale for gender analysis in control programmes.

Control programmes involve interventions to destroy or control the breeding of vectors and mass drug therapies, including immunization, information, education and communication strategies (see Chapters 2 and 8, this volume). Programme evaluations have not produced consistent reports about successes. Reasons include poor articulation of indicators for evaluation, with some projects evaluating process and others outcome. However, some qualitative studies have identified gender as a reason for limited effectiveness – even outright failure – of particular interventions (Katabarwa et al, 2001).

The control of malaria, African trypanosomiasis, dengue fever, lymphatic filariasis, leishmaniasis and onchocerciasis involve significant vector-control strategies, which are sometimes targeted at women. For example, public health messages about covering water to reduce mosquito breeding sites in the Dominican Republic were aimed at women because several studies had documented women's traditional responsibilities for the health and nutrition of their families, and household tasks related to handling water (Whiteford, 1997). However, this targeting ignored the larger outdoor containers maintained by men, which were major breeding sites that remained uncovered and exposed to *Aedes aegypti* (Whiteford, 1997).

Bed-net usage, especially bed-nets impregnated with insecticide, is a major intervention in the control of mosquito-borne diseases, such as malaria (see Kunstadter, 2003, for a summary). The nets provide a physical barrier to mosquitoes and the insecticide enhances the protection by repelling mosquitoes from the room or killing them on contact. This preventive approach is costly, however. The highest predictor of bed-net ownership is family income, with higher income groups up to three times more likely to own bed-nets (Abdulla et al, 2001). In addition, the acquisition of bed-nets is also higher if the woman in the household is employed or has access to an income (Rashed et al, 1999). In Ghana, some programmes supported by NGOs are attempting to mobilize women to form cooperatives to enable them to purchase such items.

The push towards involving communities in public health programmes has coincided to some extent with the involvement of women in projects designed to empower them by increasing their engagement in community activities, largely driven by development and poverty alleviation projects. A major drawback is that the projects discount women's household and other unremunerated work, and make recommendations that increase women's participation in control programmes, thus adding extra burdens to their overloaded schedules.

Mass chemotherapy programmes seek either to immunize the total population, which provides cohort immunity to an infectious disease, or to destroy the reservoir of infection by interrupting the life cycle of the infective agent. This area of infectious disease control appears to have paid least attention to gender discourse and, for this reason, has attracted some criticism (Vlassoff and Manderson, 1998). Vaccines and other treatments have been developed with limited or no information about gender differences in response to treatment, for example, because women were excluded from trials.

DISCUSSION

The available data on gender and tropical diseases are not balanced. Although there may not be 'hard data' to support the importance of gender in all instances, there is sufficient evidence to support a hypothesis that a gender perspective can only enhance the effectiveness of infectious disease control programmes in forests. The main limitation of the use of the gender analysis framework is its potential to become mechanistic and thereby exclude analyses of other forms of social disadvantage (Standing, 1999). However, such analyses remain very useful for assessing priorities and conceptualizing and designing research and intervention programmes (Gender and Health Group, 1999).

Questions remain to be answered on specific diseases and major gaps in the integration of gender perspectives in control programmes in forested regions. Analyses on gender and health also need to move away from the assumption that gender has the same significance in all contexts to a more empirically informed approach (Standing, 1999), based on the context in which the conditions occur. This is particularly true given the potential disease implications of climate change in the coming years (Graczyk, 2002).

Programme planning and implementation can be improved for specific diseases in several ways:

- developing or maintaining a focus on gender, with both sex-disaggregated data and systematically collected qualitative data to explain and enrich the quantitative data;
- providing a balanced account of the gender dynamics that affect the disease and its control, reporting on the multiple factors and the perspectives of both women and men;
- taking account of demographic changes, and the needs of women and men of different ages;
- providing data based on operational research to move from knowledge to practice;[2] and
- including a gender perspective in clinical trials and research on the effects of tropical diseases in forests.

CONCLUSION

Some advances have clearly been made in reducing gender inequality over the past few decades (World Bank, 2001), but gender disparities remain pervasive. This notwithstanding, as with so many aspects of life in forests the nature and extent of the disparities are largely context-dependent and vary across regions. Inappropriate generalizations about the universality of gender disadvantage are largely due to the lack of evidence in many contexts; basic sex-disaggregated data are still unavailable at many levels of health-service planning, implementation and evaluation (Corrigan, 2002; Grant, 2002; Vlassoff and Moreno, 2002). Indicators for the success of programmes that incorporate gender are poorly articulated or inappropriate, a problem that is common in interventions that require broad social change. There is a general paucity of systematic and well-documented evaluations of gender-sensitive interventions. Nevertheless, the evidence on the contribution of gender to the impacts of forest diseases makes mainstreaming gender issues intuitive. The gender perspective provides a framework that allows us to deconstruct social conditions and processes within particular environmental contexts that influence health, and develop interventions, taking account of the different needs of individuals and groups within a population.

The complex interaction of gender, health and forest diseases is clearly demonstrated by gender analysis of particular diseases. What is less obvious is the differential impact of disease on cultural, social and economic life. Much of the burden of these conditions is experienced by young and economically productive age groups. Where people are still highly reliant on primary production (as in most forests), women are responsible for up to three-quarters of the food produced annually; the figure is as high as 80 per cent in many African countries (Tinker, 1994). When home production is included, women earn 40 to 60 per cent of household income and produce about half of export crops.

So far, the gender agenda has been driven by advocacy and the need to introduce measures that would compensate for the extreme disadvantage that forest women suffer. As with most advocacy approaches, identification of the 'victim' provides the justification for a new focus for research. This approach has its shortcomings, however. Primarily, much of the gender and health literature has treated women as a homogeneous group, and most of the examples pertain to women with young children. Older women are almost totally absent in the literature – even though this group has become increasingly important as heads of households where younger parents have been lost to conflict or the HIV pandemic.

Also, the advocacy approach for women gives supremacy to gender over other critical hierarchies and factors – such as class, caste, socioeconomic status, race and ethnicity – around which societies are organized. Some assume that once gender equity and equality are achieved, other issues become easier to resolve. This is reflected, for instance, in the descriptions of the cumulative effects of the multiple layers of disadvantage (Okojie, 1994; Vlassoff, 1995; Vlassoff et al, 1996). However, the complexities arise more from an interactive rather than an additive effect, and thus the provision of opportunities (mainly for women) to improve equity as an intervention may not be sufficient to address the interaction of gender with the other social factors. For instance, health services provided to a woman who has filariasis in a society where her condition is stigmatized may improve her physical health but do little towards overcoming other social circumstances, or indeed preventing reinfection. This supports the argument for mainstreaming gender across all sectors of health and development, and highlights the advantage of comprehensive, broad-ranging programmes that address general wellbeing, basic standards of living, access to education and other principles espoused by the WHO's comprehensive Primary Health Care strategy (Walsh and Warren, 1979; 1980; Rifkin and Walt, 1986; Unger and Killingsworth, 1986; Werner and Sanders, 1997) over disease-specific or even gender-specific programmes.

Efforts to 'mainstream' the gender perspective have intensified within some international organizations and government departments since the Fourth World Conference on Women in Beijing in 1995. The goal is to move from women-specific projects to the integration of gender concerns throughout all aspects of programmes and organizations (Elson and Evers, 1998). A recent comprehensive review by CIFOR in Bogor, Indonesia, recommended that gender be integrated into all its programmes.

However, in general, the commitment has been patchy, and several organizations (including the WHO) have continued to marginalize gender from other areas of health, consulting gender experts only on a programme-specific basis.

Finally, although the gender literature has usefully highlighted points of intervention to address equity, a lack of information about the evolving relationships between men and women in societies, including men's perspectives, constrains the debate and stifles opportunities for broader social interventions. This is particularly true within a male-dominated field like forestry. In the search for equity, the female half of the population has been mobilized with little attention paid to the potential contributions that could be made by the other half of the population. Standing (1999) characterizes this as the 'impact on women' versus 'gender impact' distinction. Gender, by definition, is a comparative construct, and the description of roles and circumstances of one sex needs to be discussed with reference to the other if we are to understand and, where necessary, intervene in ways that are meaningful to both. This has been evident in some programmes that have involved community participation (Manderson et al, 1996). A study investigating the impact of the inclusion of fathers in the decision to immunize children found a significant increase in the uptake and timely completion of the immunization schedule (Brugha et al, 1996).

Vlassoff and Moreno (2002) argue that a commitment to promoting a gender perspective in health policies, planning and intervention needs to start with a better integration of women in the health-services structure at all levels, with equal opportunities (Vlassoff and Moreno, 2002). Insofar as there is interest in addressing the needs of women in forests, there is also a need for greater gender equity within the forestry profession. Gender issues should be integrated into training programmes for all health-service providers and forestry training programmes. The recognition of the importance of gender needs to be reflected in the allocation of resources for programmes. But in addition, the gender agenda needs to include a balance between advocacy, based on a need for social justice and equity, and evidence that supports cost-effective programmes so that interventions actually do improve gender equity and equality.

NOTES

1 This chapter has been revised, incorporating information from two main sources: a fuller review commissioned by the Steering Committee on Strategic Social, Economic and Behavioural Research (SEB) of the Special Programme for Research and Training in Tropical Diseases (TDR) by Allotey and Gyapong (2005) and a CIFOR study by Colfer, Sheil and Kishi (2006). The opinions expressed are those of the authors and do not necessarily reflect the views of TDR, CIFOR or their affiliated institutions.

2 Rapid changes in health systems have blurred the line between the opportunities for empowerment of vulnerable groups, particularly women, and negative factors such as increased costs of access to health care. However, the momentum created by changes in health systems provides an ideal opportunity to introduce a gender perspective and develop the structures necessary for systematic evaluation (Vlassoff and Moreno, 2002). For instance, models that integrate tropical disease programmes into maternal and child health services need to be rigorously assessed within the contexts of the settings where they are implemented.

REFERENCES

Abdulla, S., Armstrong Schellenberg, J., Nathan, R., Mukasa, O., Marchant, T., Smith, T., Tanner, M. and Lengeler, C. (2001) 'Impact on malaria morbidity of a programme supplying insecticide treated nets in children aged under 2 years in Tanzania: Community cross sectional study', *British Medical Journal* 322: 270–273

Abiose, A. (1992) 'Social and behavioral factors affecting participation in mass chemotherapy to control onchocerciasis: A pilot study in women', Rep. ID 900429, WHO TDR, Geneva

Ager, A., Carr, S., Maclachlan, M. and Kanekachilongo, B. (1996) 'Perceptions of tropical health risks in Mponda, Malawi – attributions of cause, suggested means of risk reduction and preferred treatment', *Psychology and Health* 12: 23–31

Agyepong, I. (1992) 'Women and malaria: Social, economic, cultural and behavioural determinants of malaria', in P. Wijeyaratne, E. Rathgeber and E. St-Onge (eds) *Women and Tropical Diseases*, CRDI, Ottawa, Canada, pp176–193

Allotey, P. (1995) 'The burden of illness in pregnancy in rural Ghana: A study of maternal morbidity and interventions in northern Ghana', PhD thesis, University of Western Australia, Perth, Australia

Allotey, P. and Gyapong, M. (2005) *The Gender Agenda in the Control of Tropical Diseases: A Review of Current Evidence*, Special Programme for Research and Training in Tropical Diseases, World Health Organization, Geneva, Switzerland

Allotey, P. and Reidpath, D. (2002) 'Objectivity in priority setting tools in reproductive health: Context and the DALY', *Reproductive Health Matters* 10(20): 38–46

Allotey, P. and Sundari Ravindran, T.K. (2002) 'Gender analysis in the control of malaria: the insecticide treated bed net intervention', in C. Garcia-Moreno and R. Snow (eds) *Gender Analysis and Health*, WHO, Geneva, Switzerland

Alvar, J. (1994) 'Leishmaniasis and AIDS: The Spanish example', *Parasitology Today* 10: 160–163

Anyangwe, E., Njikam, O. and Kouemeni, L. (1992) 'Urinary schistosomiasis in women: An anthropological and descriptive study of a holo-endemic focus in Cameroon', paper presented at meeting on Women and Tropical Diseases, 28–30 April, Oslo, Norway

Armijos, R.X., Weigel, M.M., Izurieta, R., Racine, J., Zurita, C., Herrera, W. and Vega, M. (1997) 'The epidemiology of cutaneous leishmaniasis in subtropical Ecuador', *Tropical Medicine and International Health* 2: 140–152

Australian Bureau of Statistics (2001) 'Australian social trends 1996: Family – family services: Principal carers and their caring roles', http://144.53.252.30/ausstats/abs@.nsf/2f762f95845417aeca25706c00834efa/154746a89736ded1ca2570ec0073d3b6!OpenDocument (alternatively, www.abs.gov.au)

Azogue, E. (1993) 'Women and congenital Chagas disease in Santa Cruz, Bolivia: Epidemiological and sociocultural aspects', *Social Science and Medicine* 37: 503–511

Bandyopadhyay, L. (1996) 'Lymphatic filariasis and the women of India', *Social Science and Medicine* 42: 1401–1410

Boatin, B., Molyneux, D.H., Hougard, J.M., Christensen, O.W., Alley, E.S., Yameogo, L., Seketeli, A. and Dadzie, K.Y. (1997) 'Patterns of epidemiology and control of onchocerciasis in West Africa', *Journal of Helminthology* 71: 91–101

Booth, M., Yuesheng, L. and Tanner, M. (1996) 'Helminth infections, morbidity indicators and schistosomiasis treatment history in three villages, Dongting Lake region, PR China', *Tropical Medicine and International Health* 1: 464–474

Bora, D. (1999) 'Epidemiology of visceral leishmaniasis in India', *National Medical Journal of India* 12: 62–68

Brabin, L. (1990) 'Sex differentials in susceptibility to lymphatic filariasis and implications for maternal child immunity', *Epidemiology and Infection* 105: 335–353

Brabin, L. and Brabin, B. (1992) 'Parasitic infections in women and their consequences', *Advances in Parasitology* 31: 1–81

Brandao-Filho, S.P., Campbell-Lendrum, D., Brito, M.E., Shaw, J.J. and Davies, C.R. (1999) 'Epidemiological surveys confirm an increasing burden of cutaneous leishmaniasis in north-east Brazil', *Transactions of the Royal Society of Tropical Medicine and Hygiene* 93: 488–494

Brieger, W.R., Ososanya, O.O., Kale, O.O., Oshiname, F.O. and Oke, G.A. (1997) 'Gender and ethnic differences in onchocercal skin disease in Oyo State, Nigeria', *Tropical Medicine and International Health* 2: 529–534

Brinkman, U.K., Kramer, P., Presthus, G.T. and Sawadogo, B. (1976) 'Transmission in utero of microfilariae of *Onchocerca volvulus*', *Bulletin of the World Health Organization* 54: 708–709

Brouwer, K.C., Ndhlovu, P.D., Wagatsuma, Y., Munatsi, A. and Shiff, C.J. (2003) 'Epidemiological assessment of *Schistosoma haematobium*-induced kidney and bladder pathology in rural Zimbabwe', *Acta Tropica* 85(3): 339–347

Brugha, R., Kevani, J. and Swan, V. (1996) 'An investigation of the role of fathers in immunization uptake', *International Journal of Epidemiology* 25: 840–845

Cancian, F.M. and Oliker, S.J. (2000) *Caring and Gender*, Sage Publications, London

Cascio, A., Luigi, G., Scarlata, F., Gramiccia, M., Giordano, S., Russo, R., Scalone, A., Camma, C. and Titone, L. (1997) 'Epidemiologic surveillance of visceral leishmaniasis in Sicily, Italy', *American Journal of Tropical Medicine and Hygiene* 57: 75–78

Chitsulo, L., Engels, D., Montresor, A. and Savioli, L. (2000) 'The global status of schistosomiasis and its control', *Acta Tropica* 77: 41–51

Cobra, C., Rigau-Perez, J.G., Kuno, G. and Vorndam, V. (1995) 'Symptoms of dengue fever in relation to host immunologic response and virus serotype, Puerto Rico, 1990–1991', *American Journal of Epidemiology* 142: 1204–1211

Cofie, P. and Adjei, S. (1999) *Promoting Gender Equity in Health*, Health Research Unit, Accra, Ghana

Colfer, C., Sheil, D. and Kishi, M. (2006) 'Forests and human health: Assessing the evidence', CIFOR Occasional Paper 45, CIFOR, Bogor, Indonesia, p111

Coreil, J., Mayard, G., Louischarles, J. and Addiss, D. (1998) 'Filarial elephantiasis among Haitian women – social context and behavioural factors in treatment', *Tropical Medicine and International Health* 3: 467–473

Corredor Arjona, A., Alvarez Moreno, C.A., Agudelo, C.A., Bueno, M., Lopez, M.C., Caceres, E., Reyes, P., Duque Beltran, S., Gualdron, L.E. and Santacruz, M.M. (1999) 'Prevalence of *Trypanosoma cruzi* and *Leishmania chagasi* infection and risk factors in a Colombian indigenous population', *Revista do Instituto de Medicina Tropical de Sao Paulo* 41: 229–234

Corrigan, O.P. (2002) 'A risky business: The detection of adverse drug reactions in clinical trials and post-marketing exercises', *Social Science and Medicine* 55: 497–507

Coura, J.R., Junqueira, A.C.V., Fernandes, O., Valente, S.A.S. and Miles, M.A. (2002) 'Emerging Chagas disease in Amazonian Brazil', *Trends in Parasitology* 18(4): 171–176

Courtright, P., Johnston, K., and Chitsulo, L. (1995) 'A new focus of onchocerciasis in Mwanza District, Malawi', *Transactions of the Royal Society of Tropical Medicine and Hygiene* 89: 34–36

da Silva Valente, S.A., de Costa Valente, V. and Neto, H.F. (1999) 'Considerations on the epidemiology and transmission of Chagas disease in the Brazilian Amazon', *Memorias do Instituto Oswaldo Cruz* 94: 395–398

de Almeida, A.B. and Freedman, D.O. (1999) 'Epidemiology and immunopathology of Bancroftian filariasis', *Microbes and Infection* 1: 1015–1022

Dias, J.C.P. (1996) 'Tropical diseases and the gender approach', *Bulletin of the Pan American Health Organization* 30: 242–260

Dreyer, G., Noroes, J. and Addiss, D. (1997) 'The silent burden of sexual disability associated with lymphatic filariasis', *Acta Tropica* 63: 57–60

Eberhard, M., Walker, E., Addiss, D. and Lammie, P. (1996) 'A survey of knowledge, attitudes and perceptions of lymphatic filariasis, elephantiasis and hydrocele among residents of an endemic area in Haiti', *American Journal of Tropical Medicine and Hygiene* 54(3): 299–303

El-Khoby, T., Galal, N., Fenwick, A., Barakat, R., El-Hawey, A., Nooman, Z., Habib, M., Abdel-Wahab, F., Gabr, N.S., Hammam, H.M., Hussein, M.H., Mikhail, N.N., Cline, B.L. and Strickland, G.T. (2000) 'The epidemiology of schistosomiasis in Egypt: Summary findings in nine governorates', *American Journal of Tropical Medicine and Hygiene* 62: 88–99

Elson, D. and Evers, B. (1998) *Sector Programme Support: The Health Sector, a Gender Aware Analysis,* Genecon Unit, University of Manchester, Manchester

Ezeh, A. (1993) 'The influence of spouses over each other's contraceptive attitudes in Ghana', *Studies in Family Planning* 24: 163–174

Fatima, N. (1991) 'The plight of rural women', in S.L. Raj (ed) *Quest for Gender Justice: A Critique of the Status of Women in India,* Satya Nilayam Publishers, Madras, India, pp12–26

Feldmeier, H., Poggensee, G. and Krantz, I. (1993) 'A synoptic inventory of needs for research on women and tropical parasitic diseases. II. Gender-related biases in the diagnosis and morbidity assessment of schistosomiasis in women', *Acta Tropica* 55: 139–169

Feldmeier, H., Poggensee, G., Krantz, I. and Helling-Giese, G. (1995) 'Female genital schistosomiasis: New challenges from a gender perspective', *Tropical and Geographical Medicine* 47: S2–15

Ferguson, P. (2000) 'Testing a drug during labour: The experiences of women who participated in a clinical trial', *Journal of Reproductive and Infant Psychology* 18: 117–131

Fonn, S. and Xaba, M. (1995) *Health Workers for Change: A Manual to Improve Quality of Care,* UNDP/World Bank/WHO Special Programme for Research and Training in Tropical Disease, Geneva, Switzerland

Fulford, A.J.C., Ouma, J.H., Kariuki, H., Thiongo, F., Klumpp, R., Sturrock, R.F. and Butterworth, A.E. (1996) 'Water contact observations in Kenyan communities endemic for schistosomiasis: Methodology and patterns of behaviour', *Parasitology* 113: 223–241

Gbakima, A. and Sahr, F. (1996) 'Filariasis in the Kaiyamba chiefdom, Moyamba district, Sierra Leone – an epidemiological and clinical study', *Public Health* 110: 169–174

Gender and Health Group (1999) *Guidelines for the Analysis of Gender and Health,* Liverpool School of Tropical Medicine, Liverpool

Graczyk, T.K. (2002)' Zoonotic infections and conservation', in A.A. Aguirre, R.S. Ostfeld, G.M. Tabor, C. House and M.C. Pearl (eds) *Conservation Medicine: Ecological Health in Practice,* Oxford University Press, Oxford, pp220–228

Grant, K. (2002) 'Gender-based analysis: Beyond the Red Queen syndrome', *Centres of Excellence for Women's Health Research Bulletin* 2: 16–19

Gratz, N.G. and Knudsen, A.B. (1997) 'The rise and spread of dengue, dengue haemorrhagic fever and its vectors, 1950–1990', *Dengue Bulletin* 21

Gwatkin, D. and Guillot, M. (1998) 'The burden of tropical diseases and the poorest and richest 20 per cent of the global population', Special Programme on Research and Training in Tropical Diseases (TDR), Geneva, Switzerland

Gyapong, J.O. (2000) 'Lymphatic filariasis in Ghana: From research to control', *Transactions of the Royal Society of Tropical Medicine and Hygiene* 94: 599–601

Gyapong, J.O., Adjei, S. and Sackey, S.O. (1996a) 'Descriptive epidemiology of lymphatic filariasis in Ghana', *Transactions of the Royal Society of Tropical Medicine and Hygiene* 90: 26–30

Gyapong, J.O., Gyapong, M. and Adjei, S. (1996b) 'The epidemiology of acute adenolymphangitis due to lymphatic filariasis in northern Ghana', *American Journal of Tropical Medicine and Hygiene* 54: 591–595

Gyapong, M., Gyapong, J., Weiss, M. and Tanner, M. (2000) 'The burden of hydrocele on men in northern Ghana', *Acta Tropica* 77: 287–294

Haaland, A. and Vlassoff, C. (2001) 'Introducing health workers for change: From transformation theory to health systems in developing countries', *Health Policy and Planning* 16: 1–16

Halstead, S.B. (1990) 'Global epidemiology of dengue hemorrhagic fever', *Southeast Asian Journal of Tropical Medicine and Public Health* 21: 636–641

Hayes, E. and Gubler, D. (1992) 'Dengue and dengue hemorrhagic fever', *Pediatric Infectious Diseases Journal* 11: 311–317

Helitzer-Allen, D., Kendall, C. and Wirima, J. (1993) 'The role of ethnographic research in malaria control: An example from Malawi', *Research in Sociology and Health Care* 10: 269–286

Hibbard, J. and Pope, B. (1983) 'Gender roles, illness orientation and use of medical services', *Social Science and Medicine* 17: 129–137

Hightower, A.W., Lammie, P.J. and Eberhard, M.L. (1993) 'Maternal filarial infection – a persistent risk factor for microfilaremia in offspring', *Parasitology Today* 9: 418–421

Hudelson, P. (1996) 'Gender differentials in tuberculosis: The role of socio-economic and cultural factors', *Tubercle and Lung Disease* 77: 391–400

Hunter, J.M. (2003) 'Inherited burden of disease: Agricultural dams and the persistence of bloody urine (*Schistosomiasis hematobium*) in the upper east region of Ghana, 1959–1997', *Social Science and Medicine* 56: 219–34

Isiugo-Abanihe, U. (1994) 'Reproductive motivation and family size preference among Nigerian men', *Studies in Family Planning* 25: 149–161

Katabarwa, M.N., Habomugisha, P., Ndyomugyeni, R. and Agunyo, S. (2001) 'Involvement of women in community directed treatment with Ivermectin for the control of onchocerciasis in Rukungiri district, Uganda: A knowledge, attitude and practise study', *Annals of Tropical Medicine and Parasitology* 95: 485–494

Kaur, V. (1997) 'Tropical diseases and women', *Clinics in Dermatology* 15: 171–178

Kazura, J.W. (1997) 'Filariasis and onchocerciasis', *Current Opinion in Infectious Diseases* 10: 341–344

Khonde, N., Pepin, J., Niyonsenga, T. and Dewals, P. (1997) 'Familial aggregation of *Trypanosoma brucei gambiense* trypanosomiasis in a very high incidence community in Zaire', *Transactions of the Royal Society of Tropical Medicine and Hygiene* 91: 521–524

Khoury, S., Saliba, E.K., Oumeish, O.Y. and Tawfig, M.R. (1996) 'Epidemiology of cutaneous leishmaniasis in Jordan: 1983–1992', *International Journal of Dermatology* 35: 566–569

Kisekka, M. (1995) 'Research for the development of the healthy women's counselling guide: Makarfi local government area, Kaduna', Rep. 931070, WHO/TDR, Geneva, Switzerland

Klaus, S.N., Frankenburg, S. and Ingber, A. (1999) 'Epidemiology of cutaneous leishmaniasis', *Clinics in Dermatology* 17: 257–260

Kunstadter, P. (2003) 'Poverty, gender and malaria', draft training module for WHO, University of California, San Francisco, CA/WHO, Geneva, Switzerland

Lampietti, J.A., Poulos, C., Cropper, M.L., Mitiku, H. and Whittington, D. (1999) 'Gender and preferences for malaria prevention in Tigray, Ethiopia', in 'Policy research report on gender and development', Working Paper Series No. 3, World Bank Development Research Group and Poverty Reduction and Economic Management Network, World Bank, Washington, DC

Liefooghe, R., Baliddawa, J.B., Kipruto, E.M., Vermeire, C. and de Munynck, A.O. (1997) 'From their own perspective: A Kenyan community's perception of tuberculosis', *Tropical Medicine and International Health* 2: 809–821

Lu, A., Valencia, L., de las Llagas, L., Aballa, L. and Postrado, L. (1988) 'Filariasis: A study of knowledge, attitudes and practices of the people of Sorsogon', Report No. 1, TDR, Geneva, Switzerland

McBride, W.J. and Bielefeldt-Ohmann, H. (2000) 'Dengue viral infections: Pathogenesis and epidemiology', *Microbes and Infection* 2: 1041–1050

McGowan, J. and Pottern, L. (2000) 'Commentary on the Women's Health Initiative', *Maturitas* 34: 109–112

Manderson, L., Mark, T. and Woelz, N. (1996) 'Women's participation in health and development projects', Tropical Health Program, ACITHN, University of Queensland, Rep. WHO/TDR/GTD/RP/96.1, World Health Organization, Geneva, Switzerland

Marty, P., Le Fichoux, Y., Pratlong, F. and Gari-Toussaint, M. (1994) 'Human visceral leishmaniasis in Alpes-Maritimes, France: Epidemiological characteristics for the period 1985–1992', *Transactions of the Royal Society of Tropical Medicine and Hygiene* 88: 33–34

Meinert, C., Gilpin, A., Unalp, A. and Dawson, C. (2000) 'Gender representation in trials', *Controlled Clinical Trials* 21: 462–475

Michael, E., Bundy, D.A. and Grenfell, B.T. (1996) 'Re-assessing the global prevalence and distribution of lymphatic filariasis', *Parasitology* 112: 409–428

Michelson, E.H. (1993) 'Adam's rib awry – women and schistosomiasis', *Social Science and Medicine* 37: 493–501

Molyneux, D.H. and Morel, C. (1998) 'Onchocerciasis and Chagas-disease control – the evolution of control via applied research through changing development scenarios', *British Medical Bulletin* 54: 327–339

Moncayo, A. (1992) 'Chagas disease: Epidemiology and prospects for interruption of transmission in the Americas', *World Health Statistics Quarterly – Rapport Trimestriel de Statistiques Sanitaires Mondiales* 45: 276–279

Moncayo, A. (1999) 'Progress towards interruption of transmission of Chagas disease', *Memorias do Instituto Oswaldo Cruz* 94: 401–404

Morrison, A. (2000) 'A woman with leprosy is in double jeopardy', *Leprosy Review* 71: 128–143

Murray, C. (1991) 'Social, economic and operational research on tuberculosis: Recent studies and some priority questions', *Bulletin of the International Union Against Tuberculosis and Lung Disease* 66: 149–156

Murray, C.J.L. and Lopez, A.D. (eds) (1996) *The Global Burden of Disease, Vol. 1,* Harvard School of Public Health, Cambridge, MA

Nakazawa, M., Ohtsuka, R., Toshio, K., Tetsuro, H., Tsuguyoshi, S., Tsukasa, I., Tomoya, A., Shigeyuki, K. and Mamoru, S. (1994). 'Differential malaria prevalence among villages of Gidra in lowland Papua New Guinea', *Tropical and Geopraphical Medicine* 46(6): 350–354

Nations, M.K. (1986) 'Epidemiological research on infectious disease: Quantitative rigor or rigormortis? Insights from ethnomedicine', in C. Janes, R. Stall and S. Gifford (eds) *Anthropology and Epidemiology,* Reidel Publishers, Dordrecht, The Netherlands, pp97–123

Nchinda, T. (1998) 'Malaria: A re-emerging disease in Africa', *Emerging Infectious Diseases* 4: 398–403

Ogunrinade, A., Boakye, D., Merriweather, A. and Unnasch, T.R. (1999) 'Distribution of blinding and nonblinding strains of *Onchocerca volvulus* in Nigeria', *The Journal of Infectious Diseases* 179: 1577–1579

Okojie, C. (1994) 'Gender inequalities of health in the Third World', *Social Science and Medicine* 39: 1237–1247

Oliveira, G. and de Pessini, M. (1997) 'The effects of leprosy on men and women: A gender study', Gender and Tropical Diseases Resource Paper No. 4, WHO, Geneva, Switzerland

Ottesen, E.A., Duke, B.O.L., Karam, M. and Behbehani, K. 1997, 'Strategies and tools for the control/elimination of lymphatic filariasis', *Bulletin of the World Health Organization* 75: 491–503

Pani, S., Balakrishnan, N., Srividya, A., Bundy, D. and Grenfell, B. (1991) 'Clinical epidemiology of Bancroftian filariasis: Effect of age and gender', *Transactions of the Royal Society of Tropical Medicine and Hygiene* 85: 260–264

Pani, S.P., Srividya, A., Krishnamoorthy, K., Das, P.K. and Dhanda, V. (1997) 'Rapid assessment procedures (RAP) for lymphatic filariasis', *National Medical Journal of India* 10: 19–22

Patz, J.A., Graczyk, T.K., Geller, N. and Vittor, A.Y. (2000) 'Effects of environmental change on emerging parasitic diseases', *International Journal for Parasitology* 30(12–13): 1395–1405

Polderman, A.M. (1995) 'Gender-specific schistosomiasis: Why?', *Tropical and Geographical Medicine* 47: S1

Potash, B. (1995) 'Women in the changing African family', in M.J. Hay and S. Stichter (eds) *African Women South of the Sahara*, Longman Group, New York, NY

Raikes, A., Shoo, R. and Brabin, L. (1992) 'Gender planned health services', *Annals of Tropical Medicine and Parasitology* 86: 19–23

Rao, S., Garole, V., Walawalkar, S., Khot, S. and Karandikar, N. (1996) 'Gender differentials in the social impact of leprosy', *Leprosy Review* 67: 190–199

Rashed, S., Johnson, H., Dongier, P., Moreau, R., Lee, C., Crépeau, R., Lambert, J., Jefremovas, V. and Schaefer, C. (1999) 'Determinants of the permethrin impregnated bednets (PIB) in the Republic of Benin: The role of women in the acquisition and utilization of PIBs', *Social Science and Medicine* 49: 993–1005

Raviglione, M.C., Snider, D.E. and Kochi, A. (1995) 'Global epidemiology of tuberculosis: Morbidity and mortality of a worldwide epidemic', *JAMA* 273: 220–226

Reidpath, D.D., Allotey, P., Kouamé, A. and Cummins, R.A. (2001) *Social, Environmental and Cultural Contexts and the Measurement of the Burden of Disease: An Exploratory Comparison in the Developed and Developing World*, Key Centre for Women's Health in Society, University of Melbourne, Melbourne, Australia

Reuben, R. (1993) 'Women and malaria: special risks and appropriate control strategy', *Social Science and Medicine* 37: 473–480

Rifkin, S. and Walt, G. (1986) 'Why health improves: Defining the issues concerning comprehensive primary health care and selective primary health care', *Social Science and Medicine* 23: 559–566

Sahoo, P.K., Geddam, J.J.B., Satapathy, A.K., Mohanty, M.C. and Ravindran, B. (2000) 'Bancroftian filariasis: Prevalence of antigenaemia and endemic normals in Orissa, India', *Transactions of the Royal Society of Tropical Medicine and Hygiene* 94: 515–517

Schiebinger, L. (1993) *Nature's Body: Gender in the Making of Modern Science*, Beacon Press, Boston, MA

Shekhar, K.C. and Huat, O.L. (1992) 'Epidemiology of dengue/dengue hemorrhagic fever in Malaysia – a retrospective epidemiological study 1973–1987, Part I: Dengue hemorrhagic fever (DHF)', *Asia Pacific Journal of Public Health* 6: 15–25

Simonsen, P.E., Meyrowitsch, D.W., Makunde, W.H. and Magnussen, P. (1995) 'Bancroftian filariasis – the pattern of microfilaraemia and clinical manifestations in three endemic communities of northeastern Tanzania', *Acta Tropica* 60: 179–187

Smith, D.H., Pepin, J. and Stich, A.H.R. (1998) 'Human African trypanosomiasis: An emerging public health crisis', *British Medical Bulletin* 54: 341–355

Standing, H. (1999) *Frameworks for Understanding Gender Inequalities and Health Sector Reform: An Analysis and Review of Policy Issues*, Global Health Equity Initiative Project on Gender and Health Equity, Harvard Center for Population and Development Studies, Cambridge, MA

Steketee, R. and Mutabingwa, T. (1999) 'Malaria in pregnant women: Research, epidemiology, policy and practice', *Annals of Tropical Medicine and Parasitology* 93: S7–S9

Steketee, R., Wirima, J., Slutsker, L., Heymann, D. and Breman, J. (1996) 'The problem of malaria and malaria control in pregnancy in sub-Saharan Africa', *American Journal of Tropical Medicine and Hygiene* 55: 2–7

Tanner, M. and Vlassoff, C. (1998) 'Treatment seeking behaviour for malaria: A typology based on endemicity and gender', *Social Science and Medicine* 46: 523–532

Terhell, A.J., Price, R., Koot, J.W.M., Abadi, K. and Yazdanbakhsh, M. (2000) 'The development of specific IgG4 and IgE in a paediatric population is influenced by filarial endemicity and gender', *Parasitology* 121: 535–543

Thakur, C.P. (2000) 'Socio-economics of visceral leishmaniasis in Bihar (India)', *Transactions of the Royal Society of Tropical Medicine and Hygiene* 94: 156–157

Thomas, D. (1997) 'Incomes, expenditures and health outcomes: Evidence on intrahousehold resource allocation', in L. Haddad, J. Hoddinot and H. Alderman (eds) *Intrahousehold Resource Allocation in Developing Countries*, Johns Hopkins, Baltimore, MD

Thomson, M.C., Obsomer, V., Dunne, M., Connor, S. and Molyneux, D. (2000) 'Satellite mapping of Loa loa prevalence in relation to ivermectin use in west and central Africa', *The Lancet* 356: 1077–1078

Thomson, M.C., Obsomer, V. Kamgno, J., Gardon, J. Wanji, S. Takougang, I. Enyong, P. Remme, J.H., Molyneux, D.H. and Boussinesq, M. (2004) 'Mapping the distribution of Loa loa in Cameroon in support of the African Programme for Onchocerciasis Control', *Filaria Journal* 3: 7

Tinker, A., Daly, P., Green, C., Saxenian, H., Lakshminarayanan, R. and Gill, K. (1994) *Women's Health and Nutrition: Making a Difference*, World Bank, Washington, DC

Triteeraprapab, S. and Songtrus, J. (1999) 'High prevalence of Bancroftian filariasis in Myanmar-migrant workers: A study in Mae Sot district, Tak province, Thailand', *Journal of the Medical Association of Thailand* 82: 735–739

Twigg, J., Atkin, K. and Perring, C. (1990) *Carers and Services: A Review of Research*, HMSO, London

Unger, J. and Killingsworth, J. (1986) 'Selective primary health care: A critical review of methods and results', *Social Science and Medicine* 22: 1001–1013

Velez, I., Hendricks, E., Roman, O. and del Pilar Agudelo, S. (1997) 'Gender and leishmaniasis in Columbia: A redefinition of existing concepts', Rep. WHO/TDR/GTD/RP/97.1, PECET, University of Anitoquia, Medellin, Colombia/WHO, Geneva, Switzerland

Vivader, R., Lafleur, B., Tong, C., Bradshaw, R. and Marts, S. (2000) 'Women subjects in NIH-funded clinical research literature: Lack of progress in both representation and analysis by sex', *Journal of Women's Health and Gender-Based Medicine* 9: 495–504

Vlassoff, C. (1995) 'Gender inequalities in health in the Third World – uncharted ground', *Social Science and Medicine* 39: 1249–1259

Vlassoff, C. (1997) 'The gender and tropical diseases task force of TDR: Achievements and challenges', *Acta Tropica* 67: 173–180

Vlassoff, C. and Bonilla, E. (1994) 'Gender-related differences in the impact of tropical diseases on women: What do we know?', *Journal of Biosocial Science* 26: 37–53

Vlassoff, C., Khot, S. and Rao, S. (1996) 'Double jeopardy: Women and leprosy in India', *World Health Statistics Quarterly – Rapport Trimestriel de Statistiques Sanitaires Mondiales* 49: 120–126

Vlassoff, C. and Manderson, L. (1998) 'Incorporating gender in the anthropology of infectious diseases', *Tropical Medicine and International Health* 3: 1011–1019

Vlassoff, C. and Moreno, C.G. (2002) 'Placing gender at the centre of health programming: Challenges and limitations', *Social Science and Medicine* 54: 1713–1723

Vlassoff, C., Weiss, M., Ovuga, E.B., Eneanya, C., Nwel, P.T., Babalola, S.S., Awedoba, A.K., Theophilus, B., Cofie, P. and Shetabi, P. (2000) 'Gender and the stigma of onchocercal skin disease in Africa', *Social Science and Medicine* 50: 1353–1368

Walsh, J.F., Molyneaux, D.H. and Birley, M.H. (1993) 'Deforestation: Effects on vector-borne disease', *Parasitology* 106: S55–S75

Walsh, J.A. and Warren, K.S. (1979) 'Selective primary health care: An interim strategy for disease control in developing countries', *New England Journal of Medicine* 301(18): 967–974

Walsh, J. and Warren, K. (1980) 'Selective primary health care: An interim strategy for disease control in developing countries', *Social Science and Medicine* 14C: 145–163

Weigel, M.M., Armijos, R.X., Racines, R.J., Zurita, C., Izurieta, R., Herrera, E. and Hinojsa, E. (1994) 'Cutaneous leishmaniasis in subtropical Ecuador: Popular perceptions, knowledge, and treatment', *Bulletin of the Pan American Health Organization* 28: 142–155

Werner, D. and Sanders, D. (1997) *Questioning the Solution: The Politics of Primary Health Care and Child Survival*, HealthWrights, Palo Alto, CA

Wernsdorfer, G. and Wernsdorfer, W. (1988) 'Social and economic aspects of malaria and its control', in I. McGregor (ed) *Malaria: Principles and Practice of Malariology*, Churchill-Livingston, Edinburgh, p1421–1471

Whiteford, L.M. (1997) 'The ethnoecology of dengue fever', *Medical Anthropology Quarterly* 11: 202–223

WHO (World Health Organization) (1995) 'Onchocerciasis and its control – report of a WHO expert committee', Rep. Series 852, WHO, Geneva, Switzerland

WHO (1997) 'Chagas disease', in *Tropical Disease Research: Progress 1995–96. 13th Programme Report of the UNDP/World Bank/WHO Special Programme for Research and Training in Tropical Diseases*, World Health Organization, Geneva, Switzerland, pp112–123

WHO (2000) 'WHO report on global surveillance of epidemic-prone infectious diseases', Rep. WHO/CDS/CSR/ISR/2000.1, Communicable Diseases Surveillance and Response (CSR), WHO, Geneva, Switzerland

WHO (2001a) 'WHO report 2001: Global tuberculosis control', Rep. WHO/CDS/TB/2001.287, Communicable Diseases, WHO, Geneva, Switzerland

WHO (2001b) 'World Health Report 2001: Mental health: new understanding, new hope', WHO, Geneva, Switzerland

Wijeyaratne, P.M., Jones Arsenault, L.K. and Murphy, C.J. (1994) 'Endemic disease and development: The leishmaniasis', *Acta Tropica* 56: 349–364

World Bank (1993) *Investing in Health: World Development Report*, Oxford University Press, New York, NY

World Bank (2001) 'Power, incentives and resources at the household level', in *Engendering Development – Through Gender Equality in Rights, Resources, and Voice*, World Bank and Oxford University Press, Washington, DC

Yumkella, F. (1996) 'Women, onchocerciasis and Ivermectin in Sierra Leone', Rep. WHO/TDR/GTD/RP/96.2, TDR, WHO, Geneva, Switzerland

8
Bat-Borne Viral Diseases

Jean Paul Gonzalez, Meriadeg Ar Gouilh, Jean-Marc Reynes and Eric Leroy

L'Homme modifie le milieu
et le milieu modifie telle partie de l'Homme
directement rattachée à la modification du milieu.

(Man [sic] changes the environment
and the environment will change that part of Man [sic]
directly linked to actual environmental change.)
(Krisnamurthi, 1964)

More than 50 million years ago, long before humankind came to light, bats were already closely associated with primary forests. They adapted to a changing environment and spread early to occupy a variety of ecological niches where they evolved and etched their natural history. Much later, humans came for a first encounter, probably in a forested environment where a tribe was hunting or gathering fruits from trees, or a clan found shelter in a cave entrance already inhabited by bats flying out at night. Human and bat domains did not naturally overlap, but their paths would cross for brief encounters.

Bats have fascinated people for centuries. Bats are present in the mythology of the peoples who populated the Americas in the pre-Columbian era. Petroglyphs in the great city of Copan (in present-day Honduras), for example, represent an anthropomorphic head of a leaf-nosed bat (Phyllostomidae) (Figure 8.1), and tribes of the Maya kingdom revered a vampire bat god of the underworld with bloodsucking servants who could turn into bats at will, a legend that perhaps inspired our comparatively modern myth of Dracula. Interestingly, Dracula appears as a vector of the Black Death, and also transmits his evil power (by saliva) to the blood, and 'the transmission of vampirism strictly mimics an infectious transmission by bite (from bats), with the (infected) bitten person becoming a vampire' (Vidal et al, 2007).

Conversely, bats in Asian culture are revered animals, sources of luck and wealth. The saying 'five bats and a peach' represents the five blessings of longevity, wealth, health, virtuousness and a peaceful end.

In this chapter, we present what scientists have learned about these unfamiliar mammals as it relates to bat-borne transmitted diseases. We analyse how, on ecological grounds, the cycle of virus transmission involving bats, humans and forested environments

Figure 8.1 *Representation of an anthropomorphic head of a leaf-nosed bat (Phyllostomidae)*

Source: Meriadeg Ar Gouilh (Chinese ink on paper)

can be understood. We describe important bat-borne viral diseases and unveil the fundamentals of emergence, including environmental and human factors and bat behaviour in forest areas (Kunz and Brock Fenton, 2003; Simmons, 2005a).

NATURAL HISTORY OF BATS

Bats are nocturnal flying mammals with both wild and peri-domestic habits. Their interactions with humans have been poorly studied. Because of their night-time behaviour, their sense of echolocation, the blood-feeding habit of some species and the colonization of caves or house attics by other species, bats have been perceived as bizarre mammals and are often mistakenly considered as potential risks to human wellbeing (Kunz and Brock Fenton, 2003).

Representing more than 20 per cent of the known mammal species, the Chiroptera order, with more than 1000 species, is one of the most diversified mammalian orders, after Rodentia (Simmons, 2005b). Chiropterans are the only mammals that have acquired powered flight. Bats have specialized forelimbs, with light skeletons that support the wing-skin membrane, the 'patagium'. They also have unique and sophisticated laryngeal echo-location, which allows them to find flying prey as well as wheel inside dark caves. This substitute for vision is well developed and is the basis for a traditional separation into two distinct groups: Microchiroptera (often insectivorous and using echolocation) and Megachiroptera (mostly frugivorous and without echolocation). Although universally accepted, this classification has been revised in light of genetics (Teeling et al, 2005).

Table 8.1 provides a succinct summary of the families, sub-families and habitats of bat species. Most of the 18 bat families are either Old World or New World. The Vespertilionidae (evening bats, mainly insectivorous) and the Emballonuridae (sheath-tailed bats, insectivorous and rarely frugivorous) families have a worldwide distribution. Recent data suggest that bats originated in Laurasia – North America and Eurasia when a single continent – in the early Eocene, a time of plant and insect diversification (Simmons, 2005a).

Feeding

Bats' feeding behaviour has implications for the health of both forests and people. Frugivorous (fructivorous) bats, or 'fruit bats', belong to Megachiroptera; some are Phyllostomidae. They locate ripe fruits by smell, chew them and disperse seeds in faeces, thus playing an important role in regeneration in tropical forests.

Nectar-eating bats are small, specialized pollinators. They collect nectar and pollen with their long tongues and locate night-blooming flowers by scent or sight.

The more than 600 species of insectivorous bats feed on insects by flying close to ground or water. Good echolocators, some can cue in on insect calls and some can anticipate the trajectory of flying prey. Some bats can adapt, reduce the intensity of their outgoing pulses, and produce higher-frequency signals not detected by insects. Small bats trap insects in their wings and sweep them into the mouth.

Carnivorous bats eat vertebrate prey (rodents, birds and other bats). Phyllostomids eat frogs and can distinguish the calls of edible and poisonous species.

The piscivorous bats of South America are noctilionids. By echolocation they detect the ripples fish make on the water's surface. They catch fish up to 12cm long with their gaff-like claws by dragging their feet through the water.

Hematophagous (sanguivorous) bats, the so-called vampire bats, like other blood-feeding animals, are a public health concern and a potential risk when blood meals are taken from humans. These specialized carnivores are limited to Central and South America, where they feed on domestic livestock. The bats have good eyesight and are heat-sensitive; they select prey whose skin area has dense surface capillaries. With their blade-like canines, they make small, painless incisions – keeping the prey's blood flowing with an anti-coagulant in their saliva – and lap up the blood. To ensure a light body for flight after eating, they urinate as feeding begins.

Migration and hibernation

Several bat species migrate locally, but Asian and African bats and Australian flying fox species can cover hundreds of kilometres a night during the breeding season. The Mexican free-tailed bat can ascend more than 3000 metres to catch tailwinds that carry it over long distances at speeds of more than 100km per hour. Flight is metabolically costly, and bats must eat and compete in cold climates. Depending on seasonal food sources, they migrate to warmer places and food sources. Bats hibernate and, by holding their breath, reduce heart rate and body temperature.

Roosting

While resting, bats hang upside-down in groups in a cave or a tree, depending on species and feeding requirements. Bats roost singly or communally (in groups of a few to millions), often with different species. They have adapted to changing environments and occupy house attics, bridges and unused buildings. In tropical biomes, pteropodids and phyllostomids roost in tents made of specific plant leaves, another example of adaptation

Table 8.1 *Chiropteran families: Main habitats and characteristics*

Family (common name): sub-families	*Habitat*
MEGACHIROPTERA (SUB-ORDER)	
Pteropodidae (fruit bats, flying foxes)	Old World fruit bats: trees of the tropical dense forests; nocturnal, hang from their feet during the day with their wings wrapped around their bodies. India, Africa, Asia, Australia.
MICROCHIROPTERA	
Emballonuridae (sac-winged bat or sheath-tailed bats)	Inhabit lowland evergreen or semi-deciduous forests, roost in territorial small colonies in hollow trees, hunt below the canopy along forest edges. Worldwide, tropical and subtropical regions.
Rhinopomatidae (mouse-tailed bats)	Treeless arid regions; roost in caves, rock clefts, wells, houses and pyramids. North and West Africa, Middle East, Pakistan, Afghanistan, Thailand.
Craseonycteridae (bumblebee bats)	Caves in previously forested area. Thailand.
Nycteridae (slit-faced bats)	Caves, hollow logs, tree branches, tunnels, human houses. Tropical forests and arid tropical regions, Africa, Southeast Asia.
Megadermatidae (false vampire)	Caves, rock crevices, buildings or trees. Tropics and subtropics, Old World (Central Africa, South Asia, Malaysia region, Philippines and Australia).
Rhinolophidae (horseshoe bats)	High and low elevations, forests and open spaces, caves, tree holes, foliage, mines, buildings. Tropical and desert biomes (Europe, Africa, Asia, northern and eastern Australia, Pacific islands). Old World leaf-nosed bat = Hipposiderinae (now promoted to a family rank).
Mormoopidae (moustached bats)	Roost in caves and tree hollows. Highly specialized echolocation system. Central and South America.
Noctilionidae (bulldog bats)	Near water, hollow trees, deep cracks in rocks. North Mexico, Central America to Paraguay and Argentina.
Phyllostomidae (New World leaf-nosed bat)	Tropical deciduous forests and rainforests, temperate rainforests. Lower elevations ranging up to 1500 metres (southwestern USA to northern Argentina, West Indies).
Mystacinidae (short-tailed bats)	Endangered by hunting in indigenous forests, roost singly or communally in hollow trees. Go into a 'torpor'. New Zealand.

Myzopodidae (Old World sucker-footed bats)	Roost head-up inside unfurled banana or *Heliconia* leaves. Suction cups at wrists and ankles. Madagascar.
Thyropteridae (disk-winged bats)	Neo-tropical rainforests. Roost head-up inside banana or *Heliconia* leaves. Madagascar.
Furipteridae (thumbless bats)	Lowland rainforest to arid western deserts (South America).
Natalidae (funnel-eared bats)	Caves and mines. New World: Mexico to Brazil, West Indies.
Molossidae (free-tailed bats)	Solitary or colonies, usually in caves. Fast-flying aerial insectivores. USA to Argentina, southern Europe, Africa, tropical and subtropical Asia and Australia.
Vespertlionidae* (vesper bats or evening bats)	Caves, mine-shafts, tunnels, tree roosts, rock crevices, buildings, etc. Worldwide distribution. Insectivorous from deserts and grasslands, roost in rock crevices or buildings, less often in caves.
Antrozoidae (pallid bats)	From western Canada to central Mexico.

*Incertae cedis

within their natural forested environment and a result of a strong association with plants and a long-term co-evolution. Flying foxes have large communal roosts; they remove leaves from the trees to improve lines of sight, or fold over large leaves for protection from insect bites, rain and predators.

Ecology and reproduction

Most bat species give birth once per year, with the young quickly acquiring autonomy. However, the parturition schedule and thus the population structure can be influenced by fertility rates related to biogeographic zone and climatic conditions. Many tropical species can give birth more than once a year.

Chiropteran groups are often variable and the populations fluctuate widely in sex ratio and age structure (Kunz, 1982). This suggests a strong mobility (daily or seasonal) of individuals and groups. Fruit-bat mobility is conditioned by food abundance: in case of food shortage, the group can move hundreds of kilometres to another feeding source.

Bats have predators and can be attacked and eaten by birds of prey (e.g., owls) or carnivores. Bat roosts are targets for snakes and hawks, as are juveniles that cannot yet fly. New World species (*Vampyrum spectrum*, *Chirotopterus auritus*) and Old World species of the *Megaderma* genus are known to prey on other bats.[1]

Bats bite when they fight each other for territory in a cave or tree canopy. However, they do not normally approach humans; bat bites to humans are usually defensive, and occur mostly when people handle sick, moribund or trapped bats. A primary mode of direct infection to humans is by infected saliva; aerosolized infected biological products are also an important basis of bat-borne disease epidemiology.

Bats and microbes and parasites

Bats carry a variety of endoparasites (such as worms, bacteria, fungi, haemoparasites and protozoans), but these pose a threat for human beings only rarely (Hill and Smith, 1984). One of the best-known such parasites is a soil fungus, *Histoplasma capsulatum*. Histoplasmosis can affect the lungs when humans enter bat roosting places and inhale airborne *Histoplasma* spores of infected soils enriched with bat droppings. Among the 23 bacteria genera recorded in Chiroptera, nine (*Salmonella typhymurium* and *anatum*, *Shigella* sp, *Yersinia pseudotuberculosis*, *Mycobacterium bovis*, *Leptospira* sp, *Bartonnella* sp, *Borrelia recurrentis*, *Coxiella burnetiihas*) are known to be pathogenic for humans, but documentation of transmission is lacking. Among bats' haemoparasites, 19 species of trypanosomes have been reported, and *Trypanosoma cruzi* from South American bats is the most dangerous for humans. Bats can host a variety of ecto-parasites, including bat fleas (Siphonaptera), bat flies (Streblidae), bat bugs (Hemiptera), ticks (Argasidae and Ixodidae) and chigger mites (Trombiculidae), which can be 'exchanged' with other mammals. In the absence of primary bat hosts, bat bugs (*Cimex adjunctus*) bite humans without transmitting any pathogen but can cause irritation, itching, burning sensations and limited skin inflammatory reactions (Reeves et al, 2005).

Bats and viruses

More than 60 species of viruses have been found associated with bats (Calisher et al, 2006). Some viruses that are known to infect bats also represent a danger to humans (e.g., lyssavirus); some others have recently emerged and are of serious concern for public health (e.g., henipaviruses); others are bat-borne viral diseases not yet totally understood (e.g., Ebola fever). See the Appendix to this chapter for a detailed table of data about virus-infected bats.

Alphaviridae family

A few members of the family have been isolated from bats. The Chikungunya virus (Hailin Zhang et al, 1989), as well as the Sinbis virus (Blackburn et al, 1982), have hardly ever been isolated from African bats. Venezuelan encephalitis virus was isolated several times from bats in the Americas, though its epidemiological significance was not understood (Wong-Chia and Scherer, 1971).

Arenaviridae family

Members of this family are strictly associated and co-evolved with their rodent host species. The Tacaribe virus has been isolated from *Artibeus jamaicensis*, a fruit bat from the rainforests of Trinidad (Downs et al, 1963). Although this virus appears closely related to the other arenavirus from the same North American phylum, it has never been isolated from another animal (or human) species anywhere. As for the other family members, Tacaribe virus has a limited distribution and a strict virus–host association, but having a bat virus reservoir is unique and its significance enigmatic.

Coronaviridae family

The severe acute respiratory syndrome (SARS) coronavirus emerged in late 2002. It was first isolated from a carnivore, the civet, but later a SARS-like coronavirus was isolated from bats, which were then identified as a natural reservoir (Chu et al, 2006). Other coronaviruses were subsequently isolated from bats (Woo et al, 2006; Wong et al, 2007).

Filoviridae family

Ebola and Marburg viruses are two unique family species. Although bats have been recently identified as a natural host and a potential virus reservoir for Ebola virus from Central Africa, other African Ebola virus subtypes from different environments have a totally unknown natural history, as does the Ebola-Reston subtype isolated from captive monkeys from the Philippines. The origin of Marburg virus in South and East Africa remains cryptic.

Flaviviridae family

These viruses are generally transmitted by arthropods – mosquitoes or ticks – but a few have been isolated only from bats without a known arthropod host, including Yokose virus from a bat in Japan in 1971, Tamana virus from *Pteronotus parnellii* in Trinidad (showing also serological evidence of infection in humans), Montana virus from *Myotis leukoencephalitis*, and Entebbe viruses from bat salivary glands. Other flaviviruses (true arboviruses) have also been isolated from bats (e.g., St Louis encephalitis). Some have been investigated for their ability to infect bats (e.g., dengue fever, West Nile fever, Japanese encephalitis viruses). Surprisingly, for some well-known arboviruses, bats could be a potential natural reservoir (Sulkin and Allen, 1974).

Hantaviridae family

Two bat species, *Eptesicus serotinus* and *Rhinolophus ferrumequinum*, have yielded evidence of hantavirus infection and have been a suspected natural reservoir of hantavirus in Korea (Kim et al, 1994). This original observation has not yet been further documented, and much more information is needed to elucidate such an association.

Paramyxoviridae family

It is believed that Australian fruit bats are the natural host for the Menangle virus and caused a rare event in which the virus 'jumped' to pigs, and was responsible for a swine outbreak in 1997 and an influenza-like illness in farm workers (Bowden et al, 2001). Two other emerging members of the family, highly pathogenic for humans, have been isolated from fruit bats: the Hendra virus in Australia (Halpin et al, 2000) and the Nipah virus in Malaysia (Wang et al, 2000). The Nipah virus represents an exemplary emerging pathogen from bats. Also, the Tioman virus was isolated in 2001 from the urine of fruit bats (*Pteropus hypomelanus*) without evidence of human infection (Chua et al, 2001).

Reoviridae family

To track the Hendra virus in bats from the Australian mainland, a large study was carried out on Tioman Island, and the Pulau virus (Pritchard et al, 2006), isolated from a bat, appears as a novel mammalian fusogenic reovirus type.

Rhabdoviridae family

The *Lyssavirus* genus includes rabies and extends worldwide. The genus contains seven distinct genotypes (six associated with bats) and four not yet classified bat lyssaviruses (King et al, 1994; Hanna et al, 2000; Favi, et al, 2002; Arguin et al, 2002; Serra-Cobo et al, 2002; see also Appendix to this chapter). These viruses are transmitted by bites or by exposure of broken skin to infected animals' saliva. Most lyssaviruses are highly pathogenic and fatal for humans and also for other mammals. Rabies in bats was first documented in Brazil in 1911. Since then, numerous bat species have been found infected by lyssaviruses and have been considered a risk for humans. From 1954 onwards, starting in Germany, lyssavirus infections in bats have been recorded in Europe, and the emerging European bat lyssavirus types have been identified in most European countries. Bat rabies has also been accidentally introduced to western Europe by smuggled wild animals. The Australian bat lyssavirus, identified in 1996, has been isolated from flying fox species, and human deaths attributed to it have been reported. Experimentally, rabies vaccine and rabies immunoglobulin protect against this lyssavirus.

Although the distribution of lyssavirus infections in bats appears ubiquitous, human infection is rare and epidemics are exceptional (Batista-da-Costa et al, 1993; Passos et al, 1998; Gupta, 2005). Despite a historical association among vampire bats, cattle and rabies in South America, bat-transmitted lyssaviruses have recently been brought to public attention mostly because of increasing global travel and human infections associated with recreational activities (Moutou et al, 2003). Vampire bats regurgitate blood to other members of the colony, thus keeping the viruses circulating in the group and enhancing the possibility that disease will be transmitted to other mammals. The virus group contains seven other distinct genotypes and four not-yet-classified bat lyssaviruses (King et al, 1994; Hanna et al, 2000; Arguin et al, 2002; Favi et al, 2002; Serra-Cobo et al, 2002; see also Appendix to this chapter). Altogether, bat rabies has a strong association or even dependence on environmental factors of physical (habitat), biological (roosting, feeding) or human (environmental disturbance) origins.

EXEMPLARY BAT-BORNE VIRAL ZOONOSES

A zoonosis is a transmitted disease that uses an animal as a reservoir. Bats can be an intermediate host, a reservoir or a vector of human pathogenic viruses and can play an important role in the emergence, transmission, spread and maintenance of viruses in nature.

SARS coronavirus

SARS, and its etiologic agent the SARS coronavirus, were discovered after an outbreak in Guangdong Province, South China, that subsequently spread worldwide. Data suggest a virus wildlife origin (Guan et al, 2003; Rest and Mindell, 2003).

The first human cases were reported in mid-November 2002, four months before the first WHO report. Formal characterization of the virus was done in April 2003 (Ksiazek et al, 2003; Drosten et al, 2003; Peiris et al, 2003; Poutanen et al, 2003; Rota et al, 2003; Poon et al, 2005; Li et al, 2005). Many viral sources were identified at the same time in several municipalities of Guangdong. The outbreak can be divided into three phases: first, the original multiple introductions of the virus into the human population, leading to the first case of human 'super-spreaders' who dispersed the virus among the Chinese population (Zhong et al, 2003); second, human-to-human transmission in urban environments, starting with a Shenzhen hospital outbreak (most of the victims were hospital staff) and ending with the imported case of the hotel 'M' in Hong Kong (Tsang et al, 2003); and third, the spread of the virus to Vietnam, Canada, Singapore and Germany – the late pandemic phase. The spread was entirely due to humans travelling locally and internationally by motorized ground and commercial air travel.

As a typical case of interspecific viral transmission, the question of virus spillover from one species to another remains central for understanding how the SARS coronavirus emerged. Below, we put forth a hypothesis for its origin and transmission.

After the virus was first isolated in a human patient, investigations led to the first isolation of the virus from wild animals and farm civets (masked palm civet, *Paguma larvata*) in Guangdong Province; consequently, civets were considered a potential virus reservoir that could infect humans. Other closely related wild animals were then investigated (Guan et al, 2003; Guan et al, 2005). The geographical origin of the SARS epidemic was near the western part of Guangzhou and Shenzhen districts in Guangdong Province, a region of rapid development since the 1970s. In these changing human and natural environments, people began to change their culinary habits, favouring the consumption of wild animals (traditionally consumed in southern China) and, particularly, the masked palm civet, a sophisticated and much-appreciated dish. Responding to the new demand, people began to farm civets instead of hunting them; the number of civet farms multiplied by a factor of 50 during the five years before the SARS outbreak (Formenty, personal communication, 2007).

Bats were apparently important in the natural life cycle of the virus (Li et al, 2005). The discovery of many SARS-like coronaviruses (see Woo et al, 2006) from southern China supports a hypothesis in which civets were a semi-domestic amplifier of the virus or an intermediate host between wildlife and humans. Multiple *Rhinolophus* species (insectivorous horseshoe bats) tested positive for the virus (Lau et al, 2006; Tang et al, 2006). Other species from the Vespertilionidae family (evening bats) – *Miniopterus magnater, Miniopterus pusillus, Pipistrellus pipistrellus, Pipistrellus abramus* – can transport other coronavirus types (Poon et al, 2006).

Bat populations provide a suitable environment for coronavirus hosting and evolution. The genetic variability of the viruses seems to be much more effective in Chiroptera than in civets or humans, suggesting that the later hosts could have inherited their virus from a wild chiropteran origin. Chiroptera share a long natural history of hosting coronaviruses, and infected bats rarely show clinical signs – a requirement for a natural candidate virus reservoir.

The coronavirus genus is usually divided into three groups hosted by a variety of animals (Siddell, 1995):

- group 1, hosted by carnivores, artiodactyls (camels, pigs, ruminants) and primates;
- group 2, hosted by artiodactyls, horses, rodents, and birds; and
- group 3, detected only in birds.

Recently, two new groups were shown to be hosted by chiropterans (Tang et al, 2006). This astonishing diversity of animal hosts is explained and sustained by the combination of molecular and ecological factors revealed by the genetic footprints underlining coronavirus evolution (Woo et al, 2006; Wang et al, 2006). (It should be mentioned that coronavirus diversity varies according to the data sets, the method of data analysis and/or authors.) It is thought that coronaviruses have a great potential for evolution because they are known to recombine genomes in nature and are also able to switch from one host to another (Rest and Mindell, 2003).

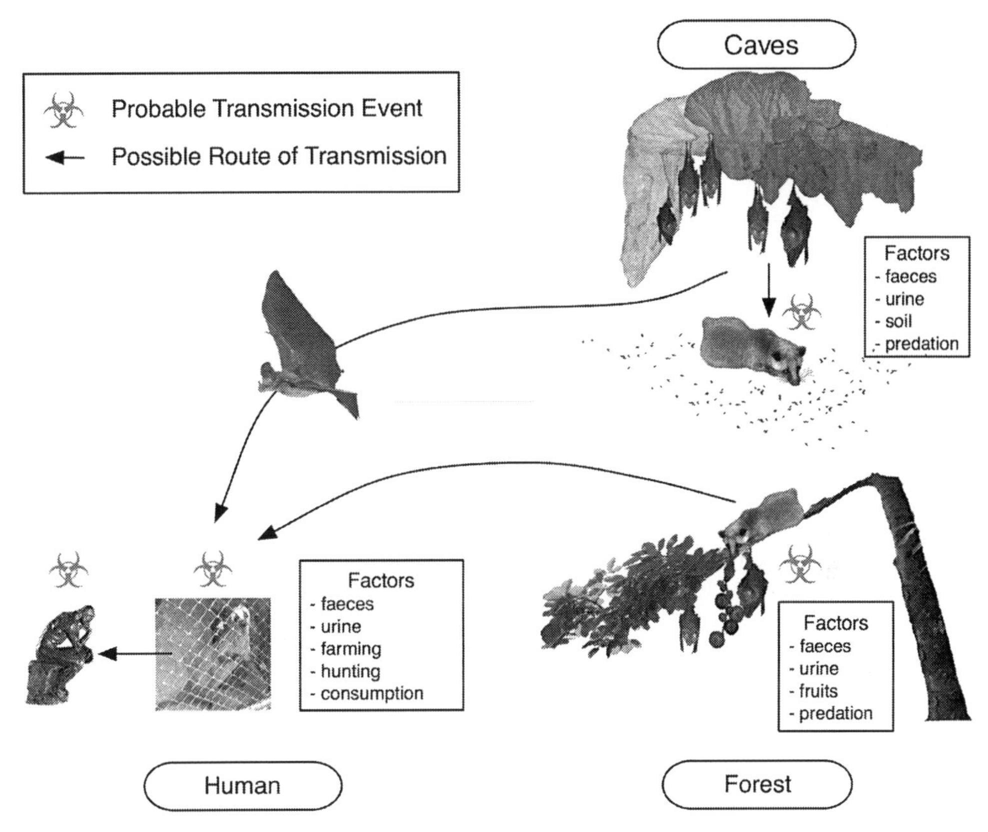

Figure 8.2 *Transmission mechanisms of SARS coronavirus*

Source: Meriadeg Ar Gouilh

Such transformations of the genetic material open a new way to evolve in relation with the host. Frequently, viruses develop modifications of the surface receptors, those that interact with host cells and determine virulence or latency. In addition, ecological factors and host species particularities – and constraints – force new patterns of virus evolution, circulation and dispersion. The nature of bat communities – large, dense colonies moving with variable frequency, roosting with other species, ranging and migrating over great distances – and the long life expectancy of individual bats are factors that can favour passage of a virus and drive its evolution.

The likelihood of such transmission from one species to another – in this case, from bats to civets – is lower than for transmission within the same genus. To jump from one host species to another, the virus has to produce enough variants to have a chance to complete the switch. Virus passage to civets could occur through contact, or licking/ingestion/inhalation of fresh faeces or urine from infected bats. Civets are mostly nocturnal, predominantly frugivorous, but also omnivorous and tree-dwelling. They could be infected by bats during night hunting or feeding in caves or trees where bats excrete biological products in great quantity (Figure 8.2). The SARS-positive civets were all farmed animals (Formenty, personal communication, 2007), and whether the human index case had a wildlife or farm origin is unknown. Here we outline the hypothesis of a natural virus cycle involving wildlife hosts.

The original coronavirus in civets was probably not sufficiently virulent and infectious to 'jump' to the first human being encountered. However, when civet density increased through intensive breeding on farms, multiple contacts between people and animals favoured the emergence of alternative human pathogenic strains. Processes of domestication and breeding, by modifying the ecological and biological characteristics of a species and its relationship with humans, can support the passage of viruses from animals to humans. And then humans, by their travel behaviour, 'extracted' the virus from the forest and disseminated it at the global level.

Ebola fever

In 1976, a double epidemic in Africa of an unknown, highly fatal viral haemorrhagic fever led to the discovery of a new virus: Ebola. Together with the Marburg virus (also of African origin, recognized in 1967), it was subsequently defined as the prototype of a new virus family, the Filoviridae (Latin *filum*, 'thread'), characterized by genetic material carried by only one thread of RNA with negative polarity (Martini and Siegert, 1971; Kiley et al, 1982). Ebola fever emerged where the Ebola River crosses the rainforest of the Democratic Republic of Congo (formerly Zaire) and also in the forest/savannah mosaic of Sudan. But the Ebola virus remained hidden for 25 years and re-emerged in Kikwit, 1000km south of where it had first appeared (Gonzalez, 1995). Moreover, the arcane natural history of the Ebola virus was not understood until the early years of the 21st century (Pourrut et al, 2005).

In an exhaustive compilation, Jens H. Kuhn provides a view of the unfathomable history of Ebola fever and Marburg disease emergence and the consequences for public health and social issues, and also for the popular imagination about deadly diseases from remote, uncharted areas: 'interest in the filoviruses has developed among the general public, in part because of novels and popular science stories and Hollywood productions that portrayed them ... and investigators used "Ebola" as a catch phrase to draw attention to their articles, many of which did not pertain to filoviruses' (Kuhn, 2008).

Moreover, after a real Ebola outbreak appeared in the news, the 1994–1995 movie *Outbreak* (by Wolfgang Petersen, based on Richard Preston's novel, *The Hot Zone*) was a huge Hollywood production:

> *The story imagines the appearance of an Ebola-type virus in the United States, which is reported to have mutated and have been transmitted by air. A military plot is woven into this story, because the military does not want to use an existing, but secret, vaccine in response to the epidemic. The two star characters were based on a husband and wife team of virologists, pioneers in the hunt for viruses of haemorrhagic fevers such as Bolivian haemorrhagic fever and Ebola fever.* (Vidal et al, 2007)

The search for the natural cycle lasted for 30 years (Table 8.2), and bats were ultimately identified as a first natural host and potential reservoir and vector (Leroy et al, 2005). All subsequent human outbreaks investigated after the resurgence of Ebola fever in Central Africa

Table 8.2 *Wild and domestic animals investigated in quest for Ebola virus hosts and reservoir*

Host type	*Habitat*	*Suspected*[1]	*Confirmed*	*Total tested*	*Origin*[1]	*Date of sampling*
INSECTIVOROUS						
Shrew sp	Sub-Sudanese savannah	N	–	7	Nzara, Sudan	1976
Shrew sp	Congolese rainforest	N	–	53	Cameroon; DRC	1979–1980
Shrew sp	Degraded rainforest	N	–	114	Kikwit, DRC	1995
Shrew sp	Rainforest	N	–	398	Taï, Côte d'Ivoire	1996–1997
Sylvisorex ollula (shrew)	Rainforest	1	–	10	CAR	1999
Insectivorous	Rainforest	N	–	46	CAR	1999
RODENTS						
Murids sp	Rainforest	N	–	131	Yambuku, DRC	1976
Murids sp	Sub-Sudanese savannah	N	–	309	Nzara, Sudan	1976
Murids sp	Rainforest	N	–	661	Cameroon; DRC	1979–1980
Arvicanthis spp (Murids)	Dry savannah	9	–	98	CAR	1979–1983
Mastomys spp (Murids)	Rainforest	2	–	91	CAR	1979–1983
Mastomys spp (Murids)	Wet savannah	10	–	265	CAR	1979–1983
Mus spp (Murids)	Rainforest	2	–	54	CAR	1979–1983
Murids sp	Degraded rainforest	N	–	1759	Kikwit, DRC	1995
Murids sp	Rainforest	N	–	283	Taï, Côte d'Ivoire	1996–1997
Murids sp	Rainforest	N	–	163	CAR	1999
Praomys spp (Murids)	Rainforest	1	–	41	CAR	1999
BATS						
Chiroptera sp	Rainforest	N	–	7	Yambuku, DRC	1976
Megachiroptera	Sub-Sudanese savannah	N	–	4	Nzara, Sudan	1976
Microchiroptera	Sub-Sudanese savannah	N	–	174	Nzara;, Sudan	1976

Table 8.2 *(cont'd)*

Host type	*Habitat*	*Suspected*[1]	*Confirmed*	*Total tested*	*Origin*[1]	*Date of sampling*
Megachiroptera	Rainforest	N	–	41	Cameroon, DRC	1979–1980
Microchiroptera	Rainforest	N	–	422	Cameroon, DRC	1979–1980
Megachiroptera sp	Degraded rainforest	N	–	125	Kikwit, DRC	1995
Microchiroptera sp	Degraded rainforest	N	–	414	Kikwit, DRC	1995
Megachiroptera sp	Rainforest	N	–	19	CAR	1999
Microchiroptera sp	Rainforest	N	–	4	CAR	1999
Chiroptera sp	Rainforest	N	–	652	Taï, Côte d'Ivoire	1996–1997
Hypsignathus monstrosus (Megachiroptera)	Rainforest	P	8	32	Gabon, RC	2002–2003
Epomops franquetti (Megachiroptera)	Rainforest	P	4	102	Gabon, RC	2002–2003
Myonycteris torquata (Megachiroptera)	Rainforest	P	4	58	Gabon, RC	2002–2003
CARNIVOROUS						
Dog	Rainforest	P	–	162	CAR	1979–1983
Unspecified carnivorous	Rainforest	N	–	27	Cameroon; DRC	1979–1980
Unspecified carnivorous	Degraded rainforest	N	–	28	Kikwit, DRC	1995
Dog	Rainforest	P	55	258	Gabon	2002
ARTIODACTYLS						
Domestic ungulates	Rainforest	N	–	13	Yambuku, DRC	1976
Artiodactyls	Rainforest	N	–	27	Cameroon; DRC	1979–1980
Artiodactyls	Degraded rainforest	N	–	22	Kikwit, DRC	1995
Cattle	Wet savannah	P	2	108	CAR	1979–1983
Pig	Rainforest	13	–	80	CAR	1979–1983

OTHER MAMMALS						
Pangolin (Pholidotes)	Rainforest	N	–	66	Cameroon; DRC	1979–1980
Pangolin (Pholidotes)	Degraded rainforest	N	–	29	Kikwit, DRC	1995
Donkey (Perissodactyles)	Rainforest	3	–	13	CAR	1979–1983
NON-HUMAN PRIMATES						
Primates	Rainforest	P	5	267	Cameroon; DRC	1979–1980
Cercopithecus spp	Rainforest	P	1	107	Cameroon; Gabon; RC	1980–2002
Papio spp	Rainforest	P	1	25	Cameroon; Gabon; RC	1980–2002
Mandrillus spp	Rainforest	P	6	215	Cameroon; Gabon; RC	1980–2002
Gorilla gorilla	Rainforest	P	3	30	Cameroon; Gabon; RC	1980–2002
Pan troglodytes	Rainforest	P	29	225	Cameroon; Gabon; RC	1980–2002
Primates	Degraded rainforest	N	–	12	Kikwit, DRC	1995
Primates	Rainforest	N	–	17	Taï, Côte d'Ivoire	1996–1997
BIRDS						
Bird	Rainforest	N	–	67	Cameroon; DRC	1979–1980
Bird	Degraded rainforest	N	–	184	Kikwit, DRC	1995
Chicken	Wet savannah	13	–	131	CAR	1979–1983

Table 8.2 *(cont'd)*

Host type	*Habitat*	*Suspected*[1]	*Confirmed*	*Total tested*	*Origin*[1]	*Date of sampling*
REPTILES AND AMPHIBIANS						
Not specified	Rainforest	N	–	5	Nzara, DRC	1976
Not specified	Rainforest	N	–	33	Cameroon; DRC	1979–1980
Not specified	Degraded rainforest	N	–	127	Kikwit, DRC	1995
Not specified	Rainforest	N	–	283	Taï, Cote d'Ivoire	1996–1997
ARTHROPODS						
Bedbugs, mosquitoes, ticks, lice, flies.	Rainforest	N	–	2318	Yambuku, DRC	1976
Mosquitoes, ticks, lice, flies	Degraded rainforest	N	–	27,843	Kikwit, DRC	1995

*Notes:*1 Ebola virus antigen-reacting antibodies (number of positive)

CAR = Central African Republic; RC = Republic of Congo; DRC = Democratic Republic of Congo

N = negative; P = positive

Sources: Arata and Johnson, 1978; Gonzalez et al, 1983; Breman et al, 1999; Formenty et al, 1999; Leirs et al, 1999; Morvan et al, 1999; Reiter et al, 1999; Leroy et al, 2004a; Leroy et al, 2004b; Allela et al, 2005; Gonzalez et al, 2005; Pourrut et al, 2005.

were associated with fatal infections of non-human primates (see Box 12.1). The recent findings on the role of bats as a potential reservoir of Ebola virus in the Congolese rainforest and the risk for enzootic extension, epizootics and epidemics through migration or feeding behaviour have also attracted attention. However, other hosts appeared to play a role in Ebola. Among wild and domestic animals tested as eventual hosts of Ebola virus, several have been suspected of harbouring Ebola antigen-reacting antibodies, though few of them appeared as a potential natural host. Lastly, only bats have been confirmed as both a potential, efficient host reservoir of the virus (experimentally, chronically shedding the virus) and a vector (associated with foraging and migration). Finally, several fruit-bat species have been targeted as potential reservoirs of Ebola virus and tested positive during human and non-human primate outbreaks (Leroy et al, 2004a, b; Gonzalez et al, 2005).

Fruit bats are very active and can easily meet a variety of hosts in their habitats, factors that satisfy the conditions of close contact and pathogenic emergence and explain their association with the consequent epizootics and, indirectly, epidemics. However, the receiving species (secondary host) must be able to multiply and transmit the virus, so natural resistance of the secondary host or fewer pathogen strains can stop resurgence. The vertebrates found infected are sensitive to infection, and once infected, these accidental hosts can transmit the virus to another accidental host in either an interspecific (primate to human) or an intraspecific (human to human) way.

When fruit bats chew or suck the pulp of fruits, their saliva can infect the fruits and their infected waste products (urine, faeces) also contaminate fruits that fall to the ground (Figure 8.3). The contaminated fruits can be consumed by mammals living in the tree canopy (particularly forest monkeys of the *Cercopithecus* genus) or by duikers or primates on the ground (gorillas, chimpanzees). It is also possible that humans could transport contaminated fruits for commercial purposes. Additionally, bats could also facilitate transmission to large monkeys by direct contact with blood or female bat placental tissue at the time of parturition.

Numerous deaths of non-human primates and monkeys from Ebola virus have occurred at the end of the dry season, when food resources are rare. Such food shortages might lead to the consumption of the same type of fruit at the same time by frugivorous animal species, such as bats and large monkeys, increasing interspecies contact. Non-human primates appear to occupy central roles as both victims and unwilling spreaders of the virus when sick animals are taken by unskilled hunters. Their relations with bats and other potential Ebola virus reservoirs need to be thoroughly investigated for controlling and preventing future epizootic and epidemics.

Because the ecology and ethology of bats as a potential reservoir of the Ebola virus are poorly documented, the chain of transmission and persistence of the virus in the natural environment are poorly understood. Some Ebola viruses with a similarly high human pathogenicity (Ebola Sudanese strain) and lesser pathogens (Ebola Reston strain from the Philippines) may be hosted by different species of chiropterans from, respectively, the Sudanese savannah and the largely degraded forests of the Philippines.

Nipah encephalitis

For more than a decade, Hendra and Nipah viruses have been emblematic as emerging pathogens for three main reasons: their unique taxonomic characters, which required the

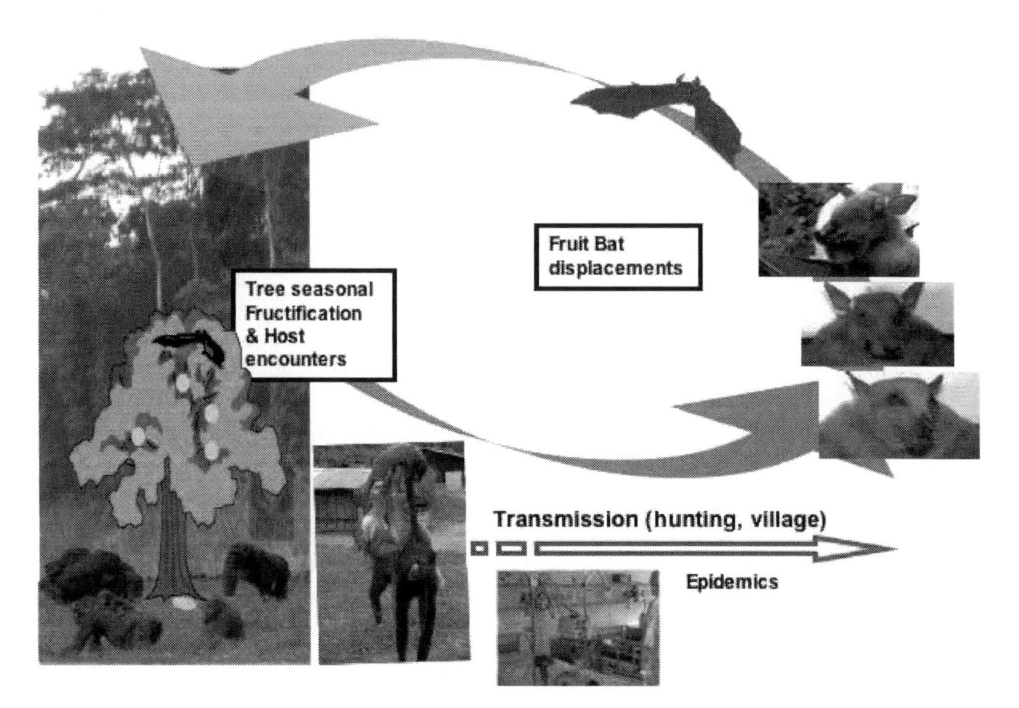

Figure 8.3 *Transmission mechanisms of Ebola virus*

creation of a new *Henipavirus* genus within the Paramyxoviridae family; their unexpected occurrence in domestic animals (horses, pigs) of the Australian continent and the Malaysian peninsula (Chua et al, 2001); and their use of Chiroptera as hosts. Since 1994, after the Hendra virus – named for a Brisbane suburb – was first isolated during an outbreak of respiratory and neurological syndromes in horses and humans, other novel viruses (Menangle, Tioman, Nipah) were isolated, mostly in fruit bats belonging to the *Pteropus* genus in Australia and Malaysia. Nipah virus appears to have a wider geographic range, sparking special public interest in better understanding its distribution, risks, hosts, reservoirs and vectors.

Nipah virus first emerged in Malaysia and Singapore in 1998–1999, when a large outbreak among swine drew immediate veterinary attention (Lam and Chua, 2002). The virus appeared to have passed by direct contact from infected pigs with respiratory syndrome to humans, but also to cats and dogs. Mostly adults developed the encephalitis syndrome (with 40 per cent fatality among pig farmers in Malaysia and abattoir workers in Singapore). In Singapore, 11 cases of Nipah virus infection (and one death) were associated with the slaughter of imported pigs from Malaysia (Table 8.3).

Four outbreaks of Nipah encephalitis followed: in Meherpur District in 2001 (69 per cent case fatality), then 150km away in Naogaon District in 2003 (67 per cent case fatality), in Goalanda and Faridpur in 2004 (75 per cent case fatality), and in Tangail in 2005 (38 per cent case fatality). During the most recent outbreak, a rickshaw driver got infected after carrying a patient to the hospital; the patient appears to have been infected by drinking juice made from dates potentially infected by bat saliva.

In 2001, after the Bengali event, an outbreak of febrile illness was observed in Siliguri, West Bengal, India, and Nipah virus etiology was retrospectively proven for the first time

Table 8.3 *Nipah bat-borne virus emergences*

Time of virus emergence or re-emergence	*Place, country of virus isolation*	*Bat as potential host and/or reservoir of virus*
September 1998	Malaysia	*Cynopterus brachyotis* *Eonycteris spelaea* *Pteropus hypomelanus* *Pteropus vampyrus* *Scotophilus kuhlii*
1999	Singapore	A pig-to-human epidemiologic chain of virus transmission (not likely a natural reservoir present in the area and involved the epidemics)
2001	Meherpur, Bangladesh	*Pteropus* sp (serology)
January–February 2001	Siliguri, India	*Pteropus* sp
2002	Cambodia	*Pteropus lylei*
2004	Faridpur and Radjabari, Bangladesh	*Pteropus giganteus*
2005	Tangail, Bangladesh	*Pteropus giganteus*
2005	Cambodia	*Pteropus lylei*
2005	Thailand	*Pteropus hypomelanus* *Pteropus vampyrus* *Pteropus lylei* *Hipposideros larvartus*

in the country. Moreover, the Indian strain was more closely related to the one from Bangladesh than to the Malaysian strain.

In 2002, the Nipah virus geographical distribution extended to bats from a Cambodian marketplace, where bats (which tested positive) were for sale as a delicacy (Olson et al, 2002). Two years later, Reynes and colleagues demonstrated the presence of a Nipah-like virus among Lyle's flying foxes (*Pteropus lylei*) (Reynes et al, 2005). These isolates were the first obtained from *Pteropus*, including *P. lylei* and *P. hypomelanus*.

Several bat species have been found to host the virus without excess mortality. However, we cannot infer that these hosts, which regularly migrate and cover a wide geographical area, can be the unique virus vector or reservoir. Moreover, bats appear as a specific host; some are infected but others from the same colony are not, which suggests that bat species distribution does not match the Nipah virus enzootic domain. Finally, virus range may be limited by environmental factors or as-yet-unidentified host (bat) subspecies (Gonzalez, 1996; Emonet et al, 2006).

Nipah virus and Hendra virus have been found within the Australasian (Australia, New Guinea and neighbouring islands) and the Indo-Malaysian ecozones (South Asian sub-continent and Southeast Asia). These biomes are separated from one another by the Wallace Line, but flying species like birds and chiropterans can cross it. Main domains of Nipah

Table 8.4 *Nipah bat-borne virus and natural host geographical distribution*

Natural host	*Known geographic distribution of Nipah virus with Chiroptera host*	*Host geographic distribution*
Cynopterus brachyotis (lesser short-nosed fruit bat)	Malaysia (from Perak to Johore province)	Sri Lanka, India, Nepal, Burma, Thailand, Cambodia, Vietnam, southern China, Malaysia, Nicobar and Andaman Islands, Borneo, Sumatra, Sulawesi, Magnole, Sanana, Sangihe Islands, Talaud Islands and adjacent small islands.
Eonycteris spelaea (lesser dawn bat)	Malaysia	India, Burma, Nepal, southern China, Thailand, Laos, Cambodia, Vietnam, western Malaysia, Borneo, Sula Islands, northern Moluccas, Sumatra, Java, Sumba, Timor and Sulawesi (Indonesia), Philippines, Andaman Islands (India)
Pteropus hypomelanus (variable flying fox)	Malaysia, Thailand	Andaman and Maldive Islands, New Guinea through Indonesia to Vietnam and Thailand and adjacent islands, Philippines.
Pteropus lylei (Lyle's flying fox)	Cambodia (Battambang Province, 80km from the Thai border), Thailand	Thailand, Vietnam, Cambodia
Pteropus giganteus (Indian flying fox)	Western Bangladesh (Faridpur District; person to person; Naogaon) Siliguri India (suspected)	Maldive Islands, India (incl. Andaman Islands), Sri Lanka, Pakistan, Bangladesh, Nepal, Burma, and Tsinghai (China) with confirmation pending
Pteropus vampyrus (large flying fox)	Thailand, Java and Sumatra, Indonesia	Vietnam, Burma, Malay Peninsula, Borneo, Philippines, Sumatra, Java, and Lesser Sunda Islands, adjacent small islands including Anak Krakatau
Pteropus rufus (Malagasy flying fox)	Madagascar	Madagascar
Eidolon dupreanum (Malagasy straw-coloured fruit bat)	Madagascar	Madagascar
Scotophilus kuhlii (lesser Asiatic yellow house bat)	Malaysia	Bangladesh, Pakistan to Taiwan, south to Sri Lanka, Burma, Cambodia, western Malaysia, Java, Bali, Nusa Tenggara (Indonesia), southeast to Philippines and Aru Islands (Indonesia)

Table 8.4 (*cont'd*)

Natural host	*Known geographic distribution of Nipah virus with Chiroptera host*	*Host geographic distribution*
Hipposideros larvatus (intermediate leaf-nosed bat)	Thailand	Northern and eastern India and Bangladesh, Yunnan, Kwangsi and Hainan (China), Burma, Thailand, Cambodia, Laos, Vietnam, western Malaysia to Sumatra, Java, Borneo, and adjacent small islands including Kangean Islands (Indonesia)

virus activity associated with bat hosts have been recorded from the Malaysian peninsula to the Indochina peninsula and the border area between western Bangladesh and India.

Bats have been found chronically infected by the Nipah virus. Transmission from bat to bat probably occurs through infected saliva and bites, and also to offspring via the placenta membrane or other infected biological products. Transmission from one person to another was suggested only during the Bangladesh outbreaks, not in the Malaysia and Singapore outbreaks.

The mode of transmission from one species to another is not totally understood but most likely requires close contact with infected biological products of an animal host. During the Malaysian and Singapore outbreaks, infected pigs were the apparent source of infection for humans, though other sources (dogs, cats) have also been suspected. In Bangladesh, humans may have been infected through exposure to bat saliva, urine or faeces.

Nipah virus appears to chronically infect several species of bat within an as-yet-unknown range. It appears to reach human populations when bats are disturbed by human beings or when people intrude in natural bat-roosting sites. The emergence of Nipah virus in Malaysia remains an enigma, but it may be that *Pteropus* of Malaysia are infected and unknown factors of emergence favour close contact between bats and pigs, or that infected *Pteropus* migrated from another domain where they were infected but not in contact with susceptible hosts.

One hypothesis regarding the sudden emergence of Nipah virus encephalitis in Malaysia involves climate change. More than a year before the outbreak, Papua New Guinea rainforests had suffered a strong and extended drought with severe forest fires that produced a thick haze over Borneo. The overcast sky lasted for months, preventing normal fruiting in the remaining trees. The fruit bats of the region, in search of food, migrated from the rainforest into the orchards of the Malaysian countryside. Close to these orchards were large pig farms. Host and vector were now together in the same place at the same time. Humans were secondarily infected by infected pigs. If the hypothesis is supported, the chain of infection occurred through infected fruits fed to the pigs (Figure 8.4). It also supposes that fruit bats from the Papua New Guinea rainforest are chronically infected with Nipah virus.

In Bangladesh, although the infected fruit can be traced to contamination by bat saliva, the fundamentals of the emergence remain a mystery and have to be found in the

Nipah Virus: A Brief History of Emergence

Oct.1998	- First cases of suspected JEV in Malaysia
Feb. 1999	- Severe cases and deaths, mostly pig farm workers
Mar. 1999	- Hendra-like virus isolated from pig, One million pigs being culled
May. 1999	- 257 cases of encephalitis in humans 40% mortality

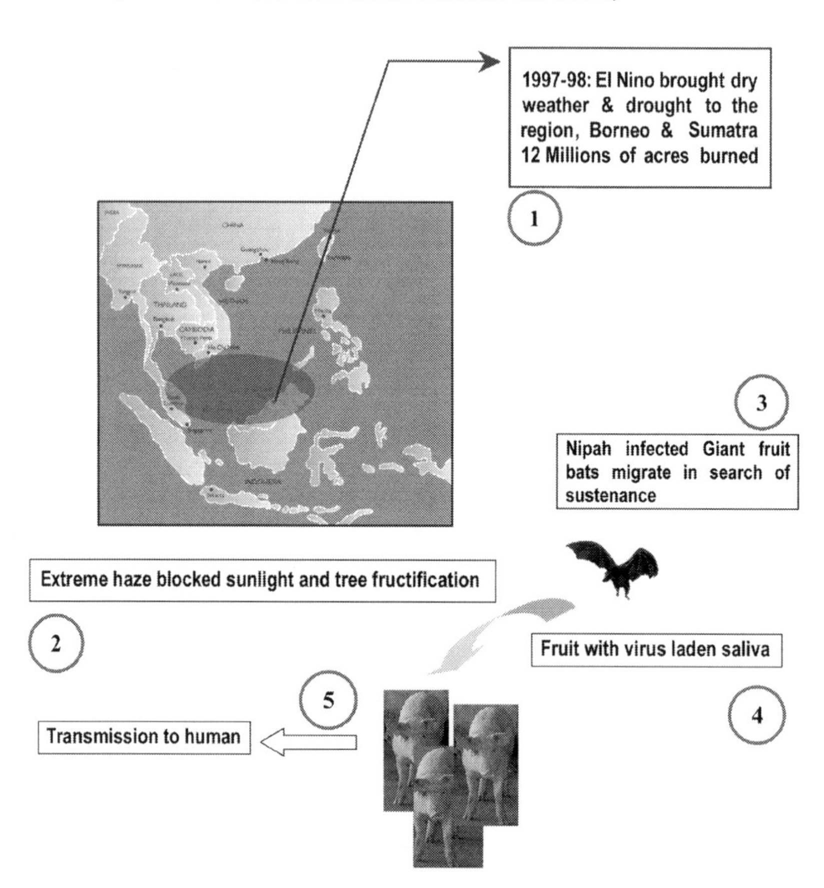

Figure 8.4 *Transmission mechanisms of Nipah virus*

ecosystem of local bat colonies and their behaviour. The Nipah virus shows us two things: the interdependence of life and environment, and the effects that one can have on the other.

Because antibodies reactive with Nipah virus were identified in local *Pteropus* bats surrounding the outbreaks, this genus may serve as the reservoir. A possible explanation for infection without an obvious domestic reservoir (pigs) may be inadvertent direct contact with bats or bat secretions. Although henipavirus did cause limited outbreaks, the virus infected a variety of host species and produced severe disease with significant mortality in humans. *Pteropus* species are distributed across Australia, Indonesia, Malaysia, the Philippines and several Pacific islands. Some species found in the Indian Ocean islands are also susceptible to henipavirus infection and are chronically infected without suffering clinical damage.

Emergence of henipavirus infections appears to be associated with human or livestock diseases. Consequently, prevention strategies are directed at disease control through good farm management practices, reducing exposure to flying foxes, and rapid disease recognition and diagnosis. A fascinating potential strategy lies in wildlife immunization through plant-derived vaccines (Yusibov et al, 2006).

CONCLUSION

Bats play crucial roles in maintaining an environmental balance, including insect population stability and ultimately reducing human exposure to several pest risks. At the same time, bats can be carriers of pathogens for humans or other animals and also victims of foreign germs associated with a changing environment or human behaviour. When it comes to public health, the potential of bats as virus carriers or efficient reservoirs is entirely related to bat life-traits: bats have long lives in the wild and can be chronically infected by human pathogens (Dobson, 2005), their immune systems have a unique potential to favour or resist the replication of viruses, and their wide ranges and ability to fly make them a powerful means for disease dispersion, extension or emergence.

In a customary state, it is rare for bats to infect humans or domestic animals and precipitate the emergence of new diseases. However, changes in the environment and human alteration of their habitat – deforestation, agricultural practices, urbanization – often displace bats from their natural contexts and favour encounters with potentially susceptible hosts. It is now a dogma that the forest and other natural environments are 'remote natural habitats where unknown disease agents exist in harmony with wild reservoir hosts' (Field, 2005). Reducing the risk of disease transmission to humans, domestic animals and other wild animals lies primarily in minimizing direct or indirect contact with bats, and improving disease recognition, epidemiology and farming biosecurity. For that, one must develop an integrated knowledge of chiropteran bio-ecology. Approaches to limiting bat-borne diseases must incorporate environmental factors and systems and, most importantly, climatic tendencies and forest health itself.

NOTE

1 See http://animaldiversity.ummz.umich.edu/site/accounts/information/Chiroptera.html#3b4691b6f3f87d21896413fe51f31ef

REFERENCES

Allela, L., Bourry, O., Pouillot, R., Délicat, A., Yaba, P., Kumulungui, B., Rouquet, P., Gonzalez, J.P. and Leroy, E.M. (2005) 'Ebola virus antibody in dogs and human risk', *Emerging Infectious Diseases* 11: 385–390

Arata, A.A. and Johnson, B. (1978) 'Approaches towards studies on potential reservoirs of viral haemorrhagic fever in southern Sudan', in S.R. Pattyn (ed) *Ebola Virus Haemorrhagic Fever*, Elsevier/Netherland Biomedical, Amsterdam, The Netherlands, pp191–202

Arguin, P., Murray, K., Miranda, M., Smith, J., Calaor, A. and Rupprecht, C. (2002) 'Serologic evidence of Lyssavirus infections among bats, the Philippines', *Emerging Infectious Diseases* 8: 258–262, http://findarticles.com/p/articles/mi_m0GVK/is_3_8/ai_84668151

Batista-da-Costa, M., Bonito, R.F. and Nishioka, S.A. (1993) 'An outbreak of vampire bat bite in a Brazilian village', *Tropical Medicine and Parasitology* 44(3): 219–220, http://cat.inist.fr/?aModele=afficheN&cpsidt=3807002

Bausch, D.G., Nichol, S.T., Muyembe-Tamfum, J.J., Borchert, M., Rollin, P.E., Sleurs, H., Campbell, P., Tshioko, F.K., Roth, C., Colebunders, R., Pirard, P., Mardel, S., Olinda, L.A., Zeller, H., Tshomba, A., Kulidri, A., Libande, M.L., Mulangu, S., Formenty, P., Grein, T., Leirs, H., Braack, L., Ksiazek, T., Zaki, S., Bowen, M.D., Smit, S.B., Leman, P.A., Burt, F.J., Kemp, A. and Swanepoel, R. (2006) 'Marburg hemorrhagic fever associated with multiple genetic lineages of virus', *The New England Journal of Medicine* 355(9): 909–919, http://content.nejm.org/cgi/content/abstract/355/9/909

Blackburn, N.K., Foggin, C.M., Searle, L. and Smith, P.N. (1982) 'Isolation of Sindbis virus from bat organs', *Central African Journal of Medicine* 28(8): 201, www.ajol.info/journal_index.php?jid=52

Bowden, T.R., Westenberg, M., Wang, L.F., Eaton, B.T. and Boyle, D.B. (2001) 'Molecular characterization of Menangle virus, a novel paramyxovirus which infects pigs, fruit bats, and humans', *Virology* 10, 283: 358–373, www.ncbi.nlm.nih.gov/sites/entrez?cmd=Retrieve&db=pubmed&list_uids=11336561

Breman, J.G., Johnson, K.M., van der Groen, G., Robbins, C.B., Szczeniowski, M.V., Ruti, K., Webb, P.A., Meier, F. and Heymann, D.L. (1999) 'A search for Ebola virus in animals in the Democratic Republic of Congo and Cameroon: 1979–1980', *Journal of Infectious Diseases* 179: Supplement 1, S139–S147

Calisher, C.H., Childs, J.E., Field, H.E., Holmes, K.V. and Schountz, T. (2006) 'Bats: Important reservoir hosts of emerging viruses', *Clinical Microbiology* 19(3): 531–545

Chu, D.K., Poon, L.L., Chan, K.H., Chen, H., Guan, Y., Yuen, K.Y. and Peiris, J.S. (2006) 'Coronaviruses in bent-winged bats (*Miniopterus* spp)', *Journal of General Virology* 87: 2461–2466

Chua, K.B., Wang, L.F., Lam, S.K., Crameri, G., Yu, M., Wise, T., Boyle, D., Hyatt, A.D. and Eaton, B.T. (2001) 'Tioman virus, a novel paramyxovirus isolated from fruit bats in Malaysia', *Virology* 283(2): 215–229

Dobson, A.P. (2005) 'What links bats to emerging infectious diseases?', *Science* 310(5748): 628–629

Downs, W.G., Anderson, C.R., Spoence, L., Aitken, T.H.G. and Greenhall, A.H. (1963) 'Tacaribe virus a new agent isolated from *Artibeus* bats and mosquitoes in Trinidad, West Indies', *The American Journal of Tropical Medicine and Hygiene* 12: 640–646, www.ajtmh.org/cgi/content/abstract/12/4/640

Drosten, C., Gunther, S., Preiser, W., van der Werf, S., Brodt, H.R., Becker, S., Rabenau, H., Panning, M., Kolesnikova, L., Fouchier, R.A., Berger, A., Burguiere, A.M., Cinatl, J., Eickmann, M., Escriou, N., Grywna, K., Kramme, S., Manuguerra, J.C., Muller, S., Rickerts, V., Sturmer, M., Vieth, S., Klenk, H.D., Osterhaus, A.D., Schmitz, H. and Doerr, H.W. (2003) 'Identification of a novel coronavirus in patients with severe acute respiratory syndrome', *The New England Journal of Medicine* 348(20): 1967–1976, www.ncbi.nlm.nih.gov/sites/entrez?cmd=Retrieve&db=PubMed&list_uids=12690091

Emonet, S., Lemasson, J.J., Gonzalez, J.P., de Lamballerie, X. and Charrel, R.N. (2006) 'Phylogeny and evolution of Old World arenaviruses', *Virology* 5350(2): 251–257

Favi, M., de Mattos, C.A., Yung, V., Chala, E., Lopez, L.R. and de Mattos, C.C. (2002) 'First case of human rabies in Chile caused by an insectivorous bat virus variant', *Emerging Infectious Diseases* 8: 79–81

Field, H. (2005) 'Emerging diseases associated with flying foxes – host management strategies', National Wildlife Rehabilitation Conference, 31 August–2 September, Gold Coast, Australia Holiday Inn, Surfers Paradise

Formenty, P., LeGuenno, B., Wyers, M., Gounon, P., Walker, F. and Boesch, C. (1999) 'Search for the Ebola virus reservoir in Taï forest, Côte d'Ivoire: 1996–1997, preliminary results', XIth International Congress of Virology, 9–13 August, Sydney, Australia

Gonzalez, J. (1995) 'Ebola, une rivière tranquille au coeur de l'Afrique', *Cahiers Santé* 5: 145–146

Gonzalez, J.P. (1996) 'Coevolution of rodent and viruses: Arenaviruses and hantaviruses', in M. Ali Ozcel (ed), *New Dimension in Parasitology, Acta Parasitolgica Turcica*, 20: Supplement 1, 617–638

Gonzalez, J.P., McCormick, J.B., Saluzzo, J.F. and Georges, A.J. (1983) 'Les fièvres hémorragiques africaines d'origine virale en République Centrafricaine', *Cahiers ORSTOM, Série Entomologie Médicale et Parasitologie*, 21: 119–130

Gonzalez, J.P., Herbreteau, V., Morvan, J. and Leroy, E.M. (2005) 'Ebola virus circulation in Africa: A balance between clinical expression and epidemiological silence', *Bulletin de la Société de Pathologie Exotique* 98(3): 210–217, www.cababstractsplus.org/google/abstract.asp?AcNo=20053198720

Guan, Y., Zheng, B.J., He, Y.Q., Liu, X.L., Zhuang, Z.X., Cheung, C.L., Luo, S.W., Li, P.H., Zhang, L.J., Guan, Y.J., Butt, K.M., Wong, K.L., Chan, K.W., Lim, W., Shortridge, K.F., Yuen, K.Y., Peiris, J.S.M. and Poon, L.L.M. (2003) 'Isolation and characterization of viruses related to the SARS coronavirus from animals in southern China', *Science* 302: 276–278

Guan, Y., Field, H., Smith, G.J.D. and Chen, H.L. (2005) 'SARS coronavirus: An animal reservoir?', in M. Peiris, L.J. Anderson, A.D.M.E. Osterhaus, K. Stohr and K.Y. Yuen (eds), *Severe Acute Respiratory Syndrome*, Blackwell Publishing, Malden, MA, pp9–83

Gupta, R. (2005) 'Recent outbreak of rabies infections in Brazil transmitted by vampire bats', *EuroSurveillance*, 10(11): E051110.3

Hailin Zhang, Z., Huafang, S., Lihua, L., Yongsin, Y., Dengyuin, Z., Zhaoxaing, L., Tianshou, Z., Wuquan, C., Zhiwei, W., Zhenming, G., Xinnian, L. and Li, J. (1989) 'Isolation of Chikungunya virus from bat in Yunnan province and serological investigations', *Chinese Journal of Virology* 5(1): 31–36

Halpin, K., Young, P.L., Field, H.E. and Mackenzie, J.S. (2000) 'Related isolation of Hendra virus from pteropid bats: A natural reservoir of Hendra virus', *Journal of General Virology* 81(Pt 8): 1927–1932

Hanna, J.N., Carney, I.K., Smith, G.A., Tannenberg, A.E., Deverill, J.E., Botha, J.A., Serafin, I.L., Harrower, B.J., Fitzpatrick, P.F. and Searle, J.W. (2000) 'Australian bat lyssavirus infection: A second human case, with a long incubation period', *Medical Journal of Australia* 172: 597–599

Hill, J. and Smith, J. (1984) *Bats: A Natural History*, University of Texas Press, Austin, TX

Kiley, M.P., Bowen, E.T.W., Eddy, G.A., Isaäcson, M., Johnson, K.M., McCormick, J.B., Murphy, F.A., Pattyn, S.R., Peters, D., Prozesky, O.W., Regnery, R.L., Simpson, D.I.H., Slenczka, W., Sureau, P., van der Groen, G., Webb, P.A. and Wulff, H. (1982) 'Filoviridae: A taxonomic home for Marburg and Ebola viruses', *Intervirology* (Basel) 18(1–2): 24–32

Kim, G.R., Lee, Y.T. and Park, C.H. (1994) 'A new natural reservoir of hantavirus: Isolation of hantaviruses from lung tissues of bats', *Archives of Virology* 134: 85–95

King, A.A., Meredith, C.D. and Thomson, G.R. (1994) 'The biology of southern African lyssavirus variants', *Current Topics in Microbiology and Immunology* 187: 267–295

Krisnamurthi, S. (1964) interview with Carlo Suarès, *Planète* 14

Ksiazek, T.G., Erdman, D., Goldsmith, C.S., Zaki, S.R., Peret, T., Emery, S., Tong, S., Urbani, C., Comer, J.A., Lim, W., Rollin, P.E., Dowell, S.F., Ling, A.E., Humphrey, C.D., Shieh, W.J., Guarner, J., Paddock, C.D., Rota, P., Fields, B., DeRisi, J., Yang, J.Y., Cox, N., Hughes, J.M., LeDuc, J.W., Bellini, W.J., Anderson, L.J. and SARS Working Group (2003) 'A novel coronavirus associated with severe acute respiratory syndrome', *New England Journal of Medicine* 348(20): 1953–1966

Kuhn, H.J. (2008) *Filoviruses: A compendium of 40 Years of Epidemiological, Clinical and Laboratory Studies*, Springer Verlag, Berlin

Kunz, T.H. (1982) *Ecology of Bats*, Boston University Press, Boston, MA

Kunz, T.H. and Brock Fenton, M. (eds) (2003) *Bat Ecology*, University of Chicago Press, Chicago, IL

Lam, S.K., Chua, K.B. (2002) 'Nipah virus encephalitis outbreak in Malaysia', *Clinical Infectious Diseases* 1(34 Supplement 2): S48–51

Lau, S.K., Woo, P.C., Yip, C.C., Tse, H., Tsoi, H.W., Cheng, V.C., Lee, P., Tang, B.S., Cheung, C.H., Lee, R.A., So, L.Y., Lau, Y.L., Chan, K.H. and Yuen, K.Y. (2006) 'Coronavirus HKU1 and other coronavirus infections in Hong Kong', *Journal of Clinical Microbiology* 44(6): 2063–2071, www.pubmedcentral.nih.gov/articlerender.fcgi?artid=1489438

Leirs, H., Mills, J.N., Krebs, J.W., Childs, J.E., Akaibe, D., Woollen, N., Ludwig, G., Peters, C.J. and Ksiazek T.G. (1999) 'Search for the Ebola virus reservoir in Kikwit, Democratic Republic of the Congo: Reflections on a vertebrate collection', *Journal of Infectious Diseases* 179: S1, S155–163

Leroy, E.M., Rouquet, P., Formenty, P., Souquière, S., Kilbourne, A., Froment, J.M., Bermejo, M., Smit, S., Karesh, W., Swanepoel, R., Zaki, S.R. and Rollin, P.E. (2004a) 'Multiple Ebola virus transmission events and rapid decline of Central African wildlife', *Science* 303: 387–390

Leroy, E.M., Telfer, P., Kumulungui, B., Yaba, P., Rouquet, P., Roques, P., Gonzalez, J.-P., Ksiazek, T.G., Rollin, P.E. and Nerrienet, E. (2004b) 'A serological survey of Ebola virus infection in Central African nonhuman primates', *Journal of Infectious Diseases* 190: 1895–1899

Leroy, E., Kumulungui, B., Pourrut, X., Rouquet, P., Yaba, Ph., Délicat, A., Paweska, J., Zaki, S.R., Rollin, P., Gonzalez, J.P. and Swanepoel, R. (2005) 'Fruit bats as a reservoir of Ebola virus', *Nature* 438(7068): 575–576

Li, W., Shi, Z., Yu, M., Ren, W., Smith, C., Epstein, J.H., Wang, H., Crameri, G., Hu, Z., Zhang, H., Zhang, J., McEachern, J., Field, H., Daszak, P., Eaton, B.T., Zhang, S. and Wang, L.F. (2005) 'Bats are natural reservoirs of SARS-like coronaviruses', *Science* 310(5748): 676–679

McCormick, J.B. (2004) 'Ebola virus ecology', editorial commentary, *Journal of Infectious Diseases* 190

Martini, G.A. and Siegert, R. (1971) *Marburg Virus Disease*, Springer-Verlag, Berlin, Germany

Morvan, J.M., Deubel, V., Gounon, P., Nakoune, E., Barriere, P., Murri, S., Perpete, O., Selekon, B., Coudrier, D., Gautier-Hion, A., Colyn, M. and Volehkov, V. (1999) 'Identification of Ebola virus sequences present as RNA or DNA in organs of terrestrial small mammals of the CAR', *Microbes Infect* 1: 1193–1201

Moutou, F., Dufour, B. and Hattenberger, A.M. (2003) 'Rapport sur la rage des Chiroptères en France métropolitaine', available at www.lpo-anjou.org/actu/chauvess/rage_rapportAFSSA2003.pdf (accessed 29 November 2007)

Olson, J.G., Rupprecht, C., Rollin, P.E., An, U.S., Niezgoda, M., Clemins, T., Walston, J. and Ksiazek, T.G. (2002) 'Antibodies to Nipah-like virus in bats (*Pteropus lylei*), Cambodia', *Emerging Infectious Diseases* 8(9): 987–988

Passos, A.D., Castro e Silva, A.A., Ferreira, A.H., Maria e Silva, J., Monteiro, M.E. and Santiago, R.C. (1998) 'Rabies epizootic in the urban area of Ribeirao Preto, Sao Paulo, Brazil', *Cad Saude Publica* 14: 735–740

Peiris, J.S., Lai, S.T., Poon, L.L., Guan, Y., Yam, L.Y., Lim, W., Nicholls, J., Yee, W.K., Yan, W.W., Cheung, M.T., Cheng, V.C., Chan, K.H., Tsang, D.N., Yung, R.W., Ng, T.K. and Yuen, K.Y., SARS Study Group (2003) 'Coronavirus as a possible cause of severe acute respiratory syndrome', *Lancet* 361(9366): 1319–1325

Poon, L.L. (2006) 'Related articles, SARS and other coronaviruses in humans and animals', *Advances in Experimental Medicine and Biology* 581: 457–462 N

Poon, L.L.M., Poon, L.L., Chu, D.K., Chan, K.H., Wong, O.K., Ellis, T.M., Leung, Y.H., Lau, S.K., Woo, P.C., Suen, K.Y., Yuen, K.Y. Guan, Y. and Peiris, J.S. (2005) 'Identification of a novel coronavirus in bats', *Journal of Virology* 79(4): 2001–2009, http://jvi.asm.org/cgi/reprint/79/4/2001.pdf

Pourrut, X., Kumulungui, B., Wittmann, T., Moussavou, G., Délicat, A., Yaba, Ph., Nkoghe, D., Gonzalez, J.P. and Leroy, E.M. (2005) 'The natural history of Ebola virus in Africa', *Microbes and Infection* 7: 1005–1014

Poutanen, S.M., Low, D.E., Henry, B., Finkelstein, S., Rose, D., Green, K., Tellier, R., Draker, R., Adachi, D., Ayers, M., Chan, A.K., Skowronski, D.M., Salit, I., Simor, A.E., Slutsky, A.S., Doyle, P.W., Krajden, M., Petric, M., Brunham, R.C. and McGeer, A.J., National Microbiology Laboratory, Canada; Canadian Severe Acute Respiratory Syndrome Study Team (2003) 'Identification of severe acute respiratory syndrome in Canada', *New England Journal of Medicine* 348(20): 1995–2005

Pritchard, L.I., Chua, K.B., Cummins, D., Hyatt, A., Crameri, G., Eaton, B.T. and Wang, L.F. (2006) 'Pulau virus; A new member of the Nelson Bay orthoreovirus species isolated from fruit bats in Malaysia', *Archives of Virology* 151(2): 229–239, www.springerlink.com/content/p76630614v738017/

Reeves, W.K., Loftis, A.D., Gore, J.A. and Dasch, G.A. (2005) 'Molecular evidence for novel *bartonella* species in *Trichobius major* (Diptera: Streblidae) and *Cimex adjunctus* (Hemiptera: Cimicidae) from two southeastern bat caves, USA', *Journal of Vector Ecology* 30(2): 339–341

Reiter, P., Turell, M., Coleman, R., Miller, B., Maupin, G., Liz, J., Kuehne, A., Barth, J., Geisbert, J., Dohm, D., Glick, J., Pecor, J., Robbins, R., Jahrling, P., Peters, C. and Ksiazek, T. (1999) 'Field investigations of an outbreak of Ebola hemorrhagic fever, Kikwit, DR Congo, 1995: Arthropod studies', *Journal of Infectious Diseases* 179: Supplement 1, S148–S154

Rest, J.S. and Mindell, D.P. (2003) 'SARS associated coronavirus has a recombinant polymerase and coronaviruses have a history of host-shifting', *Infection, Genetics and Evolution* 3: 219–225

Reynes, J.M., Counor, D., Sivuth Ong, Faure, C., Vansay Seng, Molia, S., Walston, J., Georges-Courbot, M.C., Deubel, V. and Sarthou, J.L. (2005) 'Nipah virus in Lyle's flying foxes', *Emerging Infectious Diseases* 11(7): 1042–1047

Rota, P.A., Oberste, M.S., Monroe, S.S., Nix, W.A, Campagnoli, R., Icenogle, J.P., Penaranda, S., Bankamp, B., Maher, K., Chen, M.H., Tong, S., Tamin, A., Lowe, L., Frace, M., DeRisi, J.L., Chen, Q., Wang, D., Erdman, D.D., Peret, T.C., Burns, C., Ksiazek, T.G., Rollin, P.E, Sanchez, A., Liffick, S., Holloway, B., Limor, J., McCaustland, K., Olsen-Rasmussen, M., Fouchier, R., Gunther, S., Osterhaus, A.D., Drosten, C., Pallansch, M.A., Anderson, L.J. and Bellini, W.J. (2003) 'Characterization of a novel coronavirus associated with severe acute respiratory syndrome', *Science* 300(5624): 1394–1399

Serra-Cobo, J., Amengual, B., Abellán, C. and Bourhy, H. (2002) 'European bat Lyssavirus infection in Spanish bat population', *Emerging Infectious Diseases* 8: 413–420

Siddell, S.G. (1995) 'The coronaviridae: An introduction', in S.G. Siddell (ed) *The Coronaviridae: An Introduction,* Plenum Press, New York, NY, pp1–10

Simmons, N.B. (2005a) 'An Eocene big bang for bats', *Science* 307(5709): 527–528

Simmons, N.B. (2005b) 'Chiroptera', in D.E. Wilson and D.M. Reeder (eds) *Mammal Species of the World: A Taxonomic and Geographic Reference,* 3rd edn, Johns Hopkins University Press, Baltimore, MD, pp312–529

Sulkin, S.E. and Allen, R. (1974) 'Virus infections in bats', *Monographs in Virology* 8: 1–103

Tang, X.C., Zhang, J.X., Zhang, S., Wang, P., Fan, X.H., Li, L.F., Li, G., Dong, B.Q., Liu, W., Cheung, C.L., Xu, K.M., Song, W.J., Vijaykrishna, D., Poon, L.L.M., Peiris, J.S.M., Smith, G.J.D., Chen, H. and Guan, Y. (2006) 'Prevalence and genetic diversity of coronaviruses in bats from China', *Journal of Virology* August: 7481–7490

Teeling, E.C., Springer, M.S., Madsen, O., Bates, P., O'Brien, S.J. and Murphy, W.J. (2005) 'A molecular phylogeny for bats illuminates biogeography and the fossil record', *Science* 307(5709): 580–584

Tsang, K.W., Ho, P.L., Ooi, G.C., Yee, W.K., Wang, T., Chan-Yeung, M., Lam, W.K., Seto, W.H., Yam, L.Y., Cheung, T.M., Wong, P.C., Lam, B., Ip, M.S., Chan, J., Yuen, K.Y. and Lai, K.N. (2003) 'A cluster of cases of severe acute respiratory syndrome in Hong Kong', *New England Journal of Medicine* 348(20): 1977–1985

Vidal, P., Tibayrenc, M. and Gonzalez, J.P. (2007) 'Infectious diseases and art', in M. Tibayrenc (ed) *Encyclopedia of Infectious Diseases: Modern Methodologies,* Wiley & Sons, London

Wang, L.F., Yu, M., Hansson, E., Pritchard, L.I., Shiell, B., Michalski, W.P. and Eaton, B.T. (2000) 'The exceptionally large genome of Hendra virus: Support for creation of a new genus within the family Paramyxoviridae', *Journal of Virology* 74: 9972–9979

Wang, L.F., Shi, Z., Zhang, S., Field, H., Daszak, P. and Eaton, B.T. (2006) 'Review of bats and SARS', *Emerging Infectious Diseases* 12(12): 1834–1840

Wong, S., Lau, S., Woo, P. and Yuen, K.Y. (2007) 'Bats as a continuing source of emerging infections in humans', *Reviews in Medical Virology* 17(2): 67–91

Wong-Chia, C. and Scherer, W.F. (1971) 'Isolation of the Venezuelan encephalitis virus from a frugivorous bat (*Artibeus turpis*) in Mexico' (in Spanish), *Bolitin Oficina Sanitaria Panama* 70(4): 339–343

Woo, P.C., Lau, S.K., Li, K.S., Poon, R.W., Wong, B.H., Tsoi, H.W., Yip, B.C., Huang, Y., Chan, K.H. and Yuen, K.Y. (2006) 'Molecular diversity of coronaviruses in bats', *Virology* 20, 351(1): 180–187

Yusibov, V., Rabindran, S., Commandeur, U., Twyman, R.M. and Fischer, R. (2006) 'The potential of plant virus vectors for vaccine production', *Drugs in R&D* 7(4): 203–217, www.ingentaconnect.com/content/adis/rdd/2006/00000007/00000004/art00001

Zhong, N.S., Zheng, B.J., Li, Y.M., Poon, L.L., Xie, Z.H., Chan, K.H., Li, P.H., Tan, S.Y., Chang, Q., Xie, J.P., Liu, X.Q., Xu, J., Li, D.X., Yuen, K.Y., Peiris and Guan, Y. (2003) 'Epidemiology and cause of severe acute respiratory syndrome (SARS) in Guangdong, People's Republic of China', *Lancet* 362(9393): 1353–1358

APPENDIX: TABLE OF NATURAL VIRUS-INFESTED BATS

Virus family/ genus	*Virus species[1]/ human disease[2]*	*Bats genus species*	*Name*	*Distribution bat/virus*	*Reference*
Arenaviridae	Tacaribe virus /no record (nr)	*Artibeus jamaicensis trinitatus*	Jamaica fruit-eating bat	Trinidad	i
		Artibeus lituratus palmarum	Great fruit-eating bat	Trinidad	i
Alphaviridae	Chikungunya virus/ dengue-like syndrome	*Scotophilus sp*	Asiatic yellow bat	Senegal	ii
		Rousettus leschenaultii	Leschenault's rousette	China	iii
		Hipposideros caffer	Sundevall's leaf-nosed bat	China	iii
		Chaerephon pumilus	Little free-tailed bat	China	iii
	Venezuelan Equine Encephalitis virus /encephalitis	*Carollia perspicillata*	Seba's short-tailed bat	Brazil, Colombia	iv, v, vi
		Artibeus phaeotis	Pygmy fruit-eating bat	Guatemala, Mexico, Venezuela	iii
		Desmodus rotundus	Common vampire bat	Mexico, Venezuela	iii
		Uroderma bilobatum	Common tent-making bat	Mexico, Venezuela	iii
	Sindbis virus/dengue-like syndrome	Rhinolophidae sp	Horseshoe bat	Mexico, Venezuela	
		Hipposideridae sp	Old World leaf-nosed bat	Mexico, Venezuela	vii
Bunyaviridae *Orthobunyavirus*	Catu virus/dengue-like syndrome	*Molossus molossus*[3]	Pallas' mastiff bat	Amapa	xv
	Guama virus/dengue-like syndrome	Unidentified bat	–	Southern Brazil	iii
	Nepuyo virus/dengue-like syndrome	*Artibeus jamaicensis*	Jamaican fruit-eating bat	Venezuela,	iii
		A. lituratus	Great fruit-eating bat	Guatemala	
	Mojui dos Campos virus/nr	Unidentified bat	–	Brazil	xvi
	Kaeng Khoi virus/nr	*Chaerephon plicatus*	Wrinkle-lipped free-tailed bat	Cambodia, Thailand	viii
Hantavirus	Hantaan virus/ haemorrhagic fever	*Eptesicus serotinus*	Common serotine	Korea	ix
		Rhinolophus ferrumequinum	Greater horseshoe bat		
Phlebovirus	Rift Valley Fever virus/ haemorrhagic fever	*Micropteropus pusillus*	Peters' lesser epauletted fruit bat	Guinea	x
		Hipposideros abae	Aba leaf-nosed bat		
		Hipposideros caffer	Sundevall's leaf-nosed bat		
	Toscana virus/nr	*Pipistrellus kuhlii*	Kuhl's pipistrelle	Italy	xi
Nairovirus	Bandia virus/nr	*Scotophilus nigrita*	Giant house bat	Senegal	xii
Unassigned genus	Bangui virus/nr	*Scotophilus* sp	Yellow bats	Central African Republic	xiii
		Pipistrellus sp	Pipistrelle bats		
		Tadarida sp	Free-tailed bats		
	Bhanja virus/dengue-like syndrome	Unidentified bat	–	Kyrgyzstan	xiv

Virus family/ genus	*Virus species[1]/ human disease[2]*	*Bats genus species*	*Name*	*Distribution bat/virus*	*Reference*
	Issyk-Kul virus/ dengue-like syndrome	*Nyctalus noctula*	Noctule	Kyrgyzstan	xv
		Eptesicus serotinus	Common serotine	Kyrgyzstan	
		Myotis blythii	Lesser mouse-eared myotid	Kyrgyzstan	
		Pipistrellus pipistrellus	Common pipistrelle	Tajikistan	xvi
	Kasokero virus/ dengue-like syndrome	*Rousettus aegyptiacus*	Egyptian rousette	Uganda	xvii
	Keterah virus/nr	*Scotophilus kuhlii*	Lesser Asiatic yellow house bat	Malaysia	xviii
	Yogue virus/nr	*Rousettus aegyptiacus*	Egyptian rousette	Senegal Central African Republic	xvii
Coronaviridae					
Coronavirus					
Group 1	bat-CoV HKU2	*Rhinolophus sinicus*	Chinese rufous horseshoe bat	Hong Kong SAR	xix
	bat-CoV HKU6	*Myotis ricketti*	Rickett's big-footed myotis	Hong Kong SAR	
	bat-CoV HKU7	*Miniopterus magnater*	Western long-fingered bat	Hong Kong SAR	
	bat-CoV HKU8	*Miniopterus pusillus*	Small long-fingered bat	Hong Kong SAR	
		M. magnater	Western long-fingered bat		
		M. schreibersii	Schreibers' long-fingered bat		
	bat-CoV/701/05	*Myotis ricketti*	Rickett's big-footed myotis	Anhui, Yunnan, Guangdong	
	bat-CoV/821/05	*Myotis ricketti*	Rickett's big-footed myotis	Jiangxi, Guangxi	
	bat-CoV/821/05	*Scotophilus kuhlii*	Lesser Asiatic yellow house bat	Hainan	
	bat-CoV/970/06	*Rhinolophus pearsonii*	Pearson's horseshoe bat	Shandong	
		R. ferrumequinum	Greater horseshoe bat		
	bat-CoV/A773/05	*Miniopterus schreibersii*	Schreibers' long-fingered bat	Fujian	
	bat-CoV/A011/05	*Miniopterus schreibersii*	Schreibers' long-fingered bat	Anhui, Fujian, Guangxi	
G2b	Rp3	*Rhinolophus pearsonii*	Pearson's horseshoe bat	Guangxi	xix
	Rm1 (bat-CoV/279/04)	*Rhinolophus macrotis*	Big-eared horseshoe bat	Hubei	
	Rf1 (bat-CoV/273/04)	*Rhinolophus ferrumequinum*	Greater horseshoe bat	Hubei	
	bat-SARS-CoV HKU3	*Rhinolophus sinicus*	Chinese rufous horseshoe bat	Hong Kong SAR	
	bat-CoV/A1018/06	*Rhinolophus sinicus*	Chinese rufous horseshoe bat	Shandong	
	bat-CoV/279/04	*Rhinolophus macrotis*	Big-eared horseshoe bat	Hubei	

Virus family/ genus	*Virus species[1]/ human disease[2]*	*Bats genus species*	*Name*	*Distribution bat/virus*	*Reference*
	bat-CoV/273/04	*Rhinolophus ferrumequinum*	Greater horseshoe bat	Hubei	
G2c	bat-CoV HKU4 bat-CoV HKU5 bat-CoV/133/05 bat/CoV/434/05 bat/CoV/355/05	*Tylonycteris pachypus* *Pipistrellus abramus* *Tylonycteris pachypus* *Pipistrellus pipistrellus* *Pipistrellus abramus* *Rhipicephalus ferrumequinum*	Lesser bamboo bat Japanese pipistrelle Lesser bamboo bat Japanese pipistrelle Japanese pipistrelle Greater horseshoe bat	Hong Kong SAR Hong Kong SAR Guangdong Hainan Anhui, Henan, Sichuan	xix
Filoviridae	Ebola virus subtype 'Zaire'/ haemorrhagic fever	*Epomops franqueti* *Hypsignathus monstrosus* *Myonycteris torquata*	Franquet's epauletted fruit bat Hammer-headed fruit bat Little collared fruit bat	Central Africa	xx
Flaviviridae	Bukalasa bat virus /nr	*Tadarida* sp *Chaerephon pumilus* *Mops condylurus*	Free-tailed bat Little free-tailed bat Angolan free-tailed bat	Senegal Uganda Uganda	xxi
	Carey Island virus/nr	*Cynopterus brachyotis* *Macroglossus minimus*	Lesser short-nosed fruit bat Dagger-toothed long-nosed fruit bat	Malaysia	xxii
	Dakar bat virus/fever	*Scotophilus nigrita* *Chaerephon pumilus* *Mops condylurus*	Giant house bat Little free-tailed bat Angolan free-tailed bat	Senegal Madagascar, Uganda, Senegal, Nigeria	ii
	Entebbe bat virus/nr	*Chaerephon pumilus* *Mops condylurus*	Little free-tailed bat Angolan free-tailed bat	Uganda	xxiii
	Japanese encephalitis virus /encephalitis	*Miniopterus* sp *Rhinolophus* sp *Hipposideros armiger terasensis* *Miniopterus schreibersii* *Rhinolophus cornutus*	Fingered bat Horseshoe bat Great leaf-nosed bat Schreibers' long-fingered bat Little Japanese horseshoe bat	Japan Taiwan Taiwan Japan	xxiv
	Jugra virus/nr	*Cynopterus brachyotis*	Lesser short-nosed fruit bat	Malaysia	iii
	Kyasanur Forest Disease virus/ haemorrhagic fever	*Rhinolophus rouxii*	Rufous horseshoe bat	India	xxv
	Montana myotis Leukoencephalitis virus/nr	*Myotis lucifugus*	Little brown myotis	Montana	xxvi
	Phnom Penh bat virus/nr	*Cynopterus brachyotis* *Eonycteris spelaea*	Lesser short-nosed fruit bat Lesser dawn bat	Cambodia Malaysia	xxvii

Virus family/ genus	*Virus species[1]/ human disease[2]*	*Bats genus species*	*Name*	*Distribution bat/virus*	*Reference*
	Rio Bravo virus/ dengue-like syndrome and pulmonary signs	*Eptesicus fuscus* *Molossus rufus*	Big brown bat Black mastiff bat	Texas, Trinidad California, Trinidad	xxviii
		Tadarida braziliensis mexicana	Mexican free-tailed bat	Trinidad, USA, Mexico	
	Saboya virus/nr	*Nycteris gambiensis*	Gambian slit-faced bat		xxix
	Saint Louis encephalitis virus/encephalitis	*Tadarida braziliensis mexicana*	Mexican free-tailed bat	Texas	xxx, xxxi
	Sokuluk virus/ Encephalitis	*Pipistrellus pipistrellus*	Common pipistrelle	Kyrgyzstan	xxxii
	Tamana virus/nr	*Pteronotus parnellii*	Common moustached bat	Trinidad	xxxiii
	Uganda S virus/nr	*Rousettus* sp	Flying foxes	Uganda	xxix
	West Nile virus/ encephalitis	*Roussettus leschenaultii*	Leschenault's rousette	India	xxxiv
	Yokose virus (Entebbe V-like)/nr	*Miniopterus* ssp	Long-winged bats	Japan	xxxv
Herpesviridae	Agua Preta virus/nr	*Carollia subrufa*	Grey short-tailed bat	Brazil	xvii
unassigned	A cytomegalovirus/nr	*Myotis lucifugus*	Little brown myotis	USA	
	Herpes virus/nr	Chiroptera	Indetermined genus	Central African Republic	xiii
		Eidolon helvum	Straw-coloured fruit bat	Cameroon	
	Parixa virus/nr	*Lonchophylla thomasi*	Thomas's nectar bat	Brazil	iii
Orthomyxoviridae *Influenzavirus A*	Influenza A virus/ influenza	*Nyctalus noctula*	Noctule	Kazakhstan	iii
Papillomaviridae *unassigned*	*RaPV-1/nr*	*Rousettus aegyptiacus*	Egyptian rousette	USA (captive)	xlvi
Paramyxoviridae *Rubulavirus*	Mapuera virus/nr	*Sturnira lilium*	Little yellow-shouldered bat	Brazil	xxxvi
	Menangle virus/ dengue-like syndrome	*Pteropus* spp	Flying foxes	Australia	xxxvii
	Tioman virus/nr	*Pteropus hypomelanus*	Variable flying fox	Malaysia	xxxviii
Henipavirus	Hendra virus/ encephalitis	*Pteropus poliocephalus* *Pteropus alecto* *Pteropus scapulatus* *Pteropus conspicillatus*	Gray-headed flying fox Black flying fox Little red flying fox Spectacled flying fox	Australia	xxxvi

Virus family/ genus	Virus species[1]/ human disease[2]	Bats genus species	Name	Distribution bat/virus	Reference
	Nipah virus/ Encephalitis	*Cynopterus brachyotis*	lesser short-nosed fruit bat	Malaysia (serology)	xxxix
		Eonycteris spelaea	lesser dawn bat	Malaysia (serology)	
		Pteropus hypomelanus	Variable flying fox	Malaysia	
		Pteropus vampyrus	Large flying fox[4]	Indonesia (serology),	
		Pteropus lylei	Lyle's flying fox	Cambodia, Thailand	
		Pteropus giganteus	Indian flying fox	Bangladesh (serology)	
		Hipposideros armiger	Great leaf-nosed bat	Thailand	
Picornaviridae	Juruaca virus/nr	Chiroptera	Indetermined genus		iii
Reoviridae *Orbivirus*	Ife virus/nr	Eidolon helvum	African straw-coloured fruit bat	Cameroon, Central African Republic	xl
	Japanaut virus/nr	*Syconycteris australis*	Southern blossom bat	New Guinea	iii, xxviii
	Fomede virus/nr	*Nycteris nana*	Dwarf slit-faced bat	Guinea	iii
		Nycteris gambiensis	Gambian slit-faced bat		
Orthoreovirus	Broome virus/nr	*Pteropus alecto*	Black flying fox		iii
	Nelson Bay virus/nr	*Pteropus poliocephalus*	Gray-headed flying fox	Australia	xli
	Pulau virus/nr	*Pteropus hypomelanus*	Variable flying fox	Malaysia	xlii
Rhabdoviridae *Lyssavirus*[5]	Rabies virus Fatal meningo-encephalomyelitis	*Pipistrellus hesperus*	Western pipistrelle	USA	See xliii (main hosts)
		Pipistrellus subflavus	Eastern pipistrelle		
		Myotis lucifugus	Little brown myotis		
		Tadarida brasiliensis	Brazilian free-tailed bat		
		Lasionycteris noctivagans	Silver-haired bat		
		Desmodus rotondus	Common vampire bat		
		Eptesicus fuscus	Big brown bat		
		Lasiurus borealis	Eastern red bat		
		Lasiurus cinereus	Hoary bat		
	Lagos bat virus/nr	*Eidolon helvum*	African straw-colored fruit bat	Nigeria, Senegal	xiii
		Micropteropus pusillus	Peters's lesser epauletted fruit bat	Central African Republic	
		Nycteris gambiensis	Gambian slit-faced bat	Guinea	
		Epomophorus wahlbergi	Wahlberg's epauletted fruit bat	South Africa	
	Duvenhage virus/ fatal meningo-encephalomyelitis	*Miniopterus* sp	Fingered bat	South Africa	xliv
		Nycteris thebaica	Egyptian slit-faced bat	Zimbabwe	

Virus family/ genus	Virus species[1]/ human disease[2]	Bats genus species	Name	Distribution bat/virus	Reference
	European bat Lyssavirus 1	*Eptesicus serotinus*	Common serotine	Europe	
		Pipistrellus pipistrellus	Common pipistrelle	France, Germany	
	Fatal meningo-encephalomyelitis	*Miniopterus schreibersii*	Schreibers's long-fingered bat	Spain	
		Myotis myotis	Mouse-eared myotis	Germany	
		Myotis nattereri	Natterer's myotis	Spain	
		Rhinolophus ferrumequinum	Greater horseshoe bat	Turkey	
		Rousettus aegyptiacus	Egyptian rousette	France	
		Vespertilio murinus	Particoloured bat	Ukraine	
	European bat Lyssavirus 2/ fatal meningo-encephalomyelitis	*Myotis daubentonii*	Daubenton's myotis	Switzerland	
		Myotis dasycneme	Pond myotis	Netherlands, UK	
	Australian bat Lyssavirus/ fatal meningo-encephalomyelitis	*Pteropus* sp	Flying foxes	Australia	
		Saccolaimus flaviventris	Yellow-bellied pouched bat		
	Aravan virus/nr	*Myotis blythii*	Lesser mouse-eared bat	Kyrgyzstan	
	Khujand virus/nr	*Myotis mystacinus*	Whiskered myotis	Tajikistan	iii
	Irkut virus/nr	*Murina leucogaster*	Greater tube-nosed bat	Siberia	iii
	West Caucasian bat virus/nr	*Miniopterus schreibersii*	Schreibers' long-fingered bat	Caucasus	iii
Vesiculovirus	Mount Elgon bat virus/nr	*Rhinolophus hiderbrandtii*	Hildebrandt's horseshoe bat	Kenya	
Unassigned	Gossas virus/nr	*Tadarida* sp	Free-tailed bat	Senegal	xlv
	Kern Canyon virus/nr	*Myotis yumanensis*	Yuma myotis	California	iii
	Oita 296 virus/nr	*Rhinolophus cornutus*	Little Japanese horseshoe bat	Japan	

Notes

nr = no record

1 See also the 'International Committee on Taxonomy of Viruses': www.virustaxonomyonline.com/virtax/lpext.dll?f=templates&fn=main-h.htm.

2 Human Disease relates only to human potential infection by the virus without presupposing the role of bat as host, reservoir or vector of virus, each virus will have its own specific ecology including natural cycle of transmisson.

3 Simmons NB. Order Chiroptera. In Wilson DE, Reeder DM, editors. Mammal species of the world. A taxonomic and geographic reference. 3rd ed. Baltimore: The Johns Hopkins University Press; 2005. p. 312–529.

4 Malaysian specimens of frugivorous species (*Cynopterus brachyotis* and *Eonycteris spelaea*) and insectivorous species (*S. kuhlii*) have been found carrying neutralizing antibodies to NiV. Bangladesh antibodies against NiV have been detected in the Indian flying fox, *P. giganteus*, a conspecific species with *P. vampyrus*.

5 Genotype 3. Mokola virus, known from Nigeria, Central African Republic (CAR), Zimbabwe, Cameroon, Ethiopia has been isolated from human, dog, shrew and rodents (but not from Chiroptera).

References

i Downs WG, Anderson CR, Spoence L, Aitken, THG and GreenHall AH. 1963. Tacaribe virus a new agent isolated from *Artibeus* bats and mosquitoes in Trinidad, West Indies. *Am. J. Trop. Med. & Hyg.* 12: 640–646

ii Brès P & Chambon L. 1963. Isolement à Dakar d'une souche d'arbovirus à partir des glandes salivaires de chauves-souris. *Ann. Inst. Pasteur.* 104: 701–711

iii Calisher CH, Childs JE, Field HE, Holmes KV, Schountz T. 2006. Bats: important reservoir hosts of emerging viruses. *Clin Microbiol Rev.* 19: 531–545

iv Salaun Klein Hebrard. 1974. Un nouveau virus Phnom-Penh bat, isolé au Cambodge chez une chauve souruis frugivore *Cynopterus brachyotis angulatus*, Miller 1898. *Ann Microbiol* (Paris). 125: 485–495.

v Wong-Chia C, Scherer WF. 1971. Isolation of the Venezuelan encephalitis virus from a frugivorous bat (*Artibeus turpis*) in Mexico. *Bol Oficina Sanit Panam.* 70: 339–343

vi Correa-Giropn, Calisher CH & Baer GM, Epidemic strain of Venezuelan equine encephalomyelitis virus from a vampire bat captured in Oaxaca, Mexico, 1970. *Science,* 1972, 175: 546–547

vii Blackburn NK, Foggin CM, Searle L, Smith PN. 1982. solation of Sindbis virus from bat organs. *Cent Afr J Med.* 28(8): 201

viii Osborne JC, Rupprecht CE, Olson JG, Ksiazek TG, Rollin PE, Niezgoda M, Goldsmith CS, An US, Nichol ST. Isolation of Kaeng Khoi virus from dead Chaerephon plicata bats in Cambodia. *J Gen Virol.* 2003 Oct; 84(Pt 10): 2685–9.

ix Kim GR, Lee YT, Park CH. A new natural reservoir of hantavirus: isolation of hantaviruses from lung tissues of bats. *Arch Virol.* 1994;134(1-2): 85–95.

x Boiro I, Konstaninov OK, Numerov AD. [Isolation of Rift Valley fever virus from bats in the Republic of Guinea] *Bull Soc Pathol Exot Filiales.* 1987; 80(1): 62–7

xi Verani P, Ciufolini MG, Caciolli S, Renzi A, Nicoletti L, Sabatinelli G, Bartolozzi D, Volpi G, Amaducci L, Coluzzi M, et al. Ecology of viruses isolated from sand flies in Italy and characterized of a new Phlebovirus (Arabia virus). *Am J Trop Med Hyg.* 1988 Mar; 38(2): 433–9.

xii Brès P et al. Considérations sur l'épidémiologie des arboviroses au Sénégal. *Bull Soc Path Exo,* 1969, 62: 253–259

xiii Digoutte JP & Adam F. Virus transmis par les arthropodes (arbovirus) identifies par le Centre Collaborateur oms de Reference et de Recherche sur les Arbovirus (CRORA) de l'Institut Pasteur de Dakar et autres virus du continent Africain. 2005. www.pasteur.fr/recherche/banques/CRORA/fixes/depart.htm

xiv Hubalek Z. Geographic distribution of Bhanja virus. *Folia Parasitol* (Praha), 1987, 34: 77–86

xv Lvov DK, Karas FR, Timofeev EM, Tsyrkin YM, Vargina SG, Veselovskaya OV, Osipova NZ, Grebenyuk YI, Gromashevski VL, Steblyanko SN, Fomina KB. 'Issyk-Kul' virus, a new arbovirus isolated from bats and/Argas (Carios) vespertilionis/ (Latr.,1802) in the Kirghiz S.S.R. Brief report. *Arch Gesamte Vir.*, 1973, 42: 207–209

xvi Karabatsos N. *International catalogue of arboviruses including certain other viruses of vertebrates,* 3rd edition. San Antonio, TX: American Society of Tropical Medicine and Hygiene, 1985 (www2.ncid.cdc.gov/arbocat)

xvii Kalunda M, Mukwaya LG, Mukuye A, Lule M, Sekyalo E, Wright J, Casals J. Kasokero virus: a new human pathogen from bats (*Rousettus aegyptiacus*) in Uganda. *Am J Trop Med Hyg,* 1986, 35: 387–392.

xviii Varma MG, Converse JD. Keterah virus infections in four species of Argas ticks (Ixodoidea: Argasidae). *J Med Entomol,* 1976, 13: 65–70.

xix Woo XY et al. Molecular diversity of coronaviruses in bats. 2006. *Virology.* 351; 1809: 187

xx Leroy Eric, Brice Kumulungui, Xavier Pourrut, Pierre Rouquet, Philippe Yaba, André Délicat, Jaenusz Paweska, Sherif R Zaki, Pierre Rollin, Jean-Paul Gonzalez & Robert Swanepoel 2005. Fruits bats as a reservoir of Ebola virus. *Nature.* 2005; 438(7068): 575–6

xxi Karabtsos N. Characterisation of viruses isolated from bats. *American Journal of Tropical Medicine & Hygiene,* 1969. 18(5), 893–910

xxii Miura T & Kitaoka M. Viruses isolated from bats in Japan *Journal Archives of Virology* 1977. 53(4): 281–28

xxiii Lumsdenws HR, Williams MC & Mason PJ. A virus from insectivorous bats in Uganda. *Ann. trop Med. Parasit.,* 1961, 55: 389–397

xxiv Sulkins E, Allen R, Miura T, & Toyokawak A. Studies of arthropodborne virus infections in Chiroptera : VI. Isolation of Japanese B encephalitis virus from naturally infected bats. 1960 *Amer. J. trop. Med. Hyq.,* 1970, 19: 77–87

xxv Rajagopalan PK, Paul SD & Sreenevasan MA. Isolation of Kyasanur Forest disease virus from the insectivorous bat, *Rhinolophus rouxi* and from *Ornithodoros* ticks. *Indian J. med. Res.,* 1969, 57: 805–808

xxvi Bell JF & Thomas LA. A new virus, (MML B, enzootic in bats *Myotis lucifugus)* of Montana. 1964. *Amer. J. trop. Med. Hyg.,* 13: 607–612

xxvii Salaun JJ, Klein JM & Hebrard G. Un nouveau virus, Phnom Penh bat virus, isolé au Cambodge chez un chauve souris frugivore, *Cynopterus brachyotis angulatus,* Miller, 1898. 1974. *Annales de Microbiologie* (Institut Pasteur), 125A; 485–495.

xxviii Burns KF & Farinacci J. Virus of bats antigenically related to St Louis encephalitis. *Science,* 1956, 123: 227–228

xxix Butenko AM. [Arbovirus circulation in the Republic of Guinea] *Med Parazitol* (Mosk). 1996 Apr-Jun; (2): 40–5

xxx Sulkin SE, Sims RA & Allen R. Isolation of St Louis encephalitis virus from bats *(Tadarida b. mexicana)* in Texas. *Science,* 1966, 152

xxxi Allen R, Taylor SK & Sulkin SE. Studies of arthropod-borne virus infections in Chiroptera. VIII. Evidence of natural St Louis encephalitis virus infection in bats. *Amer. J. trop. Med. Hyg.,* 1970,19: 851–859

xxxii Lvov DK, Tsyrkiny M, Karasf R, Timopheeve M, Gromashevski L, Veselovskayaoa V, Osipovan Z, Fominka B & Grebenyuki I. Sokuluk virus, a new group B arbovirus isolated from *Vespertilio pipistrellus* Schreber, 1775, bat in the kirghiz S. S. R., *Arch. ges. Virusforsch.,* 1973, 41: 170–174

xxxiii de Lamballerie X, Crochu S, Billoir F, Neyts J, de Micco P, Holmes EC, Gould EA. Genome sequence analysis of Tamana bat virus and its relationship with the genus Flavivirus. *J Gen Virol.* 2002 (Pt 10): 2443–54.

xxxiv Pauls D, Rajagolzapapn K & Sreenivasamn A. Isolation of the West Nile virus from the frugivorous bat *Rousettus leschenaulti. Indian J. med. Res.,* 1970, 58: 1169–1171

xxxv Shigeru Tajima, Tomohiko Takasaki, Shigeo Matsuno, Mikio Nakayama, Ichiro Kurane. Genetic characterization of Yokose virus, a flavivirus isolated from the bat in Japan. *Virology* 332: 38–44, 2005.

xxxvi Halpin K, Young PL, Field HE & Mackenzie JS. Isolation of Hendra virus from pteropid bats: a natural reservoir of Hendra virus *J. Gen. Virol.* 2000, 81, 1927–1932.

xxxvii Halpin K, Young PL, Field H, Mackenzie JS. Newly discovered viruses of flying foxes. *Vet Microbiol.* 1999 Aug 16; 68(1-2): 83–7.

xxxviii Chua KB, Wang LF, Lam SK, Eaton BT. Full length genome sequence of Tioman virus, a novel paramyxovirus in the genus *Rubulavirus* isolated from fruit bats in Malaysia. *Arch Virol.* 2002 Jul;147(7):1323–48.

xxxix Chua KB, Bellini WJ, Rota PA, Harcourt BH, Tamin A, Lam SK, Ksiazek TG, Rollin PE, Zaki SR, Shieh W, Goldsmith CS, Gubler DJ, Roehrig JT, Eaton B, Gould AR, Olson J, Field H, Daniels P, Ling AE, Peters CJ, Anderson LJ, Mahy BW. Nipah virus: a recently emergent deadly paramyxovirus. *Science.* 2000; 288(5470): 1432–5.

xl Kemp GE, Le Gonidec G, Karabatsos N, Rickenbach A & Copp CB. 1988. Ife: un nouvel arbovirus africain isolé chez des chauves-souris Eidolon elvum capturées au Nigeria, au Cameroun et en République Centrafricaine. *Bulletin de la Société de Pathologie Exotique,* 81: 40–48

xli Gard G, Compans RW. Structure and cytopathic effect of Nelson Bay virus. 1970. *J. Virol* 7: 100–106

xlii Pritchard LI, Chua K, Cummins B, Hyatt DA, Crameri G, Eaton BT, & Wang LF. Pulau virus; a new member of the *Nelson Bay orthoreovirus* species isolated from fruit bats in Malaysia. 2006. *Arch Virol.*; 151: 229–239

xliii Badrane H, Bahloul C, Perrin P, Tordo N. Evidence of two Lyssavirus phylogroups with distinct pathogenicity and immunogenicity. 2001. *J Virol.* 75(7): 3268–76 xliv Arai YT, Kuzmin IV, Kameoka Y, Botvinkin AD. New lyssavirus genotype from the lesser mouse-eared bat (*Myotis blythi*), Kyrghyzstan. *Emerg Infect Dis*. 2003; 9(3): 333–7

xlv Bres P. Gossas (Dak An D 401), nouvel Arbovirus non groupé. *International Catalogue of Arboviruses*, Third Edition, 1985, pp. 425–426

xlvi Rector A, Mostmans S, Van Doorslaer K, McKnight CA, Maes RK, Wise AG, Kiupel M, Van Ranst M. Genetic characterization of the first chiropteran papillomavirus, isolated from a basosquamous carcinoma in an Egyptian fruit bat: the Rousettus aegyptiacus papillomavirus type 1. 2006. *Vet Microbiol*, 117: 267–275

9

Deforestation and Malaria: Revisiting the Human Ecology Perspective

Subhrendu K. Pattanayak and Junko Yasuoka[1]

The ecological basis for disease dates at least as far back as 400B.C.E., to Hippocrates's *On Airs, Waters, and Place.* As Wilson (1995) notes, our understanding and therefore control of diseases would be inadequate without an 'ecological' perspective on the life cycles of parasitic micro-organisms and the associated infectious diseases. Smith et al (1999, p583) contend, 'many of the critical health problems in the world today cannot be solved without major improvement in environmental quality'.

In this chapter we focus on malaria because its transmission and control have clear links to ecosystem changes that result from natural resource policies on land tenure, road building and agricultural subsidies. The resulting ecosystem change has a tremendous influence on the pattern of diseases (Martens, 1998; Molyneux, 1998; Grillet, 2000), malaria in particular because mosquitoes are highly sensitive to ecosystem change. Ecological 'lenses' can improve our understanding of disease prevention, and in this chapter we articulate a particular ecological perspective that puts human behavioural change front and centre.

In the past decade, widely cited papers have drawn the connections between ecosystem change and diseases, many of which are synthesized in the 2005 Millennium Ecosystem Assessment (McMichael et al, 1998; Corvalan et al, 2005a; 2005b; Campbell-Lendrum et al, 2005; Patz et al, 2005). This renewed interest in the more distal causes of disease reflects in part the emergence of new fields such as 'sustainability science' (Kates et al, 2001) and 'biocomplexity' (Wilcox and Colwell, 2003), which argue for 'a more realistic view [requiring] a holistic perspective that incorporates social as well as physical, chemical, and biological dimensions of our planet's systems'. The resurgence also reflects the growing importance of fields of social epidemiology (e.g., Berkman and Kawachi, 2000; Oakes and Kaufman, 2006).

In joining this growing chorus, we focus on an older human ecology tradition (Wessen, 1972; MacCormack, 1984), which posits that we humans modify our natural environment, sometimes increasing disease risks, and ultimately adapt to the new disease risk environment. Two stylized yet complicating facts emerge from this viewpoint (Pattanayak et al, 2006a). First, disease prevention behaviours (including ecosystem changes that modify disease risks) respond to disease levels, suggesting a dynamic feedback between exposure and control. Second, individuals and households typically do not consider how their private actions affect public health outcomes and therefore often make socially inappropriate and sub-optimal choices. Some combination of government regulation, community norms or information and subsidies can help narrow the wedge between private and 'optimal' social behaviours. The systematic incorporation of human behaviour marks a departure from a predominantly biophysical approach, which can easily overlook the social, cultural and economic drivers that are crucial to understanding anthropogenic ecosystem disruptions and their human health impacts (McMichael, 2001; Parkes et al, 2003).

Malaria and deforestation exemplify the many links between infectious diseases and ecosystem change. We restrict ourselves to malaria not only because its transmission is clearly linked to ecological changes, but also because it is a major health concern in tropical forests (Hay et al, 2004). We focus on deforestation because it is a major development policy concern and often heralds many other land-use changes associated with malaria (Pattanayak et al, 2006c).

In the next section, we briefly review the literature on ecology of infectious diseases. We then re-introduce the human ecology perspective for better understanding the role of humans in land-use change as well as in a variety of behaviours to prevent and treat malaria. The subsequent section draws out the empirical implications of such a strategy, using our own fieldwork and secondary data sets. Finally, we conclude with a call for systematic environmental and health impact assessments that rely on interdisciplinary longitudinal studies.

MOSQUITO ECOLOGY AND MALARIA EPIDEMIOLOGY

The impacts of ecosystem change on health are diverse and long standing, but their rate and geographical range have increased markedly over the past few decades. Different kinds of environmental changes have resulted from deforestation, agricultural activities, plantations, logging, fuelwood collection, road construction, mining, hydropower development and urbanization (Walsh et al, 1993; Patz et al, 2000; 2004). The process of clearing forests and subsequent land transformation alters every element of local ecosystems, including microclimate, soil and aquatic conditions, and most significantly, the ecology of local fauna and flora. These in turn have profound impacts on the survival, density and distribution of human disease vectors and parasites (Martens, 1998; Grillet, 2000) by affecting breeding places, daily survival probability, density, biting rates, and incubation periods. Thus, the altered vector-parasite ecology modifies the transmission of vector-borne diseases such as malaria, Japanese encephalitis and filariasis (Sharma and Kondrashin, 1991; Walsh et al, 1993; see also Chapter 8, this volume).

Numerous country and area studies have described how the density and distribution of local vector species have been altered by ecosystem change, and some longitudinal studies have shown that the change in vector ecology has altered local disease incidence and prevalence (Sharma and Kondrashin, 1991; Patz et al, 2000). However, the mechanism linking ecosystem change, vector ecology and vector-borne diseases is still unclear. We draw on a paper by Yasuoka and Levins (2007) that examines the mechanisms linking deforestation, anopheline ecology and malaria epidemiology by drawing together 60 examples of changes in anopheline ecology as a consequence of deforestation and agricultural development.

Massive clearing of forests has enormous impacts on local ecosystems and human disease patterns. It alters microclimates by reducing shade, altering rainfall patterns, augmenting air movement and changing the humidity regime (Reiter, 2001). It also reduces biodiversity and increases surface water availability through the loss of topsoil and vegetation root systems that absorb rainwater (Chivian, 2002). For anopheline species that breed in shaded water bodies, deforestation can reduce breeding habitats, thus affecting their propagation. On the other hand, some environmental and climatic changes due to deforestation can facilitate the survival of other anopheline species, resulting in prolonged seasonal malaria transmission (Kondrashin et al, 1991).

As shown in Table 9.1 (drawn from Yasuoka and Levins, 2007), different land transformations have different impacts on local ecosystems and disease patterns. For example, rubber plantations increased local major malaria vectors in Malaysia and Thailand. In Malaysian hilly areas, forest clearance for rubber plantations, which started early in the 1900s, exposed the land and streams to the sun and created breeding places for *Anopheles maculatus*, which led to an increase in this species and a marked rise in the incidence and severity of malaria (Cheong, 1983). Cyclic malaria epidemics in Malaysia over 50 years are correlated with rubber replanting in response to market fluctuations (Singh and Tham, 1990). Another example is in Chantaburi, Thailand, where the land was transformed to rubber plantations and fruit trees, such as rambutan, durian and mangosteen, spurred by high markets between 1974 and 1984. The consequent ecological changes favoured *An. dirus*, which demonstrated a great ability to adapt. As a result, local malaria re-emerged, and malaria transmission was established at high levels (Rosenberg et al, 1990).

Similarly, studies on the development of irrigation systems report an increase in the density of major vectors and subsequent increase in malaria incidence. For example, irrigation schemes developed by the Mahi-Kadan Project across the River Mahi in India in 1960 had typical management problems, including over-irrigation, lack of proper drainage, weedy channels, leaking sluice gates and waterlogged fallow fields. These created extended breeding habitats for *Anopheles culicifacies*, which resulted in an increase of the vector and malaria transmission.

In some cases, different anopheline species responded differently to the same land transformation. For example, after deforestation for rice cultivation and irrigation development in Sri Lanka, *Anopheles annularis, An. barbirostris, An. culicifacies* and *An. varuna* decreased, while *An. jamesii* and *An. subpictus* increased, and *An. nigerrimus* and *An. vagus* did not change substantially (Amerasinghe et al, 1991; Konradsen et al, 2000). Not only species abundance but also species involvement in malaria transmission changed markedly during the land transformation. *An. annularis, An. culicifacies* and *An. vagus* were the main vectors during the construction phase and the first irrigation year. *An. subpictus* played a major role in the second and third years, when rice fields were fully irrigated. Throughout the process, *An. culicifacies* demonstrated continuous involvement in malaria transmission.

Table 9.1 *Ecosystem change and malaria*

Deforestation/ agricultural development	*Country/ region*	*Density decrease*		*Density increase*		*Increased human contact*	
		Species	*Malaria*	*Species*	*Malaria*	*Species*	*Malaria*
Deforestation	Thailand	*An. dirus*	–				
	Nepal	*An. minimus*		*An. fluviatilis*			
	India			*An. fluviatilis*			
	Sri Lanka	*An. barbirostris*		*An. annularis*	+		
				An. jamesii	+		
				An. nigerrimus	+		
				An. subpictus	+		
				An. peditaeniatus	?		
	Sahel, Africa	*An. funestus*					
Land exploitation/ pollution	Mediterranean	*An. labranchiae*	–				
		An. sacharovi	–				
		An. superpictus	–				
Cacao plantation	Trinidad			*An. bellator*			
Cassava	Thailand	*An. dirus*	–	*An. minimus*	+		
	Thailand	*An. dirus*	–				
Sugar cane	Thailand	*An. dirus*		*An. minimus*	+		

Coffee plantation + irrigation dams + tree crops	India	*An. fluviatilis*	–		
	Thailand			*An. minimus*	+
Tea plantation	Sri Lanka			*An. culicifacies*	
Rubber + Fruits + Orchards	Malaysia			*An. maculatus*	+
	Thailand			*An. dirus*	+
	Thailand			*An. dirus*	
	Thailand			*An. dirus*	+
Rice	China			*An. sinensis*	
	Malaysia	*An. umbrosus*	–	*An. campestris*	+
	Indonesia			*An. aconitus*	+
	Southeast Asia	*An. dirus*			
	Nepal	*An. fluviatilis*		*An. culicifacies*	+
	Sri Lanka	*An. annularis* *An. barbirostris* *An. culicifacies* *An. varuna*		*An. jamesii* *An. subpictus*	
	Africa			*An. funestus* *An. gambiae*	
Rice + maize	Thailand	*An. dirus*		*An. minimus*	
Irrigation system	India			*An. culicifacies*	+
	Afghanistan	*An. superpictus*		*An. pulcherrimus*	+

Table 9.1 *(cont'd)*

Deforestation/ agricultural development	*Country/ region*	*Density decrease*		*Density increase*		*Increased human contact*	
		Species	*Malaria*	*Species*	*Malaria*	*Species*	*Malaria*
	Africa			*An. arabiensis*	+		
				An. gambiae	+		
	Sahara			*An. gambiae*	+		
	Guyana	*An. darlingi*		*An. aquasalis*	+		
Hydropower dam	Sri Lanka			*An. culicifacies*	+		
Clearing of mangroves/ swamps for fish ponds or mining	Malaysia			*An. sundaicus*	+		
	Indonesia			*An. sundaicus*	+		
	Indonesia			*An. sundaicus*			
Mining	Thailand					*An. dirus*	
+ settlement	Amazon			*An. darlingi*	+		
Settlements + urbanization or highway construction	Amazon	*An. darlingi*					
	Indonesia					*An. balabacensis*	+
	Indonesia	*An. balabacensis*					
		An. leucosphyrus					
	India			*An. stephensi*	+		

Source: Yasuoka and Levins, 2007

Other cases demonstrated species replacement. Land use such as cassava and sugar-cane plantations, which need little water, provide little shade and often create unfavourable environments for anophelines, especially those which require shade. In Thailand, the transformation from forest to cassava or sugar-cane cultivations eliminated shady breeding habitats for the primary vector species, *Anopheles dirus*, but created widespread breeding grounds for *An. minimus*, which has greater sun preference and was the predominant species throughout the year. Consequently, malaria transmission among resettled cultivators rose (Prothero, 1999).

That same kind of land transformation can create totally different malaria situations, depending on locality and ecological characteristics of local vector species. For example, deforestation followed by development of coffee plantations in southeastern Thailand favoured the breeding of *Anopheles minimus* and made the previously malaria-free region hyperendemic (Suvannadabba, 1991). On the contrary, in Karnataka, India, large-scale deforestation for coffee plantations reduced seepages, which were the principal breeding sites for *An. fliviatilis*, a vector responsible for hyperendemic malaria in the region. As a result, this vector population completely collapsed, and malaria disappeared from the area (Karla, 1991).

Deforestation for mine development not only creates breeding sites but also significantly increases human contact with vectors. Where settlement and mining activities took place in the Amazon, *Anopheles darlingi* increased because of the increase in breeding sites, including borrow pits after road or settlement construction, drains and opencast mine workings. As a result, malaria, which was present in the Amazon's indigenous population, was spread to immigrants and miners (Conn et al, 2002; see also Chapter 11 of this book).

In summary, the changes in anopheline density and malaria incidence are both varied and complex, depending on the kind of land transformation, ecological characteristics of local mosquitoes, and altered human behaviour (to be discussed further). We summarize the major findings thus:

- Some anopheline species were directly affected by deforestation or subsequent land use, some preferred or could adapt to the new environmental conditions, and some invaded or replaced other species in the process of development and cultivation.
- Malaria incidence fluctuated according to different stages of development, changes in vector density and altered human contact patterns with vectors.
- More mosquitoes (density or variety) were neither a necessary nor a sufficient condition for increases in malaria incidence. In fact, inverse relationships between vector abundance and disease incidence have been reported from different regions (Ijumba and Lindsay, 2001; Amerasinghe, 2003), presumably because of human adaptations.

In general, a complex set of macro-economic, demographic, policy and behavioural factors underlie the ecosystem changes and land transformations that influence mosquito ecology and malaria epidemiology (Sharma and Kondrashin, 1991; Molyneux, 1998). We turn to these considerations in some detail next.

REVISITING THE HUMAN ECOLOGY PERSPECTIVE

Insecticide-treated bed-nets and indoor residual spraying of insecticides are the predominant vector control tools today. But if ecosystem changes affect mosquito density and activity, and possibly malaria incidence, then environmental management (e.g., vegetation management, modification of river boundaries, drainage of swamps, reduction of standing water, oil application) could reduce malaria incidence and also counter the increasing resistance of mosquitoes to insecticides and of the pathogen to anti-malarials (Lindsay and Birley, 2004). Keiser et al's (2005) review of 24 environmental management studies suggests that environmental management can reduce the malaria risk ratio by 88 per cent (compared with 79.5 per cent for human habitation modifications, for example).

Furthermore, if environmental changes are indeed modifiable behavioural causes, it should be possible to induce these behaviours. Yasuoka et al (2006a) conducted a 20-week pilot education programme to improve community knowledge and mosquito control with participatory and non-chemical approaches in Sri Lanka. They evaluated programme effectiveness using before-and-after surveys in two intervention and two control villages, and found that the participatory education programme led to improved knowledge of mosquito ecology and disease epidemiology, changes in agricultural practices and an increase in environmentally sound measures for mosquito control and disease prevention. The success of the intervention was attributed to three 'human ecology' characteristics: community-based education that enhanced residents' understanding of the mosquito-borne disease problems in their community; a participatory approach that allowed participants to gain hands-on experience, using non-chemical measures that decreased environmental and health risks in residential areas and paddy fields; and an approach that required no costly or extensive instruments. This community-based approach suppressed the density of adult *Anopheles* in the monsoon season, though little impact was detected on *Culex* and *Aedes* densities (Yasuoka et al, 2006b).

Vegetation and water management, however, are just one class of human behaviours that affect the transmission and control of malaria. The links between ecosystem change, vector ecology and disease epidemiology all depend critically on human density, gender ratio, the immigration of non-immune people, and knowledge, attitudes and practices, primarily because they alter the pattern and frequency of human contact with vectors. Furthermore, a recent special colloquium of the International Society of Ecosystem Health (Patz et al, 2004) suggests that malaria can be exacerbated by a broad array of land-use drivers and underlying human behavioural factors beyond changes to the biophysical environment. These include movement of agriculture, urbanization, and populations, pathogens and trade. Deforestation features prominently in this review and is closely linked to many of these mechanisms.

Pattanayak et al (2006b) underscore this behavioural aspect of malaria control and present four reasons why it is important to understand the role of deforestation from a policy and planning perspective:

1 Deforestation is not merely the exogenous removal of forest cover. It is the beginning of an entire chain of activities – including forest clearing, farming, irrigation, livestock

and non-timber forest product collection – that affect vector habitats as well as exposure and transmission.

2 Deforestation is an integral part of life and the landscape in many parts of the world with high malaria rates (Wilson, 2001; Donohue, 2003). Consequently, sustainable forest management is becoming an important policy goal as donor agencies and local policy-makers take a more integrated view of people in the landscape. The resulting changes in land cover, as well as changes in how people interact with the forest, have implications for malaria. Thus, conservation policies aimed at slowing deforestation will affect malaria (Walsh et al, 1993; Taylor, 1997).
3 Millions of rural households depend directly on a wide variety of forest products and services (Byron and Arnold, 1999). By lowering local people's natural wealth, deforestation can reduce household capacity to invest in health care and pay for malaria prevention and treatment. At the same time, deforestation may increase the wealth of other households, which will then be better able to avoid and cure malaria.
4 Deforestation and malaria are central elements of the vicious cycle of poverty in rural areas of developing countries. In simplistic terms, malaria could be considered to 'cause' deforestation because it can make people poorer, and poverty can cause deforestation under some conditions. In reality, the linkages are more complex and site-specific.

These ideas lead us to a human ecology framework for understanding the links between deforestation and malaria. Human ecology involves the study of human–environment interactions and explicitly traverses boundaries between 'nature and culture' and 'environment and society' (Parkes et al, 2003). Others have labelled these perspectives 'environmental health' or 'ecology and health' (Aron and Patz, 2001). As Parkes et al (2003) clarify, ultimately all these fields converge on three themes:

1 integrated approaches to research and policy;
2 methodological acknowledgement of the synergies between the social and biophysical environments; and
3 incorporation of core ecosystem principles into research and practice.

Specific to malaria, we need to shift our view of humans as passive or constant factors in malaria epidemiology to a view in which humans change epidemiological patterns (Wessen, 1972; MacCormack, 1984). The centrality of human behaviour is confirmed by the number of instances in which human behaviours show up in Figure 9.1 in this chapter and in the Patz et al (2004) review.

EMPIRICS OF HUMAN ECOLOGY: APPROACH AND EVIDENCE

In this section we examine the importance of human behaviours in malaria transmission and control and the 'active', dynamic aspects of human behavioural response. Omitting behavioural responses from any analysis of malaria and ecosystem change would result in a classic case of confounding, because human behaviour is (i) correlated with the outcome and the risk factor; (ii) not necessarily in the causal chain; and (iii) very likely to be unbalanced across the different levels of risks. As such, behavioural confounders can mimic the risk factor and mask the ecological relationship we are attempting to discover.

What does this mean in practical terms? If, for example, we are using cross-sectional or time-series variation in data on deforestation and malaria only, we will face what is an 'omitted variable' problem in statistics and econometrics. This problem leads to biased inferences and inconsistent estimates of policy parameters because the real cause – for example, the in-migration of susceptible subpopulations – is an omitted variable. A related and possibly more pernicious issue is that of endogeneity or reverse causality (or simultaneity). Consider an example from Sawyer (1993) to better understand this bias: high rates of malaria can encourage forms of land use in which men work as day labourers (in logging or ranching), allowing their wives and children to live in towns with relatively lower threats of malaria, rather than establishing family farms. In such a situation it is often difficult to disentangle the causal role of deforestation in malaria transmission.

To further investigate the empirical implications of these behavioural or human ecology models, we offer two simple tests. First, we compare a simple regression model of malaria and deforestation (naïve model) with a model including linear behavioural controls (linear controls model). Second, we compare the same naïve model with one in which the behavioural factors are used as determinants of deforestation, or the 'endogenous' risk exposure. Behaviour in this case is an instrument for the deforestation risk (the instrumental variable model). Economic theory provides one basis for identifying variables that can explain deforestation and thus serve as instruments (Sills and Pattanayak, 2006).

Arguably, the naïve model is a bit of a straw man, but it allows us to investigate the importance of a human ecology strategy. We conduct these evaluations at three scales: a micro-analysis of child malaria and community deforestation (a case from Indonesia), a meso-analysis of regional malaria and regional deforestation (a case from Brazil), and a macro-analysis of national malaria and deforestation. Data limitations preclude the use of accurate behavioural indicators and force us to use proxy variables. If the measurement error (because of the use of proxy variables) is of the classical variety – i.e., uncorrelated with the regression error – then we would face an attenuation bias (which would make the correlation seem smaller than it is). Data weakness is not the main problem here. Instead, we would argue that the paucity of good data is ultimately because of inadequate attention to the human ecology perspective in empirical analysis, both statistical estimation and numerical simulation. Thus, our analysis should be considered preliminary but illustrative of the overarching human ecology approach proposed here.

Macro-analysis using global data from 120 countries

In this case study, we examine the macro-level correlation of malaria and forest using a global data set. Pattanayak et al (2006c) describe the combination of data from five sources to produce a global malaria data set and use it to examine how disease prevention behaviours respond to disease levels. The WHO's Global Health Atlas provides data on a range of malaria variables, including the number of cases, for up to 195 countries from 1990 to 2004. The World Development Reports provide data on forest cover in 1990 and the annual rate of increase from 1990 to 2000. We obtain behavioural proxies from three other sources. First, data from the 2001 World Development Report provide measures of economic conditions (per capita gross domestic product, GDP) and social conditions (adult literacy rates, educational enrolment rates and life expectancy). Second, Kaufmann et al (2003) provide data on political stability, voice and accountability, and control of corruption. We also include a malaria ecology index to capture vector ecology and climatic factors (Kiszewski et al, 2004). This index combines climatic factors (e.g., rainfall and temperature), the presence of different mosquito vectors, and the human biting rates of these vectors to proxy for mosquito transmission. This index captures the ecological conditions with the strongest influence on the intensity of malaria prevalence and can therefore predict the actual and potential stability of transmission. Descriptive statistics and other details of the data compilation and synthesis are included in Pattanayak et al (2006a).

Our major variable is the number of malaria cases in a country in the 1996–2000 period. The variables (malaria cases, malaria ecology index and GDP index) are converted into natural logarithms to reduce scale differences, improve linearity and pull in outliers. Median regression methods are used to remove the influence of outliers. Results of the three models – naïve, linear controls and instrumental variable – are presented in columns 2, 3 and 4 of Table 9.2 (Panel 1). We report the coefficient on the deforestation variable, the probability value (p value) associated with this coefficient and the overall significance of the model. The regression coefficient reflects the size and sign of the correlation with malaria incidence. The p value reflects the statistical significance of the correlation (i.e., less than 0.1 is suggestive of a relationship).

The naïve model is statistically significant and explains about 41 per cent of the variation. We also find confirmation of our major hypothesis: annual rate of forest cover increase (during the 1990–2000 period) is negatively correlated with malaria incidence in the 1996–2000 period, and more deforestation is positively correlated with higher levels of malaria.

The linear controls model (where we account for potential confounding due to GDP, school enrolment, voice and accountability, and stability of the governmental institutions) is also statistically significant and explains about 52 per cent of the variation in the malaria cases. We also find that deforestation is positively correlated with malaria, except now the correlation is twice as big.

Finally, the instrumental variable model is also significant and explains about 54 per cent of the variation. In this model, first, behavioural variables are used to predict deforestation, and then the predicted deforestation is used to explain malaria. Again we see that the deforestation variable is positively correlated with malaria, but now the coefficient is almost four times as big as the naïve model, providing statistically significant evidence of a much stronger correlation between the disease and exposure change due to deforestation.

Table 9.2 *Empirics of 'human ecology' modelling of malaria and deforestation links*

MACRO	*Naïve model*	*Linear controls model*	*Instrumental variable model*
annual forest increase	–0.049	–0.089	–0.168
p value	(0.065)	(0.013)	(0.038)
ecology controls	yes	yes	yes
behavioural controls	no	yes	as instrumental variables
RSq	0.407	0.519	0.540
MESO			
deforestation	7.89e-07	2.25e-06	6.99e-06
p value	(0.047)	(0.000)	(0.004)
ecology controls	yes	yes	yes
behavioural controls	no	yes	as instrumental variables
RSq	0.461	0.555	0.235
MICRO			
log (primary forests)	–0.062	–0.163	–0.382
p value	(0.497)	(0.106)	(0.046)
log (secondary forests)	0.234	0.401	0.609
p value	(0.006)	(0.000)	(0.008)
ecology controls	yes	yes	yes
behavioural controls	no	yes	as instrumental variables
Pseudo RSq	0.055	0.153	0.153

Meso-analysis (regional) using the case of 480 Brazilian micro-regions

In this case study, we examine the hypothesis regarding the regional-level correlation of malaria and forest cover. We use a cross-sectional data set of approximately 490 Brazilian micro-regions. The malaria data come from the DATASUS website and are reported in terms of 1000 inhabitants, representing hospital morbidity over the 1992–2000 period (see Chapter 11, this volume). Climate is represented by long-run temperature and rainfall (averaged over several years) in the 490 micro-regions, based on weather stations (approximately one per micro-region). Census data – on housing, population, education levels, income, medical care (proxied by number of doctors and hospital beds) and infrastructure (percentage of households connected to water, sanitation, and all-weather roads) – are for 1991. Forest cover and vegetation data of the same vintage and protected-area data are from two Brazilian data agencies (IPEA and INPE, respectively). Pattanayak et al (2006b) present additional detail on the compilation and use of these data in analysis.

Using the same structure as the previous case study, major results are as follows. The naïve model (including some ecological controls for weather) is statistically significant and explains 46 per cent of the variation. First, we see that micro-regions with higher forest cover have lower rates of malaria, all things considered. Second, we find that micro-regions with higher deforestation (in the 1985–1995 time period) have greater rates of malaria.

The linear controls model (where we account for potential confounding due to demographics, income, infrastructure and institutions, such as protected areas) is statistically significant and explains 56 per cent of the variation. First, we find that micro-regions with higher deforestation have greater rates of malaria, with a correlation that is significantly larger than the naïve model coefficients (almost twice as large). Second, micro-regions in the Amazon with conservation units have lower malaria rates for a given level of deforestation.

Finally, the instrumental variable model uses several regional factors – presence of protected area, distance to highway and to state capital, population, size and location of the micro-region – as instruments for deforestation in the micro-regions. The overall model is significant. Now the deforestation coefficient is almost three times as big as for the linear controls model. The results are consistent across the three models (i.e., deforestation is correlated with more malaria), but the sizes of the estimated coefficient are much larger (three- to six-fold) in the models that include proxies for human behaviour.

Micro-analysis using data on 340 children from Flores, Indonesia

Malaria is highly contextual, with incidence and transmission depending on local conditions, perturbations and catastrophes. Thus, household or community-level multi-factor research is perhaps best suited to incorporate the diversity and heterogeneity of the ecological, epidemiological and economic phenomena surrounding malaria. This case study examines the evidence on whether deforestation causes child malaria in Ruteng Park on Flores Islands in eastern Indonesia.

The data for this analysis are drawn from a 1996 household survey in the Manggarai district of Flores around a protected area, Ruteng Park, that was established to conserve biodiversity. The survey and accompanying secondary data collection generated household data on wealth, housing quality and number of adults, as well as individual data on age, sex, occupation, education and disease history during the 12 months prior to the survey. Geographic information system (GIS) data are used to combine environmental statistics, including the amount and extent of primary and secondary forest cover at the village level, with the survey data and secondary data on public infrastructure, such as subregional health-care facilities. The sample includes approximately 340 children under the age of five. Given the binary nature of the data on malaria in children under the age of five, we estimate and report probit models of child malaria. Pattanayak et al (2005) include details.

Starting with the naïve model, we find that the overall model is significant and, this being micro-data, explains only about 6 per cent of the variation. We find that the extent of protected (primary) forest cover is not statistically related to malaria, whereas the extent of disturbed (secondary) forest is positively correlated with malaria rates.

The linear controls model accounts for potential confounding due to various individual, household and village characteristics. The overall model is significant, now explaining about 15 per cent of the variability in the malaria data. As in the naïve model, the extent of disturbed forests is positively correlated with malaria (although the coefficient is twice as big as before). Most interesting, we now confirm our central hypothesis – that the extent of protected forests is indeed negatively correlated with malaria incidence.

Finally, the instrumental variable model uses community-level factors – distance to highway, population, village size, elevation and rainfall – as instruments for protected and disturbed forest cover around the villages. The overall model is highly significant. Most crucially, now the coefficients are almost three times as big as in the linear controls model. Malaria in small children is highly positively correlated with the extent of disturbed forests and negatively correlated with the extent of protected forests.

Concluding thoughts

Vector-borne diseases such as malaria wreak havoc on the lives of many millions of people in poor, tropical countries, partly because these regions are exposed to deforestation, livestock rearing, irrigated farming, road construction and dam building: environmental changes that encourage vector abundance and disease transmission. We argue that it is critical to focus on deforestation because it is the beginning of an entire chain of activities that affect malaria risks; it can trigger human behavioural changes through accompanying increases or decreases in wealth; it can lock communities into a vicious cycle of poverty, illness and environmental degradation; and it is an integral part of the landscape and therefore of donor agencies' and policy-makers' focus. Recognizing that deforestation often precedes many other land-use changes (particularly conversion to agriculture) and taking deforestation as a starting point allow us to look at the impact of other elements in the matrix of transformations. As such, it serves as a broad indicator of change in the ecology of infectious disease paradigm.

That leads us to recommend a human ecology that focuses on the role of humans in land-use change as well as in a variety of behaviours to prevent and treat malaria, from sleeping under nets and taking prophylaxis to seeking medical care and following the drug regimen. We then review the implications of this framework change for empirical research and application, both in data collection and analysis, and in inference.

The empirical case studies draw attention to the role of socioeconomic determinants of malaria and the importance of including behavioural variables in empirical models of malaria incidence and prevalence. They illustrate how omitting behavioural factors from the analysis can lead to erroneous and biased interpretations regarding the nature of ecosystem changes and disease transmission: the size, sign and statistical significance of regression coefficients can be wrong. In general, they highlight different elements of human-induced ecosystem change, disease outcomes, and economic causes and consequences.

What we have not discussed is the inherent dynamics of coupled natural and social systems. In a recent paper, for example, Pattanayak et al (2006a) analyse global and micro-data to show that malaria prevention behaviours depend on malaria prevalence. They find that households and countries engage in greater degrees of prevention if they face high rates of malaria, and fewer prevention behaviours if they confront low rates of malaria. That is, the causal arrow can also flow in the other direction (see the dotted arrow in Figure 9.1, typically missing from most assessments). The logical feedback and dynamic between prevention and prevalence suggest that it is insufficient and inappropriate to model and consider socioeconomic behaviours outside the malaria infection and transmission process. Behaviour and its determinants are part and parcel of the ecology and epidemiology and must be built into analysis and planning.

In fact, it is safe to say that many of these findings hold for the general class of vector-borne infectious diseases, such as dengue, leishmaniasis, hantavirus pulmonary syndrome, schistosomiasis, filariasis, Lyme disease, onchocerciasis and loiasis. Space

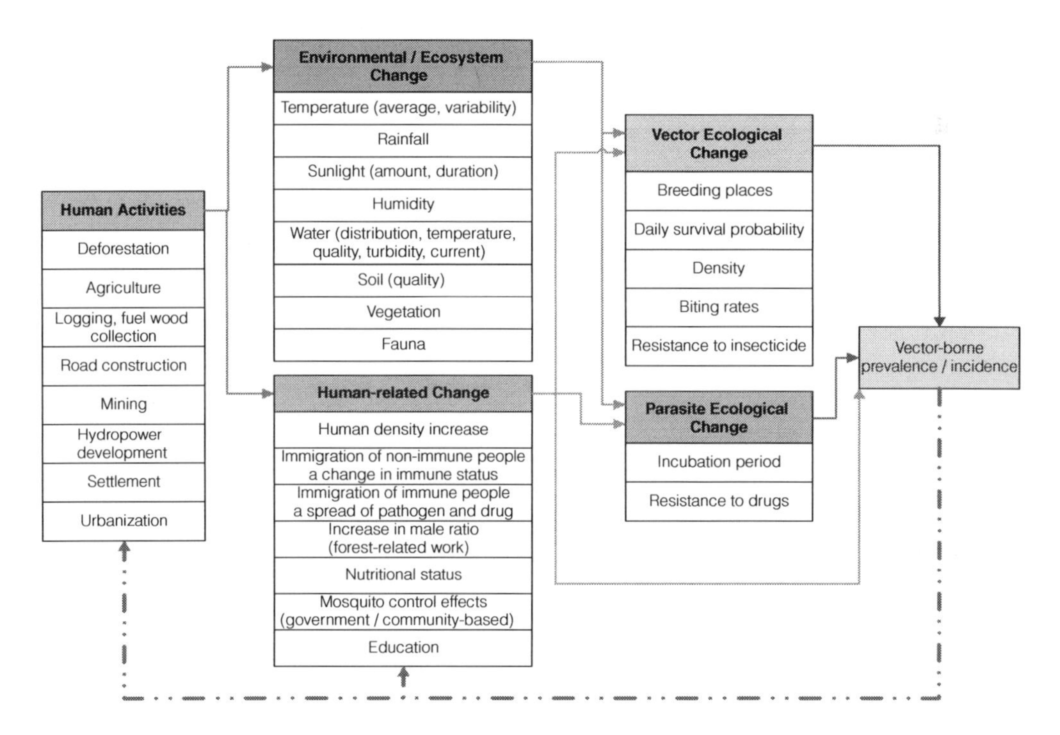

Figure 9.1 *Ecosystem change and malaria*

limitations preclude a comprehensive discussion of these diseases (for additional details, see Wilson, 2001, and Tables 4 and 5 in Colfer et al, 2006; and Chapters 5, 7 and 8, this volume). As suggested in Figure 9.1, ecosystem changes influence the emergence and proliferation of these diseases by altering the ecological balance and context within which disease hosts or vectors and parasites breed, develop and transmit diseases (Patz et al, 2000). For example, deforestation is often followed by water resources development and livestock management, which open up numerous possibilities for disease risks.

Moreover, the simultaneity between prevalence and prevention, discussed previously by Pattanayak et al (2006a), only hints at the dynamic inherent in coupled natural and social systems. As Hammer (1993) suggests, in the case of malaria, very little is known about the interrelated dynamics of ecosystem changes, vector density and infectivity, development of immunity and resistance (to pesticides and drugs) and human response. Wiemer's (1987) case of schistosomiasis in China and Gersovitz and Hammer's (2005) model of malaria prevention and treatment are early attempts to examine these dynamics through mathematical simulations. Much more conceptual work is needed before ecosystem change dynamics can be incorporated into such models. Empirical research must test hypotheses about the nature and magnitude of these relationships and generate statistical parameters that can then be used for policy scenario analysis.

In the interim, however, the human ecology approach to public health can take root and thrive through the conduct of systematic economic and health impact assessments of forest policies. Such evaluations need to be interdisciplinary longitudinal studies that take the following issues into account:

- It is impossible to design and implement a rigorous study and make credible inferences without a clear understanding of the policy scenario. Specificity of the policy scenario – be it a project at a site, a programme that includes a collection of projects, or a national or region-wide policy – allows the analyst to understand the mechanism of disease transmission and economic impacts in terms of 'modifiable causes'.
- With a clear scenario, it is then possible to design rigorous evaluations to infer 'causal policy impacts'. These are typically through randomized assignment of the programme or a quasi-experimental design that includes data collection in programme and control sites during various stages of implementation, including baseline and endline data.
- The credibility of the resulting evaluation will ultimately ride on the quality of the data and the rigour and care taken in data analysis. For a study of this type, outcome variables include indicators of health, wealth and the environment. Extent of forest cover and forest condition are among the major explanatory variables. Other explanatory variables include socioeconomic, demographic, environmental, health and public health policy indicators. The challenge in empirical work is to identify robust measures of these variables and separate independent and dependent variables. The multiple channels for feedback between malaria, deforestation and poverty suggest that these variables would be dependent variables in some specifications, and independent variables in others.

- Although researchers can employ an array of sophisticated techniques to remedy defects in available data, clearly 'prevention' in the form of careful data collection is superior to 'cure' in the form of ad hoc statistical fixes. Longitudinal data sets – and particularly panel data sets – are the key to at least three critical issues in the types of research proposed here: heterogeneity, endogeneity and dynamics or mobility (Ezzati et al, 2005). Ideally, data should be collected at several scales, ranging from individual-level health and demographic data, to household-level economic information, to community- and regional-level environmental statistics and policy factors.

The human ecology approach proposed in this chapter can be used for at least two practical purposes (Pattanayak et al, 2006c). First, it can help organize the conceptual links between coupled natural and socioeconomic systems and serve as a platform for generating testable hypothesis and policy parameters. Such efforts are critical for understanding the ecological, entomological, epidemiological and economic aspects of deforestation and malaria, and their behavioural underpinnings. Second, it will be vital for building decision analysis and scenario simulation tools (Kramer et al, 2006), which rely on estimated parameters, for formulating integrated strategies that cut across health, environment and economic sectors to address the broad idea of ecosystem change and disease control. Scenario simulation can, for example, inform the design of surveillance and monitoring frameworks to detect changes in the environment, vector density, human migration and behaviour, and incidence of diseases – both to contain vector-borne diseases and prevent epidemics.

NOTE

1 Address correspondence to Subhrendu K. Pattanayak, Fellow and Senior Economist in Public Health and Environment, RTI International and Research Associate Professor at North Carolina State University, subhrendu@rti.org or Tel (919) 541-7355. Junko Yasuoka is an Assistant Professor, Department of International Community Health, Graduate School of Medicine, the University of Tokyo. Many ideas reflected in this paper are based on discussions with Erin Sills, Keith Alger, Gene Brantley, Kelly Jones, Christine Poulos, Jonathan Patz, Montira Pongsiri, Andy Speilman and George Van Houtven. We are grateful to Carol Colfer for her encouragement and, most of all, patience with the development of this chapter. This chapter was completed while Pattanayak was a Visiting Scholar at the University of California, Berkeley.

REFERENCES

Amerasinghe, F.P.A. (2003) 'Irrigation and mosquito-borne diseases', *Journal of Parasitology Special Edition: Selected Papers of the 10th International Congress of Parasitology*

Amerasinghe, F.P., Amerasinghe, P.H., Peiris, J.S.M. and Wirtz, R. (1991) 'Anopheline ecology and malaria infection during the irrigation development of an area of the Mahaweli project, Sri Lanka', *American Journal of Tropical Medicine and Hygiene* 45: 226–235

Aron, J.L. and Patz, J.A. (eds) (2001) *Ecosystem Change and Public Health: A Global Perspective*, Johns Hopkins University Press, Baltimore, MD

Berkman, L.F. and Kawachi, I. (eds) (2000) *Social Epidemiology*, Oxford University Press, New York, NY

Byron, N. and Arnold, M. (1999) 'What futures for the people of the tropical forests?', *World Development* 27(5): 789–805

Campbell-Lendrum, D., Molyneux, D., Amerasinghe, F., Davies, C., Fletcher, E., Schofield, C., Hougard, J-M., Polson, K. and Sinkins, S. (2005) 'Ecosystems and vector-borne disease control', in *Ecosystems and Human Well-being: Policy Responses, Volume 3*, Millennium Ecosystem Assessment, World Health Organization, Geneva, Switzerland, pp353–372

Cheong, W.H. (1983) 'Vectors of filariasis in Malaysia', in J.W. Mak (ed) *Filariasis*, Bulletin No. 19, Institute for Medical Research, Kuala Lumpur, Malaysia, pp37–44

Chivian, E. (ed) (2002) *Biodiversity: Its Importance to Human Health, Interim Executive Summary*, Harvard Medical School, Boston, MA

Colfer, C.J.P., Sheil, D. and Kishi, M. (2006) 'Forests and human health: Assessing the evidence', Occasional Paper 45, CIFOR, Bogor, Indonesia

Conn, J.E., Wilkerson, R.C., Segura, M.N., de Souza, R.T., Schlichting, C.D., Wirtz, R.A. and Póvoa, M.M. (2002) 'Emergence of a new neotropical malaria vector facilitated by human migration and changes in land use', *American Journal of Tropical Medicine and Hygiene* 66: 18–22

Corvalan, C., Hales, S., McMichael, A., Butler, C., Campbell-Lendrum, D., Confalonieri, U., Leitner, K., Lewis, N., Patz, J., Polson, K., Scheraga, J., Woodward, A. and Younes, M. (2005a) 'Ecosystems and human well-being: Human health synthesis', Millennium Ecosystem Assessment, World Health Organization, Geneva, Switzerland,

Corvalan, C., Hales, S., Woodward, A., Campbell-Lendrum, D., Ebi, K., De Avila Pires, F., Soskolne, C.L., Butler, C., Githeko, A., Lindgren, E. and Parkes, M. (2005b) 'Consequences and options for human health', in *Ecosystems and Human Well-being: Policy Responses*, Volume 3, Millennium Ecosystem Assessment, World Health Organization, Geneva, Switzerland, pp467–486

Donohue, M. (2003) 'Causes and health consequences of environmental degradation and social injustice', *Social Science and Medicine* 56: 573–587

Ezzati, M., Utzinger, J., Cairncross, S., Cohen, A.J. and Singer, B.H. (2005) 'Environmental risks in the developing world: Exposure indicators for evaluating interventions, programmes, and policies', *Journal of Epidemiology and Community Health* 59: 15–22

Gersovitz, M. and Hammer, J. (2005) 'Tax/subsidy policies toward vector-borne infectious diseases', *Journal of Public Economics* 89(4): 647–674

Grillet, M.E. (2000) 'Factors associated with distribution of *Anopheles aquasalis* and *Anopheles oswaldoi* (Diptera: Culicidae) in a malarious area, northeastern Venezuela', *Journal of Medical Entomology* 37: 231–238

Hammer, J. (1993) 'The economics of malaria control', *The World Bank Research Observer* 8(1): 1–22

Hay, S.I., Guerra, C.A., Tatem, A.J., Noor, A.M. and Snow, R.W. (2004) 'The global distribution and population at risk of malaria: Past, present and future', *Lancet Infectious Diseases* 4: 327–336

Ijumba, J.N. and Lindsay, S.W. (2001) 'Impact of irrigation on malaria in Africa: Paddies paradox', *Medical and Veterinary Entomology* 15: 1–11

Karla, N.L. (1991) 'Forest malaria vectors in India: Ecological characteristics and epidemiological implications', in V.P. Sharma and A.V. Kondrashin (eds) *Forest Malaria in Southeast Asia*, WHO/MRC, New Delhi, India, pp93–114

Kates, R.W., Clark, W.C., Corell, R., Hall, J.M., Jaeger, C.C., Lowe, I., McCarthy, J.J., Schellnhuber, H.J., Bolin, B. Dickson, N.M., Faucheux, S., Gallopin, G.G., Grubler, A.,

Huntley, B., Jager, J., Jodha, N.S., Kasperson, R.E., Mabogunje, A., Matson, P., and Mooney, H. (2001) 'Sustainability science', *Science* 292(5517): 641–642

Kaufmann, D., Kraay, A. and Mastruzzi, H. (2003) *Governance Matters III: Governance Indicators for 1996–2002,* World Bank, Washington, DC

Keiser, J., Singer, B.H. and Utzinger, J. (2005) 'Reducing the burden of malaria in different eco-epidemiological settings with environmental management: A systematic review', *Lancet Infectious Diseases* 5: 695–708

Kiszewski, A., Mellinger, A., Spielman, A., Malaney, P., Sachs, S.E. and Sachs, J. (2004) 'A global index of the stability of malaria transmission', *American Journal of Tropical Medicine and Hygiene* 70(5): 486–498

Kondrashin, A.V., Jung, R.K. and Akiyama, J. (1991) 'Ecological aspects of forest malaria in Southeast Asia', in V.P. Sharma and A.V. Kondrashin (eds) *Forest Malaria in Southeast Asia,* WHO/MRC, New Delhi, India, pp1–28

Konradsen, F., Amerasinghe, F.P., van der Hoek, W. and Amerasinghe, P.H. (eds) (2000) *Malaria in Sri Lanka, Current Knowledge on Transmission and Control,* International Water Management Institute, Colombo, Sri Lanka

Kramer, R.A., Dickinson, K.L., Fowler, V.G., Miranda, M.L., Mutero, C.M., Saterson, K.A. and Weiner, J.B. (2006) 'Decision analysis as an integrative tool for improved malaria control policy making', Working Paper, Duke University, Durham, NC

Lindsay, S.W. and Birley, M. (2004) 'Rural development and malaria control in sub-Saharan Africa', *EcoHealth* 1: 129–137

MacCormack, C.P. (1984) 'Human ecology and behaviour in malaria control in tropical Africa', *Bulletin of the World Health Organization* 62 Supplement S: 81–87

McMichael, A.J. (2001) *Human Frontiers, Environments and Disease: Past Patterns, Uncertain Futures,* Cambridge University Press, Cambridge

McMichael, A., Patz, J. and S. Krovats (1998) 'Impacts of global environmental change on future health and health care in tropical countries', *British Medical Bulletin* 54(2): 475–488

Martens, P. (1998) *Health and Climate Change: Modelling the Impacts of Global Warming and Ozone Depletion,* Earthscan, London

Molyneux, D.H. (1998) 'Vector-borne parasitic diseases – an overview of recent changes', *International Journal for Parasitology* 28: 927–934

Oakes, J.M. and Kaufman, J.S. (eds) (2006) *Methods in Social Epidemiology,* John Wiley & Sons, San Francisco, CA

Parkes, M., Panelli, R. and Weinstein, P. (2003) 'Converging paradigms for environmental health theory and practice', *Environmental Health Perspectives* 111(5): 669–675

Pattanayak, S.K., Corey, C.G., Lau, Y.F. and Kramer, R. (2005) 'Conservation and health: A microeconomic study of forest protection and child malaria in Flores, Indonesia', RTI Working Paper, Research Triangle Institute, NC

Pattanayak, S.K, Poulos, C., Jones, K., Yang, J.-C. and Van Houtven, G. (2006a) 'Economics of environmental epidemiology', RTI Working Paper, Research Triangle Park, NC

Pattanayak, S.K., Ross, M., Timmins, C., Depro, B., Jones, K. and Alger, K. (2006b) 'Climate change, human health, and biodiversity conservation', paper presented at US EPA conference, Multidisciplinary Approach to Examining the Links between Biodiversity and Human Health, September, Washington, DC

Pattanayak, S.K., Dickinson, K., Corey, C., Sills, E.O., Murray, B.C. and Kramer, R. (2006c) 'Deforestation, malaria, and poverty: A call for transdisciplinary research to design cross-sectoral policies', *Sustainability: Science, Practice and Policy* 2(2): 1–12

Patz, J.A., Graczyk, T.K., Geller, N. and Vittor, A.Y. (2000) 'Effects of environmental change on emerging parasitic diseases', *International Journal for Parasitology* 30: 1395–1405

Patz, J.A., Daszak. P., Tabor, G.M., Aguirre, A.A., Pearl, M., Epstein, J., Wolfe, D.N., Kilpatrick, A.M., Foufopoulos, J., Molyneux, D., Bradley, D., and the Working Group on Land Use Change and Disease Emergence (2004) 'Unhealthy landscapes: Policy recommendations on land use change and infectious disease emergence', *Environmental Health Perspectives* 112: 1092–1098

Patz, J., Confalonieri, U.E.C., Amerasinghe, F.P., Chua, K.B., Daszak, P., Hyatt, A.D., Molyneux, D., Thomson, M., Yameogo, L., Malecela-Lazaro, M., Vasconcelos, P., Rubio-Palis, Y., Campbell-Lendrum, D., Jaenisch, T., Mahamat, H., Mutero, C., Waltner-Toews, D. and Whiteman, C. (2005) 'Human health: Ecosystem regulation of infectious diseases', in *Ecosystems and Human Well-being: Current State and Trends, Vol. 1*, Millennium Ecosystem Assessment, World Health Organization, Geneva, Switzerland, pp391–415

Prothero, R.M. (1999) 'Malaria, forests and people in Southeast Asia', *Singapore Journal of Tropical Geography* 20: 76–85

Reiter, P. (2001) 'Climate change and mosquito-borne disease', *Environmental Health Perspectives* 109(S1): 141–161

Rosenberg, R., Andre, R.G. and Somchit, L. (1990) 'Highly efficient dry season transmission in malaria in Thailand', *Transactions of the Royal Society of Tropical Medicine and Hygiene* 84: 22–28

Sawyer, D. (1993) 'Economic and social consequences of malaria in new colonization projects in Brazil', *Social Science and Medicine* 37(9): 1131–1136

Sharma, V.P. and Kondrashin, A.V. (eds) (1991) *Forest Malaria in Southeast Asia,* WHO/MRC, New Delhi, India

Sills, E.O. and Pattanayak, S.K. (2006) 'Tropical tradeoffs: An economics perspective on tropical deforestation', in S. Spray and M. Moran (eds), *Tropical Deforestation*, Rowman and Littlefield Publishers, Lanham, MD, pp104–128

Singh, Y.P. and Tham, A. (1990) 'Case history of malaria control through the application of environmental management in Malaysia', WHO/WBC/88.960, World Health Organization, Geneva, Switzerland,

Smith, K.R., Corvalán, C.F. and Kjellstrom, T. (1999) 'How much global ill health is attributable to environmental factors?' *Epidemiology* 10: 573–584

Suvannadabba, S. (1991) 'Deforestation for agriculture and its impact on malaria in southern Thailand', in V.P. Sharma and A.V. Kondrashin (eds) *Forest Malaria in Southeast Asia*, WHO/MRC, New Delhi, India, pp221–226

Taylor, D. (1997) 'Seeing the forests for more than the trees', *Environmental Health Perspectives* 105: 1186–1191

Walsh, J.F., Molyneux, D.H. and Birley, M.H. (1993) 'Deforestation: Effects on vector-borne disease', *Parasitology* 106 (supplement): 55–75

Wessen, A.F. (1972) 'Human ecology and malaria', *American Journal of Tropical Medicine and Hygiene* 21(1): 658–662

Wiemer, C. (1987) 'Optimal disease control through combined use of preventive and curative measures', *Journal of Development Economics* 25: 301–319

Wilcox, B.A. and Colwell, R.R. (2003) 'Emerging and reemerging infectious diseases: Biocomplexity as an interdisciplinary paradigm', *EcoHealth* 2: 244–257

Wilson, M.E. (1995) 'Infectious diseases: An ecological perspective', *British Medical Journal* 311(7021): 1681–1684

Wilson, M.L. (2001) 'Ecology and infectious disease', in J. Aron and J.A. Patz (eds) *Ecosystem Change and Public Health*, Johns Hopkins University Press, Baltimore, MD, pp285–291

Yasuoka, J. and Levins, R. (2007) 'Impact of deforestation and agricultural development on anopheline ecology and malaria epidemiology', *American Journal of Tropical Medicine and Hygiene* 76(3): 450–460

Yasuoka, J., Mangione, T.W., Spielman, A. and Levins, R. (2006a) 'Impact of education on knowledge, agricultural practices, and community actions for mosquito control and mosquito-borne disease prevention in rice ecosystems in Sri Lanka', *American Journal of Tropical Medicine and Hygiene* 74(6): 1034–1042

Yasuoka, J., Levins, R., Mangione, T.W. and Spielman, A. (2006b) 'Community-based rice ecosystem management for suppressing vector anophelines in Sri Lanka', *Transactions of the Royal Society of Tropical Medicine and Hygiene* 100(11): 995–1006

PART II – THEMATIC AND REGIONAL HEALTH SLICES

10

The Subversive Links between HIV/AIDS and the Forest Sector

Pascal Lopez

In 2006, 25 years after scientists reported the first clinical evidence of what would later become known as acquired immune deficiency syndrome, or AIDS, the disease is still deadly. AIDS killed 2.8 million people in 2005, mainly adults in the prime of life (UNAIDS, 2006a). Currently available anti-retroviral therapies can only slow down the mechanism by which the human immunodeficiency virus (HIV) multiplies itself and thereby delay the outbreak of AIDS; they can neither reverse the deadly course of the infection nor prevent further infections. However, both modern and traditional medicines – mainly derived from plant resources – are able, to a certain extent, to effectively treat AIDS symptoms and some of the typical opportunistic diseases of AIDS. In many areas HIV is predominantly transmitted through (unprotected) sexual intercourse. Because the virus exploits 'one of the most complex areas of human life: our sexual relationships' (UNAIDS, 2005), which are subject to many taboos, discretions and traditions, it is difficult though not impossible to effectively fight against its further spread.

HIV/AIDS is most widespread in sub-Saharan Africa, which is home to almost two-thirds of all HIV-infected people, though it has only just over 10 per cent of the world's population. The majority of the sub-Saharan countries are experiencing generalized epidemics, with HIV prevalence (expressed as a percentage of the population of 15 to 49 year olds, those presumed sexually active) above 1 per cent (UNAIDS, 2004a). In a generalized epidemic, the virus spreads beyond the high-risk groups (such as sex workers, their clients and injecting drug users) into and through the general population, rendering it difficult to target people and groups. In 38 African countries, the AIDS epidemic has caused a decline of life expectancy at birth, now estimated at 47 years, 5.7 years lower than it would have been in the absence of AIDS (UN Population Division, 2003).

According to the estimations of UNAIDS (2005), the Joint United Nations Programme on HIV/AIDS, the average levels of HIV prevalence 'are lower in East Africa (6 per cent) and West and Central Africa (4.5 per cent), and much lower in North Africa (under 0.1 per cent) while Southern Africa is most severely affected, with more than 16 per cent of its adult population HIV-positive'. In eastern and southern Africa, recent publications (e.g., ABCG, 2002; Mauambeta, 2003; Erskine, 2004; Swallow, 2004) indicate considerable impacts of HIV/AIDS on affected households and institutions related to natural resources, including forests and other plant resources, their management and the personnel in the forest sector. As these natural resources become critical safety nets for rural households in times of crisis, it is almost an inevitable consequence that they are especially important when HIV/AIDS strikes rural and poor households. Poverty plays a central role in the HIV/AIDS epidemic. It is poverty that creates the conditions that increase vulnerability to HIV, and at the same time 'AIDS tends to affect the poor more heavily than other population groups' (UNAIDS, 2006a).

HIV/AIDS AND THE FOREST

The links between the HIV/AIDS epidemic and forest resources still remain – at the international level – largely unnoticed, leading to still-insufficient acceptance of and support for HIV/AIDS prevention measures within the forest sector. Until now, the forest sector, including informal activities of local forest users, has usually not been associated with a 'health issue', a view reflected in the attitudes of many people and institutions involved in the forest sector, particularly those operating outside the heavily affected countries. The officials and project managers for natural resource management and rural development have not considered HIV/AIDS and human health to be their concern and instead leave it to the health departments, whose staff have not considered the natural environment of importance in their work. Consequently, at the end of the day, no one has taken responsibility or action.

However, recognition that the HIV/AIDS epidemic is affecting how people use forests and other natural resources, and that people and resources in these sectors are victims of the epidemic, is coming from initiatives by the United Nations Food and Agriculture Organization (FAO), the United States headquarters of the World Wide Fund for Nature (WWF-US), Oxfam International, the African Development Bank, and German and US development agencies: Gesellschaft für Technische Zusammenarbeit (GTZ), Kreditanstalt für Wiederaufbau and the United States Agency for International Development (USAID). These organizations increasingly incorporate actions to fight HIV/AIDS within their natural resource management projects, programmes and institutions.

The links between HIV/AIDS and the forest sector may seem bizarre at first sight. People form the connection. The HIV/AIDS epidemic affects forest resources by affecting the behaviour of resource users and managers: infection with HIV and the development of AIDS can lead to stigma, illness and finally death, thereby changing the needs, behaviour and living conditions of infected and affected persons, households and institutions.

HIV/AIDS alters the physical and psychological conditions of a person, leading to socioeconomic change, with direct implications for their personal environment (household, family, workplace, etc). This alteration is marked by extraordinary medical needs and household expenditures, the loss of jobs and income, and the decline of productivity at the individual and household level. Forest resources (from dense forests to woodlands and savannahs to trees outside forests) can serve these new or extended needs to a certain extent and mitigate some of the impacts by providing food, medicinal plants and many other products, and by functioning as a source of cash income.

Staff of public and private forest administrations and enterprises are not excluded from the HIV/AIDS epidemic, and within certain sub-sectors even show an elevated vulnerability to HIV. Several regions and countries within sub-Saharan Africa have already experienced the effects of accelerated HIV/AIDS infections on their staff. The reduced workforce reduces their capacity to carry out their mandates to supervise and manage forests.

Here, we examine the interactions between forest resources and HIV/AIDS-affected persons, households and forest institutions, looking at the role of forest resources, particularly for HIV/AIDS affected households; and the observed and potential impacts of the aforementioned stakeholders on forest resources. The conclusion will be a strong plea for action against HIV/AIDS in the forest sector and a closer integration of it with the public health sector. Because of the epidemic's magnitude on the African continent, this article focuses primarily on the situation in sub-Saharan Africa; three boxes present complementary examples from Zimbabwe, the Dominican Republic and Indonesia.

FOREST RESOURCES FOR MITIGATING HIV/AIDS IMPACTS

The influence of the HIV/AIDS epidemic on the forest sector and other natural resources has rarely been investigated and is little understood. Only in very recent years have comprehensive studies been published and initiatives launched – based on existing models of natural resources management – that analyse how the HIV/AIDS epidemic influences the use and management of natural resources, such as forests, agricultural land and protected areas (e.g., UNAIDS, 1999; Erskine, 2004; Barany et al, 2005; Hunter et al, 2005). In addition, there are sporadic newspaper articles, workshop proceedings and personal communications that take up the theme. Major findings are presented here.

Forests and trees as sources of medicinal plants

Medicinal plants are among the most important non-wood forest products in Africa (Walter, 2001). Within the forest ecosystem, 'herbal' or 'traditional' medicines come from tree roots, bark, leaves and fruits, with a large share of medicinal plants in sub-Saharan Africa coming from forest and tree-based sources (see Chapter 3, this volume). For example, Erskine (2004) found that in KwaZulu-Natal, South Africa, 'almost 50 per cent of the plant material traded for medicinal purposes comes from the forest biome', and the World Agroforestry Center (Swallow, 2004) estimates that two-thirds of all medicinal plants in Africa are woody perennials. Thus forest and tree resources, if sustainably harvested, can support the health sector by providing medicine to large sections of the population in developing countries.

For people with HIV/AIDS, 'medicinal plants assist in supporting the strengths of the immune system ... [and] improve appetite to help arrest the worrying tendency of loss of appetite among some people living with HIV/AIDS, combating some AIDS opportunistic infections, especially certain digestive, respiratory, skin and mouth problems' (Gari, 2004). Traditional health care and several medicinal plants have proven helpful in the treatment of symptoms related to AIDS and are now widely accepted (Bodeker et al, 2000a; Naur, 2001, Shenton, 2004). Misleading information on the effectiveness of specific medicinal plants, however, may also cause illusion among patients, who may collect or purchase and use plants that lack actual therapeutic effects.

BOX 10.1 HIV/AIDS AS A CHALLENGE FOR SUSTAINABLE NATURAL RESOURCE MANAGEMENT: EVIDENCE FROM ZIMBABWE

Alexander Fröde

With an estimated HIV prevalence of 20 to 25 per cent, Zimbabwe keeps its position as one of the countries in the world hit hardest by the epidemic (UNAIDS, 2006a). Like most other African peoples, Zimbabweans have a long tradition of medicinal plant use, with traditional remedies prescribed by traditional healers, sold by collectors in urban markets and collected by the rural population.

The recent growth in the market for herbal remedies and traditional medicines in Zimbabwe is probably fuelled by the HIV/AIDS epidemic (Mander and Le Breton, 2006). Many of the plants used for people with HIV or AIDS are the same as those used for treating sexually transmitted diseases and include plants known as 'immune boosters', such as white syringa (*Kirkia acuminata*), weeping wattle (*Peltophorum africanum*), sausage tree (*Kigelia africana*), lemon bush (*Lippia javanica*) and wild custard-apple (*Annona senegalensis*). Several other plants are used to treat opportunistic diseases, such as the pepper-bark tree (*Warburgia salutaris*) for influenza and pneumonia.

Experience has shown that most traditional healers and herbalists in Zimbabwe are careful harvesters, who try to ensure sustainability by leaving a sufficient quantity of plants for natural regeneration. However, the urgency of HIV/AIDS, the prospects for substantial profit and the country's economic crises have led to a steady increase in the number of collectors and the harvested amounts. Traditional healers interviewed by the author and his colleagues in Zimbabwe's eastern highlands and the Matobo area in the south of the country gave evidence that some plants are now locally extinct or seriously threatened.

The insecure tenure rights and the competition between modern and traditional laws and practices complicate the regulation of harvesting. Furthermore, surveys have shown that many of the 'new' collectors employ harvesting methods that are destructive to plant individuals and populations (Nyambuya et al, 2004). For instance, harvesting the bark of *Warburgia salutaris* in complete rings prevents the circulation of essential fluids and girdles the tree, killing it.

The threat to sustainability is even higher for plants said to be a cure for HIV. With anti-retrovirals and other HIV medications out of the reach of many, people are desperate to believe promises of remedies against the virus. In the late 1990s, the 'African potato', *Hypoxis hemerocallidea*, a perennial with a thick tuber with yellow flesh, which had been used for centuries in traditional health care, was widely promoted in southern Africa as a cure. In Zimbabwe and in other countries of the region, *H. hemerocallidea* and related *Hypoxis* species subsequently became subject to high collecting pressure, which has led to partial extinction in the wild (UNEP-WCMC, 2003).

Since 2003, the horseradish tree, *Moringa olifeira*, a succulent with a squat, swollen stem, has gained a reputation as a cure against HIV/AIDS. Moringa is an easily grown, multi-purpose tree that can help boost the immune system of people living with HIV or AIDS, though it is not a cure. In its area of distribution in the east of Zimbabwe in the Zambezi basin, it is extensively collected. Now its increasing cultivation all over the country ensures a sufficient supply for domestic and commercial uses and limits the ecosystem effects. This tree may exemplify the expedient use of medicinal plants for treatment of HIV/AIDS without compromising the resource base, and in addition, it offers an opportunity for economic benefits from local, small-scale commercialization.

Among the more than 25 million HIV-infected people in sub-Saharan Africa, the demand for proven medicinal plants and traditional medicine is high and rising, especially because this is practically the only health-care system to which the overwhelming majority of Africans have access (Shenton, 2004). HIV/AIDS has triggered an increasing demand for medicinal plants. Not only are rural people falling back on the traditional health-care system but also people from urban centres, adding to the growing trade in medicinal plants. Dold and Cocks (2002), for example, report from the Eastern Cape Province, South Africa, that the use of medicinal plants and traditional medicine has increased and is predicted to further increase, mainly because of the ubiquity of AIDS. This assessment is complemented by findings of two country studies, from Malawi and Mozambique, commissioned by the FAO forest department (Barany et al, 2005; FAO, 2005a), showing both a decrease in available medicinal plant resources ('species scarcity') and an increase in people using them.

Forests and trees as sources of food

Proper nutrition plays a central role for HIV-infected people in slowing down the development of AIDS and extending the life expectancy and productivity of those who have developed AIDS symptoms. An important function of forests and trees is the provision of food, either through plant components (leaves, mushrooms, flowers and fruits, roots and tubers) or forest-dependent faunal species, from insects to large mammals (see Chapter 4, this volume). Often 'forest foods' complement food derived from farming, livestock husbandry or simply purchased food. However, the negative socioeconomic impacts of HIV/AIDS (loss of jobs, money, reduced productivity) interfere with these nutritional needs because they generally decrease access to food. Again forests and tree resources, when sustainably harvested, have the potential to be important components of the food supply, in both quality and quantity, especially for poor rural households. During food shortages, forest-based foods serve as 'buffers', helping families meet dietary needs (Lipper, 2000). Access to natural resources typically renders forest and tree-based foods a crucial option for HIV/AIDS-affected (rural) households by reducing expenditures. The replacement of cultivated food items with collected food (indigenous and wild vegetables) is an important response of HIV/AIDS-affected rural households in sub-Saharan Africa (UNAIDS, 1999). An increase in poaching of wildlife has also been reported from Malawian and South African protected areas within regions of elevated HIV/AIDS prevalence (Mauambeta, 2003; Erskine, 2004). Economic misery, especially in rich ecosystems such as protected areas, often increases anthropogenic pressure (hunting, harvesting of foods, wood, etc) on the resources on which local poor people depend (Lopez et al, 2004). The situation gets even worse when institutions responsible for the environment lose their capacity to protect it because of high HIV/AIDS rates among staff.

Forest resources as sources of income

HIV/AIDS can cause financial hardship for AIDS-affected persons in several ways. The AIDS-related illness of a productive household member in an agrarian production system means the loss of his or her labour, inducing a decline or even total loss of cash income. The latter can be the result of lower productivity and output of cash crops or the loss of salaries because of periodic or permanent absenteeism (since employment contracts and social insurance, insofar as they exist in developing countries, usually do not include the continuation of salary payments). At the same time, people living with HIV/AIDS face new expenditures related to medical care, hygiene and dietary needs. To handle these varied pressures, AIDS-affected households can reduce expenditures or generate income from other sources. The particular mitigating strategies of households are manifold and can vary from begging, to selling assets (e.g., livestock and household goods such as bicycles and radios), to decreasing spending on education (UNAIDS, 1999). According to the same source, 'coping strategies not requiring any cash were most frequently adopted' by sub-Saharan households, where income diversification was a common and important strategy. It seems probable that AIDS-affected households with access to natural resources fall back on them for generating income. A literature review showed that in South Africa, the 'ad hoc trade in non-timber forest products is a common emergency net' (Shackleton and Shackleton, 2004). As forests are by nature multi-purpose resources, the range of income-generating activities based on forest and tree resources is correspondingly high. Examples of such activities implemented by AIDS-affected households are the collection or production and selling of: firewood (UNAIDS, 1999; FASAZ and FAO, 2003; Kowero et al, 2004; Barany et al, 2005); charcoal (FAO, 2005a); herbal remedies; wild foods such as mopane worms (*Imbrasia belina*), bark products, fruits and mushrooms; mats and baskets; and brewed beverages and cooked food (all examples taken from Barany et al, 2005).

Basically, the whole range of forest-based products is available for the generation of income. Market demand or potential and access, investments (time, labour, equipment) and the experience of the resource users will finally influence their decisions about what to choose. The sustainable collection and trade of medicinal plants is a suitable option because of the growing market demand. Furthermore, investments are not high and the collection of medicinal plants is traditional among rural households in certain regions.

INTERACTIONS OF HIV/AIDS AT INSTITUTIONAL LEVELS

With more than 25 million people infected by HIV in sub-Saharan Africa – the majority of them in their most productive years – and massive effects on national economies, it is no surprise that HIV/AIDS also threatens the forest sector. The sector is already strained: 'drastic staff reductions ... have dealt debilitating blows to public forest administration capacities in most countries during the last two decades, with further staff loss due to the HIV/AIDS scourge' (Ndinga and Owino, 2004). The subregional report of the Forestry Outlook Study for Africa (ADB, EC and FAO, 2003) states plainly that for southern Africa, 'the high level of HIV/AIDS infection will have significant impacts on forestry. Most important, HIV/AIDS will reduce the ability of ... governments to allocate resources for tree growing and forest management. It has also reduced the number of technical and professional forestry staff.'

Figures from research undertaken in the forest industry by the South African Institute of Natural Resources (2005) indicate HIV infection rates among staff in five regions of 28 to 48 per cent (in 2003), whereas the national prevalence of people living with HIV is estimated at 21.5 per cent (UNAIDS, 2004b). Similar findings were reported from a meeting in 2006 between GTZ and representatives of the Cameroonian timber industry, 'Groupement de la filière bois du Cameroun' (GFBC and GTZ, 2006). The rate of HIV infections of some workforces among its member enterprises is substantially above the average national rate of 6.9 per cent (UNAIDS, 2004b). Though no further reports or data are available, individual forest enterprises and forest administrations appear to face similar rates in countries of generalized HIV prevalence. This means the temporary or permanent loss of colleagues, their death, attendance at funerals, mourning, lost knowledge, disorientation and increasing paralysis of the institution. On the administrative level, the loss of staff means loss of human resources (knowledge, experience, continuity) and loss of investments in training of staff as well as high staff turnover and expenditures in recruiting new employees. Indeed, it is often difficult for institutions to document AIDS-related impacts and deaths because of stigma. There needs to be a genuine institutional commitment to show the costs and redirection of funds required to address the effects of AIDS on personnel.

Given the existing conditions, often characterized by already-insufficient staffing, lack of adequately trained foresters and strained budgets, and the need for technology advances in the African forest sector, it is obvious that the HIV/AIDS epidemic is further undermining the sector's ability to implement its programmes and reach its objectives. In contrast to private enterprises, public forest administration generally has slower reactions to sudden changes, a further argument for prevention rather than crisis management.

Obviously, the effects of HIV/AIDS among staff go beyond the practical level. As a reaction to the spreading constraints and frightening future scenarios, several governments in sub-Saharan Africa have acknowledged the HIV/AIDS epidemic as a menace to the development of the national forest sector. For example, the governments of Tanzania and the Republic of South Africa have included actions against HIV/AIDS in their national forest programmes for 2001–2010, where addressing the 'declining number of male and female forest staff due to HIV/AIDS' is of 'high priority', just as is – for example – 'ineffective forest management' (Ministry of Natural Resources and Tourism, 2001). In the South African Framework for the National Forest Programme, addressing HIV/AIDS in the forestry sector is among the major national policies and strategies (Department of Water Affairs and Forestry, 2005).

Analysing what makes the forest sector vulnerable to HIV/AIDS is a necessary exercise before any coherent preventive action can be taken. There are two conditions that render the forest sector particularly vulnerable.

First, employment in the forest sector often means working in remote, rural areas with reduced access to health services and information. It also demands mobility from its staff. Changing forest operation sites, field missions such as supervision and control, extension services, conferences and workshops frequently separate staff from their families. This increases the likelihood of multi-partner sexuality. These conditions, characteristic of many employment situations within the forest sector, contribute to the vulnerability of staff through physical and social conditions that favour the spread of HIV (UNAIDS, 2004a).

Second, the less interest an institution or a sector shows in HIV/AIDS, the harder it is to communicate prevention messages and take action. Ignorance, apathy and uncertainty then dominate the attitudes of staff and executive personnel towards HIV/AIDS-prevention. As a result, people working within this atmosphere may receive insufficient information about how to protect themselves. The consequence is high vulnerability.

BOX 10.2 THE CONTRIBUTION OF ENVIRONMENTAL DEGRADATION TO THE HIV/AIDS EPIDEMIC IN THE DOMINICAN REPUBLIC

Maria Nanette Roble

Access to health care, education and behavioural factors have all been widely attributed to the HIV/AIDS epidemic. Environmental degradation, however, which has been shown to increase migration, poverty (Myers, 2002) and levels of disease (Patz et al, 2004), may have an important contribution, too (Tan et al, 2003). It is postulated that the active inclusion of environmental factors within public health policies could offset the potential consequences of existing land degradation on the HIV/AIDS epidemic in the Dominican Republic in the Caribbean, the 'second most-affected region in the world after Africa', with an HIV prevalence of 1.6 per cent (UNAIDS, 2006a).

Many Dominicans base their livelihood on natural resources, and between 1920 and 1981, the Dominican Republic experienced rapid deforestation because of increasing population, migration and poverty (Garcia and Roersch, 1996; Brothers, 1997). Deforestation and unsustainable land-use practices have been linked to increased exposure to public health risks, particularly in developing countries. Pollution of water resources, depletion of topsoil, erosion and the loss of habitat richness all contribute to an increase in the incidence rates of disease and the emergence of zoonotic pathogens (Patz et al, 2004). Within this setting, HIV/AIDS evolved as an important public health problem in the Dominican Republic, which has a generalized HIV prevalence of approximately 1.1 per cent (UNAIDS, 2006a).

Current public health policies on HIV/AIDS in the Dominican Republic focus mainly on factors known to have clear and direct associations with infection, morbidity and mortality: high-risk behaviour and access to health care (COPRESIDA, 2002; UNAIDS, 2004a). However, HIV/AIDS has been also associated with a lack of adequate nutrition and water, poverty, migration and exposure to pathogens. These factors have been linked to a lack of infrastructure, inadequate resources, cultural and social factors, and environmental degradation (COPRESIDA, 2002; Patz et al, 2004). Though efforts have begun to offset the impacts of HIV/AIDS in the country (COPRESIDA, 2002; UNAIDS, 2006b), rates of infection, morbidity and mortality remain high (UNAIDS, 2006c) and point to a need to broaden the scope of HIV/AIDS projects and policies beyond the known factors.

In the Dominican Republic, the poorest sectors of the population depend directly on the natural environment. They migrate internally, are the most malnourished, lack adequate water resources, and have higher rates of exposure to disease pathogens (COPRESIDA, 2002; Urena et al, 2003). All of these factors, which have varying levels of association with environmental degradation and land-use patterns, may also play an important role in the impact of the HIV/AIDS epidemic. Malnutrition has been shown to lower immune response and complicate HIV/AIDS infections (Ambrus and Ambrus, 2004). Populations that rely on low-productivity subsistence farming become malnourished because of the incremental lack of yields from plots cultivated unsustainably (Myers, 2002). Dominican subsistence farmers are generally the most impoverished populations in the country (Ortiz, 2002). Subsistence farmers migrate from their homes in search of arable lands and become more impoverished (Myers, 2002). Migration is also a factor affecting the spread of HIV in the Dominican Republic (Brewer et al, 1998; COPRESIDA, 2002; Tan et al, 2003).

Deforestation, which is often a consequence of subsistence farming by migrating populations and unsustainable land-use practices, such as slash-and-burn agriculture, contributes to the pollution of water-courses, often rendering fish and wildlife unfit for human consumption and decreasing a hydrological ecosystem's ability to breakdown or absorb potentially harmful organic substances, thus increasing levels of water-borne pathogens. Habitats favourable to zoonotic pathogens are created, increasing incidence of disease in local populations (Patz et al, 2004); tuberculosis is a known opportunistic disease of AIDS, and the interaction of malaria with HIV has been shown to bring on the onset of AIDS in some individuals (Russel, 2004). All forms of disease further strain the immune systems of HIV-infected individuals, increasing the risk of development of AIDS (Ambrus and Ambrus, 2004) or provoking its outbreak.

Public health professionals that focus on migration, malnutrition, inadequate water resources and pathogens search for associated variables that can be changed or modified. Political alliances with neighbouring countries, outreach to migrants and refugees, mass education, counselling and condom distribution all have varying levels of impact on the HIV/AIDS epidemic and these efforts require continued support. The emerging understanding that poverty is a risk factor affecting HIV/AIDS, and the epidemiological difficulties in providing answers to the problem of 'poverty', point towards the need for creative solutions. The natural environment, a resource that is generally more accessible to the poor, should be conserved and maintained to offset the effects of impoverished lands on increased disease and human poverty, and in its turn contribute to the reduction of the impact of the HIV/AIDS epidemic.

HIV/AIDS-INDUCED EFFECTS ON THE FOREST SECTOR

Because forest resources play a vital role in people's livelihoods and also in business and national economies, they are subject to management strategies by individuals, private enterprises and public forest administrations. The effects of the HIV/AIDS epidemic on individuals and institutions induce a change in their forest management strategies. This change takes place in the global context of still-decreasing tropical forest area (FAO, 2005b) and the 'elusive goal in many regions of the world, but particularly so in tropical areas' to implement sustainable forest management (Contreras-Hermosilla, 1999).

According to Hunter et al (2005), AIDS-affected rural households alter their strategies for managing natural resources by changing the type of resource, its use, harvesting strategies and the quantity of consumption or collection. The use of forest resources for income generation and provision of medicinal plants induces mainly a change in quantities; the need for cash intensifies harvesting and a change of resource type (new resources are tapped). Both changes can become critical if harvesting quantities and techniques are unsustainable. The common use of forest resources as a safety net makes the application of medium- or long-term investments in forest resource management difficult, if not impossible. 'Management' of natural resources in times of crisis is limited to mere harvesting of forest products, with few inputs to make value-added products. The cultivation of plants used for medicinal purposes is exceptional: it 'is nearly exclusively wild harvested' (Swallow, 2004). A look at the non-timber forest products literature, especially for medicinal plants, reveals that harvesting quantities and techniques may be unsustainable. A regional study on the medicinal plant trade from South Africa states that 'present harvesting is indiscriminate, destructive, and unsustainable for many species, particularly those harvested from Afromontane Forest' (Dold and Cocks, 2002). The authors conclude that '93 per cent of the species traded are harvested unsustainably, as they are either entirely or partially removed, resulting in the death of the plant' (see Chapter 3, this volume). Other studies that have investigated the trade in medicinal plants used to treat AIDS symptoms also conclude that harvesting methods and quantities are not sustainable (Bodeker et al, 2000b; Erskine, 2004; Swallow, 2004). Probably the most prominent example of an over-used medicinal plant is *Prunus africana*, whose bark is the basis for a medicine used for treating enlarged prostate glands. After its 'international discovery', demand rose to such an extent that unsustainable harvesting methods, combined with habitat loss, resulted in the risk of extinction of this tree species, which once covered montane tropical forests of Africa and Madagascar. Though research has been carried out for many years to find and implement sustainable harvesting and domestication methods, *P. africana* is still listed in Appendix 2 of CITES as a species that is threatened by extinction unless it is subject to strict trade regulation (Dawson et al, 2000).

The effects of unsustainable management of forest and tree resources, triggered by the HIV/AIDS epidemic, are becoming obvious. Barany et al (2005) report that according to herbalists in Malawi and Mozambique, medicinal plants, including species used in the treatment of AIDS-related illnesses, are becoming less available. Species scarcity is apparent in the increasing collecting distances and unavailability of certain species in given areas. Mauambeta (2003) reports (also within the context of the HIV/AIDS epidemic in Malawi) dwindling forest and forest-dependent resources because of the increased poaching of wildlife, deforestation due to the production of charcoal, and increasing selective cutting of timber to meet the demands of a 'growing coffin industry'. Many researchers indicate that the degradation of forest resources is still anecdotal and call for further research.

In addition to all the potential effects by direct and local forest resource users, the indirect effects on forest resources due to reduced management capacities and lack of trained forest staff have to be taken into account. What has been described as the diminishing ability of the sector to implement its programmes at a national level, can be felt at the forest management level in the reduced support for implementing sustainable forest management and conservation efforts and reduced staff for supervising the management and exploitation of forests by industries and local forest users. This development takes place in an environment of chronic labour shortages of professional foresters, where staffing is already limited by insufficient training (ADB, EC and FAO, 2003; Temu et al, 2005).

The implication of these deficits is a higher risk of forest resource degradation. 'With the repeated and long-term stress that accompanies high HIV/AIDS it is likely that tree and forest resources will be over-exploited, reducing long-term production potential'

(Swallow, 2004). Yet forest and tree-based resources are essential for rural people to meet their direct needs, and functioning forest ecosystems provide a wider range of indirect health services, such as stable local climatic conditions, clean drinking water and soil conservation. Hence, major degradations of functioning natural ecosystems are linked with the rise of unfavourable health conditions and the promotion of disease: 'Unhealthy landscapes' is the term that describes these linkages (Patz et al, 2004).

BOX 10.3 LOCAL FOREST GOVERNANCE IN EAST KALIMANTAN, INDONESIA, IN THE CONTEXT OF HIV/AIDS

Godwin Limberg

The following information is based on observations from 2000 to 2006 in the Malinau and Kutai Barat districts, East Kalimantan. No firm data exist on the regional prevalence of HIV, but HIV infections have been reported (Anonymous, 2006c), and anecdotal information suggests that sexually transmitted infections are widespread. The national HIV prevalence for Indonesia is 0.1 per cent.

In 1999, the Indonesian government embarked on a decentralization policy that included handing over forest management to district governments. Even before the decentralization law (22/1999) was fully implemented in 2001, district governments in forested areas were already issuing permits for small-scale logging. Under these systems, for the first time in Indonesian history indigenous communities obtained substantial fees from the logging companies operating in customary territories.

In Malinau and Kutai Barat districts, more than 650 permits were issued, covering some 100,000 ha of forests (Bagian Ekonomi, 2002; Casson and Obidzinski, 2002; BPS, 2005). This surge in logging activities created a hectic, Wild West atmosphere: local elites were striking lucrative deals in the tens and even hundreds of millions of rupiahs, people were using any means (including road blockades and demonstrations) to obtain benefits, and a sector rapidly developed to 'assist' indigenous communities and their representatives in spending their new wealth on a range of goods and services, including, of course, entertainment.

In the early stages, influential people from the communities were invited by the logging companies to the larger cities, such as Samarinda and Tarakan, to discuss contractual propositions and to 'relax'. Once people got used to this lifestyle (including the consumption of alcoholic beverages and commercial sex) and the money really started flowing, the entertainment business moved closer to the sources of money: Sendawar and Malinau City, the district capitals. In Kutai Barat, for example, where fees were higher and the population bigger, cafes and karaoke bars eventually opened closer to logging hotspots. The conditions were ideal for the transmission of HIV.

For a long time, the national government in Indonesia and the provincial government of East Kalimantan denied that HIV/AIDS was a threat. It was only in late 2005 that more attention began to be paid to HIV/AIDS locally. No signs of an HIV/AIDS epidemic are visible yet, but HIV infections have been detected in the region. The mayor of Tarakan, for example, has pointed out that the new HIV infections are not just among commercial sex workers (Anonymous, 2006a), which indicates a generalization of the epidemic. The local authorities and civil society are still at the beginning of a debate on how to deal with the problem. Two approaches can be observed in East Kalimantan. Some people are progressive and try to address the problem, although they are still looking for appropriate action and strategies (including regulation, making condoms available near commercial sex zones, awareness campaigns, voluntary counselling and testing centres). Others continue to deny there is a problem, asserting that the number of HIV infections has not increased (Anonymous, 2006b) and that more people die in car accidents than from HIV infection (Anonymous, 2006d), or threaten to confiscate condom dispensers (Anonymous, 2006c). The highly religious character of society means that public campaigns concerning HIV/AIDS transmission and its association with sexual behaviour are considered inappropriate.

Looking through the HIV/AIDS lens at local forest governance, we can draw several conclusions. Forestry by itself is certainly not the cause of the spread of HIV/AIDS, but related developments can foster it as income for the local population (through logging fee payments) combines with changing behaviour and opportunities to increase the chance of high-risk behaviour (the use of condoms is not common). Moreover, forest operations are often conducted by trained staff who move from one logging site to another, a pattern known to increase high-risk behaviour. Figures from Central Africa show that mobile workers in the logging industry have a significantly above-average HIV prevalence.

The sociocultural and political environment is not yet suitable to support a comprehensive fight against HIV/AIDS because of the lack of awareness among local leaders. Given the lack of information, the people at high risk are simply unaware of the dangers.

To meet the challenge of fighting HIV/AIDS at this local level, it is not enough to act within the forest sector alone. Policy support and advocacy from civil society are needed to build a comprehensive, multi-sectoral strategy against HIV/AIDS. However, within its sphere of action, stakeholders in the forest sector can each play a part, by targeting proven but also innovative measures against HIV/AIDS at forest workers and the local populations that have spatial or contractual relationships with the logging operations.

OUTLOOK AND CONCLUSIONS

With the persistent spread of HIV/AIDS in sub-Saharan Africa, even more widespread and dire effects are likely to be felt in the forest sector. We would like to emphasize the rationale for the forest sector's involvement in HIV/AIDS-related actions.

- The workers of the forest sector, especially employees of private companies and public forest administrations, bear risks of increased vulnerability to HIV infection. This imposes on employers economic, legal and moral obligations to protect their employees from work-related vulnerability.
- The importance of forest-based resources and services for mitigating health and socioeconomic-related impacts, especially for poor HIV/AIDS-affected households, necessitates that the forest sector take real action to develop and implement sustainable management strategies.
- Comprehensive national strategies against HIV/AIDS are based on a multi-sectoral approach. That means that every sector within a country has a role in the fight against HIV/AIDS. The forest sector, possibly jointly with environment, agricultural or rural development partners, is no exception and must identify where and how it can contribute.

The good news is that in different countries, sectors and social strata, the spread of HIV can be substantially reduced, if not stopped. This has been made possible by the development and implementation of tailor-made prevention strategies and support by high-ranking politicians and cultural leaders. Classifying actions against HIV/AIDS as a priority within high-profile forest sector documents, as is the case in Tanzania and South Africa, represents a clear and necessary commitment by the political leaders. The most prominent examples of successful practice are Brazil, Uganda and Thailand, which have established multi-sectoral strategies to fight against HIV/AIDS and have achieved nationwide and sector-wide reductions in HIV prevalence. Even in Haiti, which suffers from conflict and poverty and has the region's highest HIV prevalence, the percentage of HIV-infected pregnant women was reduced by half from 1993 to 2003–2004 (UNAIDS, 2006a). Well-adapted prevention strategies for the forest sector have a similarly good chance of being successful.

Decision-makers should not be afraid of additional costs or workloads or any lack of competences. Implementing actions against HIV/AIDS demands neither that forestry staff become public health specialists, nor that additional financial resources be allocated, since many HIV/AIDS and health-specific programmes, projects and organizations already provide financial and technical support to stakeholders in other fields who want to implement HIV/AIDS strategies.

Successful strategies for fighting against HIV/AIDS range from tools for sensitizing specific target groups and developing HIV/AIDS workplace policies to complex, national multi-sectoral strategies. Technical support is available to adapt these strategies and tools to specific conditions. Moreover, an increasing number of countries have elaborated multi-sectoral strategies against HIV/AIDS at national, regional and local levels. The integration of forest-sector stakeholders into these strategies is possible, since many coordination units are in place to assist this process and can also facilitate connections to financial and technical support.

Mainstreaming actions against HIV/AIDS within an institution does not undermine its mandate. Mainstreaming is a process that enables institutions to overcome or reduce the causes and consequences of HIV/AIDS in their work more effectively and sustainably. This requires that they analyse and then adapt or improve their activities and workplaces with regard to HIV/AIDS. Small changes in working routines can have large positive impacts on the vulnerability of personnel.

The sustainable management and conservation of forest products and services are already core activities of the forest sector. Such resource use for mitigating HIV/AIDS impacts does not need special management: it is only the application and use of the resource that alters. This is an excellent example of the comparative advantages a non-health sector has in the fight against HIV/AIDS; it is only the forestry and closely related sectors that have the competence and mandate to manage these resources.

Some forest stakeholders are already taking action, as the examples given in this chapter have shown. Many others still have to be sensitized to understand the implications of HIV/AIDS and the forest sector before they support or agree to take action. Research and fact-based information that show the potential as well as the documented negative and positive impacts of HIV/AIDS on the environmental sector, including forests and other natural resources, are necessary. Such information has to be provided in a form that will allow managers and policy-makers to take action in both natural resources management and HIV/AIDS-prevention strategies. This can be facilitated through an active professional discussion and systematic information exchange between stakeholders in the environmental, forestry and health sectors. With this in mind, we conclude this chapter with a call for action.

REFERENCES

ABCG (Africa Biodiversity Collaborative Group) (2002) 'HIV/AIDS and natural resource management linkages', Nairobi, Kenya

ADB, EC and FAO (2003) 'Forestry outlook study for Africa – subregional report southern Africa', ftp://ftp.fao.org/docrep/fao/005/y8672e/y8672e00.pdf (accessed 13 July 2006)

Ambrus, J.L. Sr and Ambrus, J.L. Jr (2004) 'Nutrition and acquired immunodeficiency syndrome', *Experimental Biology and Medicine* 229: 865

Anonymous (2006a) 'Bayi juga Terkena AIDS – Dua Pengidap Meninggal, Samarinda Tambah 7 Orang' ('Baby also contracts AIDS – Two people living with HIV/AIDS pass away, 7 new cases in Samarinda'), *Kaltim Post*, 25 February

Anonymous (2006b) 'Belum Ada Perubahan Jumlah Penderita – Kutim Relatif Aman HIV/AIDS' ('No change yet in number of patients – Kutai Timur relatively safe from HIV/AIDS'), *Kaltim Post*, 2 January

Anonymous (2006c) 'Bila Ada, Kami Kembalikan – Pemkot – DPRD Tolak Mesin Kondom' ('If there are any, we will return them – city council and city assembly reject condom dispenser'), *Kaltim Post*, 24 January

Anonymous (2006d) 'Tiga Bulan 20 Nyawa Melayang di Jalan – Lakanlantas Lebih Dahsyat dari AIDS' ('Three months 20 souls killed on the road – Traffic accidents more horrifying than AIDS'), *Kaltim Post*, 24 April

Bagian Ekonomi (2002) 'Izin Pemungutan dan Pemanfaatan Kayu IPPK' ('[List of] small scale timber harvesting permit holders'), Bagian Ekonomi, Malinau, Indonesia,18 July

Barany, M., Holding Anyonge, C., Kayambazinthu, D. and Mumba, R. (2005) 'Miombo woodlands and HIV/AIDS interactions: Malawi Country Report', Forestry Policy and Institutions Working Paper 6, FAO, Rome, Italy

Bodeker, G., Kabatesi, D., King, R. and Homsy, J. (2000a) 'A regional task force on traditional medicine and AIDS', *The Lancet* 355: 1284

Bodeker, G., Burford, G., Gemmill, B., Kabatesi, D. and Rukangira, E. (2000b) 'Traditional medicine and HIV/AIDS in Africa', Report from the International Conference on Medicinal Plants, Traditional Medicine and Local Communities in Africa, a parallel session of the Fifth Conference of the Parties to the Convention of Biological Diversity, Nairobi, Kenya 16–19 May

BPS (Badan Pusat Statistik) (2005) 'Kabupaten Malinau dalam Angka 2004' ('2004 Malinau District Statistics'), BPS, Malinau, Indonesia

Brewer, T.H., Hasban, J., Ryan, C.A., Hawes, S.E., Martinez, S., Sanchez, J., de Lister, M.B., Costanzo, J., Lopez, J. and Holmes, K.K. (1998) 'Migration, ethnicity and environment: HIV risk factors for women on the sugar cane plantations of the Dominican Republic', *AIDS* 12: 1879–1887

Brothers, T. (1997) 'Deforestation in the Dominican Republic: A village-level view', *Environmental Conservation* 24: 213–223

Casson, A. and Obidzinski, K. (2002) 'From new order to regional autonomy: Shifting dynamics of "Illegal" logging in Kalimantan, Indonesia', *World Development* 30: 2133–2151
Contreras-Hermosilla, A. (1999) 'Towards sustainable forest management: An examination of the technical, economic and institutional feasibility of improving management of the global forest estate', Working Paper FAO/FPIRS/01, FAO, Rome, Italy
COPRESIDA (Consejo Presidencial de SIDA, Dominican Republic) (2002) Personal communication during a Columbia University-sponsored practicum
Dawson, I., Were, J. and Lengkeek, A. (2000) 'Conservation of *Prunus africana*, an over-exploited African medicinal tree', *Forest Genetic Resources* 28: 27–33
Department of Water Affairs and Forestry (2005) *Framework for the National Forestry Programme (NFP) in South Africa*, Department of Water Affairs and Forestry, Pretoria, South Africa
Dold, A.P. and Cocks, K.L. (2002) 'The trade in medical plants in the Eastern Cape Province, South Africa', *South African Journal of Science* 98: 589–597
Erskine, S. (2004) 'Red ribbons and green issues: How HIV/AIDS affects the way we conserve our natural environment', University of KwaZulu-Natal, Durban, South Africa
FAO (2005a) 'Miombo woodlands and HIV/AIDS interactions – Mozambique Country Report', FAO, Rome, Italy, www.fao.org/docrep/008/j5251e/j5251e00.HTM (accessed 4 October 2006)
FAO (2005b) *State of the World's Forests*, FAO, Rome, Italy
FASAZ (Farming Systems Association of Zambia) and FAO (2003) *Interlinkages Between HIV/AIDS, Agricultural Production and Food Security: Southern Province, Zambia*, FAO, Rome, Italy
Garcia, R. and Roersch, C. (1996) 'Politica de manejo y utilizacion de los recursos floristicos en la Republica Dominicana', *Journal of Ethnopharmacology* 51: 147–160
Gari, J.A. (2004) *Plant Diversity, Sustainable Rural Livelihoods and the HIV/AIDS Crisis*, UNDP and FAO, Bangkok, Thailand and Rome, Italy
GFBC and GTZ (2006) Compte rendu de la 1ère réunion entre GFBC et GTZ sur la problématique VIH/SIDA au secteur forestier', unpublished project report, GFBC and GTZ, Yaoundé, Cameroon
Hunter, L.M., Twine, W. and Johnson, A. (2005) 'Population dynamics and the environment: Examining the natural resource context of the African HIV/AIDS pandemic', working paper at the Institute of Behavioral Science (IBS), University of Colorado, Boulder, CO
Institute of Natural Resources (2005) *Pilot State of the Forest Report – A Pilot Report to Test the National Criteria and Indicators*, Institute of Natural Resources, Scottsville, South Africa
Kowero, G., Njuki, J. and Nair, C.T.S. (2004) 'What shapes forestry in Africa?', report for the project 'Lessons Learnt on Sustainable Forest Management in Africa', www.afornet.org/images/pdfs/What%20shapes%20forestry%20in%20Africa.pdf (accessed 11 July 2007)
Lipper, L. (2000) 'Forest degradation and food security', *Unasylva* 202, FAO, Rome, Italy, www.fao.org/docrep/X7273e/x7273e05.htm#P0_0 (accessed 11 July 2007)
Lopez, P., Bergmann, U., Dresrüsse, P., Fröde, A., Hoppe, M. and Rotzinger, S. (2004) 'VIH/SIDA: Un nouveau défi pour la gestion des Aires Protégées à Madagascar – l'intégration des mesures contre le VIH/SIDA dans le travail du Parc National Ankarafantsika', SLE/Humboldt-Universität Berlin
Mander, M. and Le Breton, G. (2006) 'Overview of the Medicinal Plants Industry in Southern Africa', in N. Diederichs (ed) *Commercialising Medicinal Plants. A Southern African Guide,* Sun Press, Stellenbosch, South Africa, pp1–8
Mauambeta, D.D.C. (2003) 'HIV/AIDS mainstreaming in conservation: The case of Wildlife and Environmental Society of Malawi', Malawi Wildlife and Environmental Society, Imbe, Malawi

Ministry of Natural Resources and Tourism (2001) 'National Forest Programme in Tanzania 2001–2010', Forestry and Beekeeping Division, Ministry of Natural Resources and Tourism, Dar Es Salam, Tanzania

Myers, N. (2002) 'Environmental refugees: A growing phenomenon of the 21st century', *Philosophical Transactions of the Royal Society* B 357: 609–613

Naur, M. (2001) 'HIV/AIDS: Traditional healers, community self-assessment, and empowerment', IK Notes 37, Africa Region's Knowledge and Learning Centre, World Bank, Washington, DC

Ndinga, A. and Owino, F. (2004) 'Study on forest administration and related institutional arrangements', KSLA, AFORNET and FAO, Rome, Italy

Nyambuya, O., Mujuru, F. and Khumalo, S.G. (2004) 'Traditional medicinal plants utilisation', report for Matabeleland South, SAFIRE (Southern Alliance for Indigenous Resources), Harare, Zimbabwe

Ortiz, D.T. (2002) 'Universidad Autonoma de Santo Domuingo', Columbia University-sponsored practicum, Dominican Republic

Patz, J.A., Daszak, P., Tabor, G.M., Aguirre, A.A., Pearl, M., Epstein, J., Wolfe, N.D., Kilpatrick, A.M., Foufopoulos, J., Molyneux, D., Bradley, D.J. and Members of the Working Group on Land Use Change and Disease Emergence (2004) 'Unhealthy landscapes: Policy recommendations on land use change and infectious disease emergence', *Environmental Health Perspectives* 112: 1092–1098

Russel, S. (2004) 'The economic burden of illness for households in developing countries: A review of studies focusing on malaria, tuberculosis and human immunodeficiency virus/acquired immunodeficiency virus', *American Journal of Tropical Medicine and Hygiene* 71(2 supplements): 147–155

Shackleton, C. and Shackleton, S. (2004) 'The importance of non-timber forest products in rural livelihood security and as safety nets: A review of evidence from South Africa', *South African Journal of Science* 100: 658–664

Shenton, M. (2004) 'AIDS and traditional health care in Africa: The role of traditional healers in prevention strategies and treatment options', in S. Twarog and P. Kapoor (eds) *Protecting and Promoting Traditional Knowledge: Systems, National Experience and International Dimensions*, United Nations Conference on Trade and Development, New York, NY and Geneva, Switzerland, pp21–24

Swallow, B. (2004) 'Overview of links between HIV/AIDS and agroforestry', in B. Swallow, P. Thangata, S. Rao and F. Kwesiga (eds) *Proceedings of the Workshop on Agroforestry Responses to HIV/AIDS in East and Southern Africa*, World Agroforestry Centre Office, Gigiri, Nairobi, Kenya

Tan, D., Upshur, R. and Ford, N. (2003) *Global Plagues and the Global Fund: Challenges in the Fight against HIV, TB and Malaria*, BioMed Central International Health and Human Rights 3, London

Temu, A., Rudebjer, P., Kiyiapi, J. and van Lierop, P. (2005) 'Forestry education in sub-Saharan Africa and Southeast Asia: Trends, myths and realities', FOP Working Paper, FAO, ANAFE and SEANAFE, Rome, Italy

UNAIDS (1999) *A Review of Household and Community Responses to the HIV/AIDS Epidemic in the Rural Areas of Sub-Saharan Africa*, UNAIDS, Geneva, Switzerland

UNAIDS (2004a) *2004 Report on the Global AIDS Epidemic: Fourth Global Report*, UNAIDS, Geneva, Switzerland

UNAIDS (2004b) 'UNAIDS/WHO Epidemiological Fact Sheets on HIV/AIDS and Sexually Transmitted Infections, 2004 Update', UNAIDS, UNICEF and WHO

UNAIDS (2005) *AIDS in Africa: Three Scenarios to 2025*, UNAIDS, Geneva, Switzerland

UNAIDS (2006a) *2006 Report on the Global AIDS Epidemic*, UNAIDS, Geneva, Switzerland

UNAIDS (2006b) 'Uniting the world against AIDS – Caribbean', www.unaids.org/en/Regions_Countries/Regions/Caribbean.asp (accessed 19 October 2006)

UNAIDS (2006c) 'Uniting the world against AIDS – Dominican Republic', www.unaids.org/en/Regions_Countries/Countries/Dominican_Republic.asp (accessed 20 October 2006)

UNEP-WCMC (2003) *Traditional Lifestyles and Biodiversity Use – Regional Report: Africa*, WCMC, Cambridge

UN Population Division (2003) *The HIV/AIDS Epidemic and its Social and Economic Implications*, United Nations, New York, NY

Urena, F.I., Duarte, I., de Moya, E.A., Pérez-Then, E., Hasbùn, J. and Tapia, M. (2003) 'Primera Parte – Analysis de la situacion la respuesta al VIH/SIDA en Republica Dominicana: Informe final (Diciembre de 1998)', Universidad Autonomia de Republica Dominicana, Practica, Dominican Republic

Walter, S. (2001) *Non-wood Forest Products in Africa: A Regional and National Overview*, FAO, Rome, Italy

11

Forest Disturbance and Health Risks to the Yanomami

Gale Goodwin Gómez

The Yanomami live deep in the Amazon rainforest on both sides of the border between Venezuela and Brazil. With an estimated total population of 26,000 people, their traditional territory extends over 192,000km^2 of tropical forests located to the west of the Guiana Shield (Milliken et al, 1999). Prior to the 1940s, the Yanomami in Brazil had little contact with the outside world, and thus, the danger of contracting exogenous diseases was minimal. Until the 1960s, permanent contact with outsiders was limited to religious missions and government outposts established by the Indian Protection Service.

Today, however, any disturbance of the forest can have serious repercussions on the health and welfare of these remote, indigenous people. This case study on the health situation of the Yanomami Indians in Brazil from the mid-1970s to the 1990s reveals the link between forest disturbance and their deteriorating health. A significant increase in the incidence of malaria, in particular in the state of Roraima (Figure 11.1), closely parallels that state's substantial population growth between 1962 and 1990 (Figure 11.2). Both increases reflect the results of development initiatives in the region, including highway construction, colonization and mineral exploration, intended to 'vivify' the international border zone (Ramos, 1995).

The secret creation of the northern borderlands militarization project (Projeto Calha Norte) in 1985, during the first year of the civilian government, involved the construction of four military outposts and seven airfields in Yanomami territory (Ramos, 1995). The presence of indigenous peoples – in particular, the remote and autonomous Yanomami – along the border regions was considered a threat to national security. No action was taken to protect the Yanomami or to control the arrival of non-Indians on their traditional lands. In fact, during that time the Governor of Roraima, who was the former President of the National Indian Foundation (FUNAI), Romero Jucá Filho, 'openly encouraged the free entrance of *garimpeiros* (gold prospectors) into Yanomami lands' (Ramos, 1995, p63; English translation by the author). Dramatic spikes in Figures 11.1 and 11.2 coincide with a massive influx of these gold prospectors into Yanomami territory during the gold rush of 1987–1990.

The arrival of thousands of placer miners and wildcat prospectors in Roraima from impoverished areas of Brazil caused the destruction of their environment in ways that disrupted traditional indigenous subsistence activities, while creating conditions favourable to breeding *Anopheles* mosquitoes, the vector of the malaria parasite (see Chapter 9, this volume). At the same time, *garimpeiros* brought highly contagious diseases, including respiratory infections, tuberculosis and drug-resistant strains of malaria, to previously healthy, self-sufficient Yanomami communities.

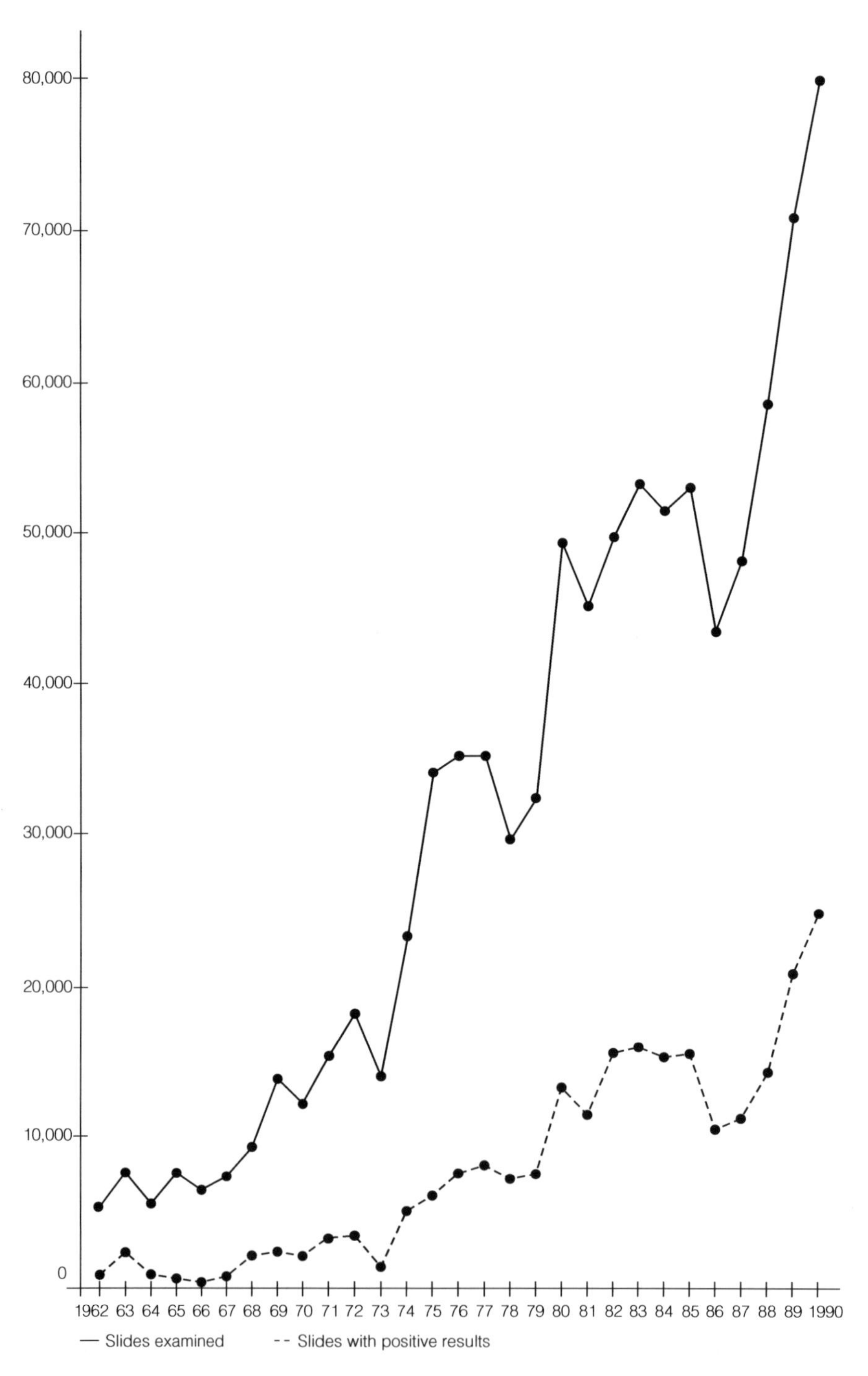

Figure 11.1 *Incidence of malaria in Roraima, 1962–1990*

Source: FNS/RR (1991a)

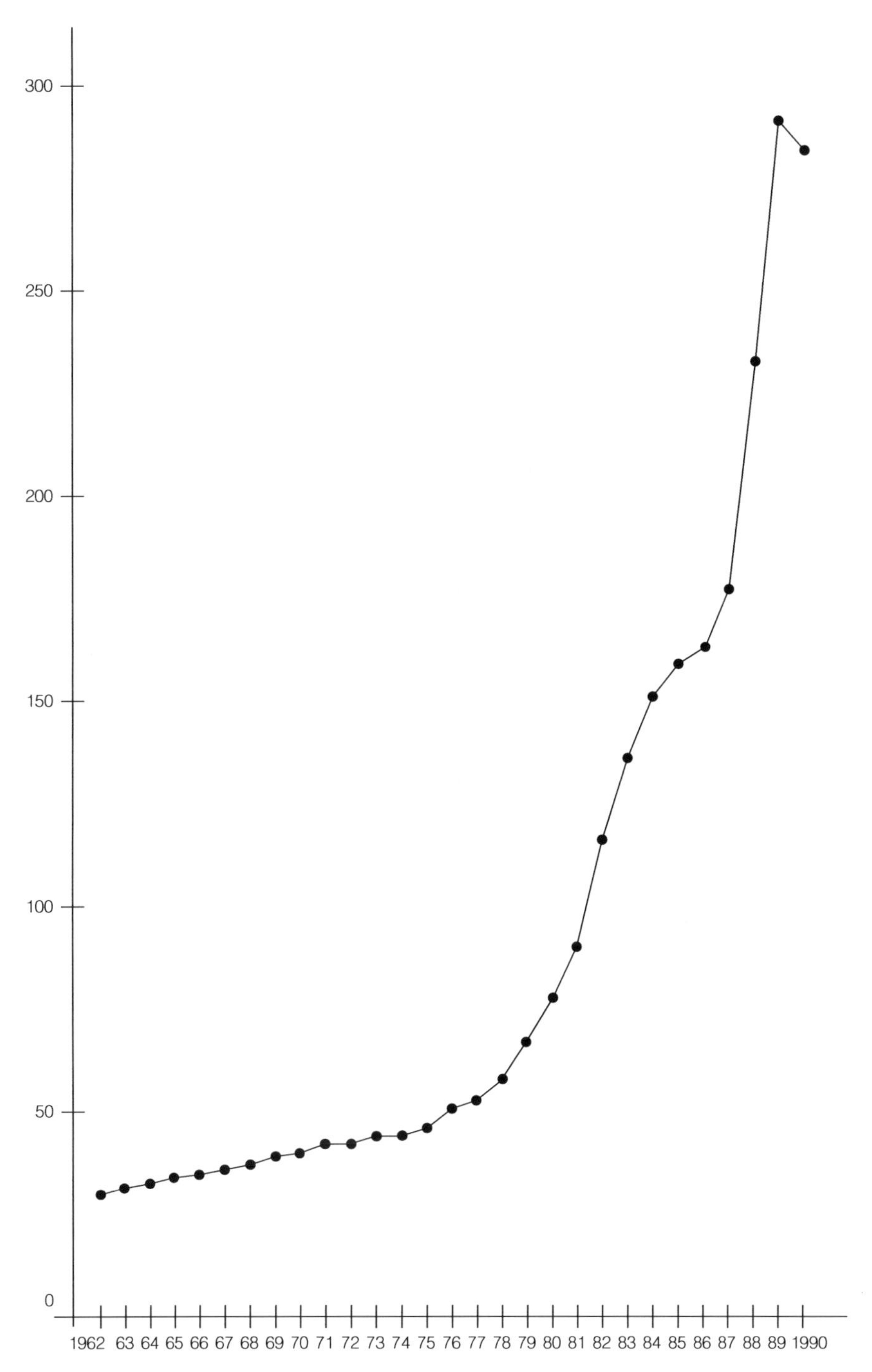

Figure 11.2 *Population of Roraima, 1962–1990*

Source: FNS/RR (1991b)

The Brazilian Yanomami, estimated by the National Health Foundation (FUNASA) to have a population of 12,500 people, live in 188 villages in Roraima and Amazonas states (Albert, 2006). An additional 13,000 Yanomami live across the border in the Venezuelan states of Bolivar and Amazonas. Although this chapter focuses on the Yanomami living in Brazil, the health problems, like the gold prospectors, are not limited by the international boundary. Outbreaks of malaria and epidemics of communicable diseases easily spread from one region to the next. Consequently, the health of the Yanomami in Venezuela can be affected by epidemics among the Brazilian Yanomami, and vice versa.

BOX 11.1 MALARIA IN AMAZONIA

Manuel Cesario

Historically, four phases of malaria can be distinguished worldwide:

1 Beginning in 1820, the *mala aria* of the Italian swamps was treated with quinine. The agent (*Plasmodium* spp) was identified by Laveram in 1890, and the mechanism of transmission was described by Ross in 1897.
2 From the 19th century until the mid-20th century, efforts were made to reduce populations of *Anopheles* mosquitoes through drainage and landfills and the use of larvicides and insecticides. Mosquito netting was also used.
3 From the mid-20th century until the 1990s, the dream of eradicating malaria worldwide was inspired by the initial success of DDT and new therapeutics, such as cloroquine, primaquine, amodiaquine and mefloquine.
4 Over the past 25 years, we have seen less emphasis on the biological aspects of the disease cycle and an increasing concern with the socioeconomic, environmental and political factors, since both *Anopheles* and *Plasmodium* species are becoming increasingly resistant to chemicals.

About 90 per cent of the malaria cases in South America are found in Amazonia. The Pan-American Health Organization finds that on average, 45 per cent of inhabitants are exposed; Brazil is in first place, with 93.9 per cent of its population at risk. From 1980 to 2003, Brazilian Amazonia was responsible for 94.8 to 99.7 per cent of all the country's cases of malaria.

In the early 1940s, Brazil had 6 million cases of malaria annually (more than 10 per cent of the population), partially because of the introduction of *Anopheles gambiae* from Africa, which was later eradicated by the use of Piretro and Green-Paris (both abandoned because of their high toxicity). In the 1960s, the Campaign for the Eradication of Malaria used DDT in the entire southern and southeastern region, in almost all of the northeast, and in part of the central western region. On the basis of environmental and socioeconomic characteristics, it was assumed that eradication would take a long time in Amazonia but could be achieved more quickly in the remainder of the country. But the number of cases of malaria registered per year in Brazil grew, from 50,000 in the 1970s to almost 600,000 at the end of the 1980s, when the country accounted for 10 per cent of all the cases in the world, apart from Africa, and then surpassed 600,000 cases per year at the end of the 20th century. This situation prompted the Brazilian Ministry of Health to create a new plan that relied on early diagnosis and treatment and surveillance systems. The programme reduced the incidence of malaria in Brazilian Amazonia by 41 per cent, from 31.9 cases per thousand inhabitants in 1999 to 18.8 in 2001; it also reduced malaria-related hospitalizations by 69.2 per cent and deaths from malaria by 54.7 per cent. For political reasons, the programme was discontinued in early 2003.

Between 2001 and 2005 the incidence of malaria increased significantly in the states of Acre, Rôndonia, Amazonas and Roraima with a smaller increase in Amapa. Only in Pará, Maranhão and Tocantins has the rate declined. An increase in malaria is observed in the region as a whole, with a concentration of cases in northern and western Amazonia, particularly in Acre, with a 153 per cent increase in the number of new cases from 2003 to 2004, and a 63 per cent increase from 2004 to 2005.

Numerous studies have demonstrated that malaria follows railroads and highways. Two massive migrations, which exposed people who had no natural immunity to the 'green hell', coincided with the first two great epidemics of malaria in Amazonia. At the end of the 19th century, the rubber-tapping industry (*seringalismo*) drew many migrant workers from northeastern Brazil to western Amazonia. Then, to build the Madeira–Mamore Railroad, known as the 'Devil's Railway', 20,000 workers came from many parts of the world, and more than 6000 of them lost their lives to malaria and other tropical diseases.

A third wave, with an increase from 50,000 to 600,000 annual cases of malaria in the 1970s and 1980s, coincided with the 'Brazilian miracle' and construction of the Trans-Amazonian and Cuiaba–Santarem highways, countless hydroelectric dams, and settlement and land-tenure projects that caused internal migrations and ecosystem change.

Today, western Amazonia is subject to a new wave of migrant workers and unprecedented environmental changes attributable to infrastructure mega-projects. Three hydroelectric dams in the Madeira River watershed are increasing the mobility of people, widening water surfaces and modifying the hydrologic regimen of a major tributary in the Amazonian basin. Highways are being improved to allow the export of grain to promising markets of the Pacific, transforming an isolated region into a corridor for the transportation of people and products between Brazil and its Andean neighbours. Malaria is certain to follow.

Manuel Cesario provides a global and historical context for the high incidence of malaria in the Amazon region, which accounts for 90 per cent of all cases in South America. He points out that in efforts to eradicate the disease over the past 25 years, there is 'an increasing concern with the socioeconomic, environmental and political factors'. This chapter on the Yanomami in northern Brazil highlights these concerns and serves as a present-day example – following the rubber boom in western Amazonia and the construction of the Madeira–Mamore Railroad between Bolivia and Brazil a century ago – of massive migrations into the forest that coincided with great epidemics of malaria in Amazonia. The case of the Yanomami illustrates the linkage between the surge in economic development in Brazil during the 1970s and 1980s and the dramatic increase in malaria that Cesario attributes to 'internal migrations and ecosystem change'. Thus, past, present and future health risks to the Yanomami are best understood within this wider context.

CULTURAL BACKGROUND: RELATIONSHIP TO FOREST

The Yanomami are hunter-horticulturists who have maintained their traditional lifestyle of hunting game, gathering wild foods and forest resources, and cultivating swidden gardens.

Important staple crops are bitter and sweet manioc and several varieties of bananas; other common cultivars in Yanomami gardens include yams, potatoes, papaya, sugar cane and tobacco. The traditional communal roundhouse (called a *yano* or *shapono*) and the temporary forest structure *(naa nahi)* used by hunters and travellers are constructed from a variety of trees, vines and other wild plant species chosen for specific properties, such as strength, hardness, flexibility and availability (Milliken et al, 1999). The success of a Yanomami hunter depends not only on his intimate knowledge of the habits of game animals (including tapir, peccaries, deer, paca, agouti, tortoises, several types of monkeys, and numerous bird species) but also a profound understanding of rainforest ecology and the interactions between the plant and animal species found there (Milliken et al, 1999). Every aspect of daily life among the Yanomami reflects generations of accumulated knowledge and depends on an in-depth understanding of their tropical forest environment.

The extent to which one large area of tropical forest is necessary for the survival of the Yanomami people and their culture was clearly explained in the 1978 proposal and justification for the Yanomami Indian Park in northern Brazil (Committee for the Creation of the Yanomami Park, 1979). The arguments presented then still hold; the tropical rainforest ecosystem with all its associated flora and fauna remains integrated into the daily lives and traditional beliefs and practices of the Yanomami today. In 1978, Brazilian anthropologist Alcida Rita Ramos estimated that an area of 707km^2 of forest would be required to provide adequate subsistence for a typical Yanomami village of 30 to 100 people (Ramos, 1979). This includes 900m^2 per person for gardening sites around the village, and 'in addition, a concentric and much more extensive area is used in the acquisition of indispensably necessary resources of a sparse and/or random distribution, such as game animals, fish and forest products' (Ramos, 1979, p103). Their swidden agriculture requires that every two years or so the Yanomami relocate their gardens and, with less frequency, their villages.

Furthermore, the Yanomami may be characterized as periodically nomadic because they spend some months during the dry season each year hunting game and gathering seasonal wild foods, often while travelling to visit kin in distant villages. A significant amount of travel between villages is the norm and 'fundamental to the maintenance of the social dynamics and cohesion of Yanomami communities' (Ramos, 1979, p102). Social relations and ritual obligations create wide networks of kin and trade partners, and these networks are physically represented throughout the forest by a complex of interwoven trails between villages. Thus, for the Yanomami, 'the concept of territory ... cannot be thought of as just the site and immediate surroundings of the villages without completely distorting the understanding of Yanomami life and culture' (Ramos, 1979, p104). The Yanomami are truly people of the forest and disturbances to their environment have profound effects on all aspects of their lives (cf. Chapters 13 and 16, this volume).

HIGHWAY CONSTRUCTION AND EARLY EPIDEMICS

When development projects arrived in the territory of the Brazilian Yanomami more than 30 years ago, exploitation of natural resources was the priority, and the presence of indigenous inhabitants in the forest was of little or no concern to the military government. At that time the myth of the 'empty' Amazon was motivating colonization schemes, and a belief that indigenous peoples were obstacles to development was voiced by many, including the governor of the Territory of Roraima, Colonel Fernando Ramos Pereira. In an interview in 1975 in the *Jornal de Brasília*, Pereira stated that in his opinion, 'an area such as that cannot afford the luxury of half a dozen Indian villages holding up development' (Taylor, 1979, p49). He was referring to the Surucucús highlands, an area of potentially rich mineral deposits, but also the historical centre of the Yanomami homeland, which at that time had an estimated indigenous population of 4500.

The beginning of the construction of the Northern Perimeter Highway in 1973 through the southeastern part of Yanomami territory in Roraima brought a wave of epidemics that devastated the villages located near the highway. The National Indian Foundation 'did nothing to try to prevent or minimize the harmful effects of the road building. No teams were sent ahead of the highway workers to attempt a vaccination campaign that would protect the Indians against such lethal diseases as measles, whooping cough, tuberculosis and the common cold, before the arrival of the workers' (Ramos, 1979, p7). Of those living near the construction areas, 22 per cent died in the first two years, principally from respiratory diseases contracted from the workers. The ensuing social breakdown of these communities led the survivors to prostitution and begging along the highway. Even after the highway construction was abandoned in 1976, the effects of contact continued to kill Yanomami in peripheral areas. In 1977, for example, a measles epidemic killed half the population of four communities in the Upper Catrimani region (Ramos, 1979).

Another event that stimulated exploitation of the forest and, consequently, had a dramatic effect on the Yanomami was the publication of the results of an aerial survey of the Amazon, carried out in 1975 by the Radar Amazonia Project, which produced satellite photographs of the Amazon basin, indicating the location of potential mineral deposits. This, of course, brought immediate attention to the development potential of Amazonia and attracted mineral prospectors as well as large mining companies to the area. Since the discovery in 1975 of large deposits of cassiterite (tin ore) in the Serra de Surucucús region in the centre of Yanomami territory, illegal invasions by wildcat prospectors have remained a continuous threat (Gomes, 1986). Subsequently, uranium, gold, diamonds and titanium were also discovered on Yanomami lands. Invasions into the territory by *garimpeiros* began in earnest in 1976 and culminated in the gold rush of 1987.

GOLD RUSH YEARS, 1987–1990

The immediate impact of this gold rush was devastating to the four Yanomami communities at Paapiú, located at the centre of gold-mining activities. The low incidence of serious disease prior to the invasion (1984–1985) contrasts sharply with the statistics after 1987 (Figure 11.3). As many as 91 per cent of the inhabitants in some communities in the Paapiú region contracted malaria, with *Plasmodium falciparum* (the most lethal type of malaria parasite) predominating. In addition, serious clinical complications, such as

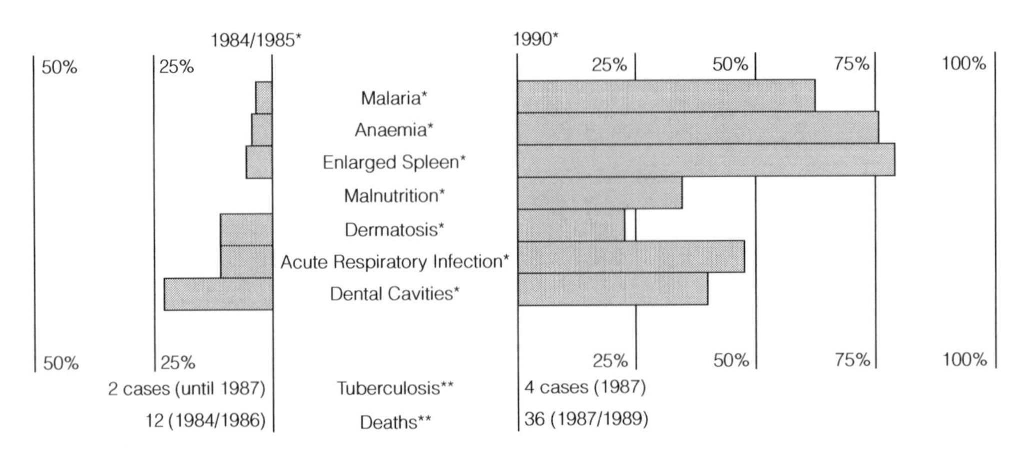

Figure 11.3 *The relative incidence of disease in four villages in the Paapiú region*

Source: * D. Levy-Bruhl (1985)
** Ação pela Cidadania (1990, p31)

malnutrition, ensued in 84 per cent of these cases. Even communities located far from mining centres, such as Demini in Amazonas State, had high rates of malaria infection – 40 per cent of those examined – probably a result of the custom of inter-village visiting among the Yanomami (CCPY, 1990).

The impact of gold-mining activities on the environment in the Amazon region is characterized by 'localized deforestation, biodiversity reduction, game depletion and displacement, mercury and other pollution, river and stream bank destruction and siltation, and decline and degradation of fisheries' (Sponsel, 1997, p103). By 1987, an estimated 45,000 miners (CCPY, 1990) were in Yanomami territory, extracting gold using hydraulic pumps and hoses and leaving huge water-filled craters in areas that they had deforested, often destroying indigenous garden sites in the process. Not only were these craters of standing water ideal breeding grounds for *Anopheles* mosquitoes, but such craters in the forest also disrupted Yanomami trails and communicative networks, isolating local communities from one another (Ramos, 1995). Fresh water for drinking, cooking and bathing disappeared as local streams and rivers were contaminated with mercury, polluted by human waste and mining run-off, and muddied to such an extent that fish could not survive.

Hundreds of aircraft flew daily into 110 clandestine airstrips to provide the miners with supplies and transportation (Ação pela Cidadania, 1990). Game animals that had not already been killed for meat by the miners were scared away by the constant noise of planes, helicopters, machinery and firearms, leaving villagers without a major source of dietary protein. Every Yanomami meal depends on hunting or fishing, produce harvested from carefully cultivated gardens, and seasonal wild fruits and nuts (cf. Chapter 4, this volume). Disruption of these normal subsistence activities not only causes malnutrition but also affects the social wellbeing of the villagers by diminishing their ability to care for and provide for one another. The deforestation, water and noise pollution, and scarcity of game that resulted during the gold rush were compounded by the breakdown of caregiving roles within communities. In addition to the high rate of infection by exogenous diseases, common secondary effects of malaria infection – anaemia and fatigue – prevented adults in many Yanomami communities from fulfilling their roles as caregivers and food providers. The activities of the gold prospectors located in or near indigenous communities disrupted every aspect of the daily lives of the Yanomami villagers, thus provoking the collapse of social caregiving along with traditional subsistence practices and putting malnourished villagers at even greater risk of disease and death (see Chapter 10, this volume, for similar effects of HIV/AIDS in southern Africa).

A comparison of the location of Yanomami villages in 1985 (Figure 11.4), and the location of clandestine airstrips used by the *garimpeiros* until 1990 (Figure 11.5) illustrates the

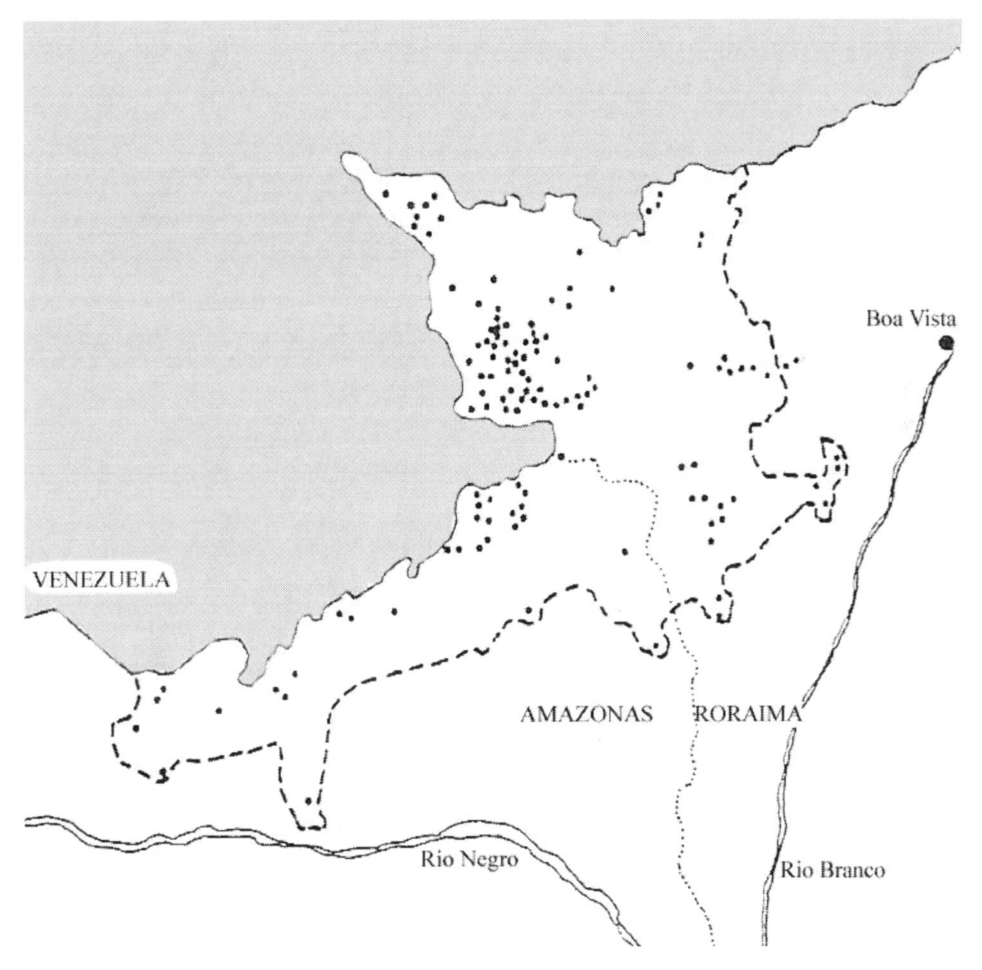

Figure 11.4 *Location of Yanomami villages, 1985*

Source: Ação pela Cidadania (1990, p16)

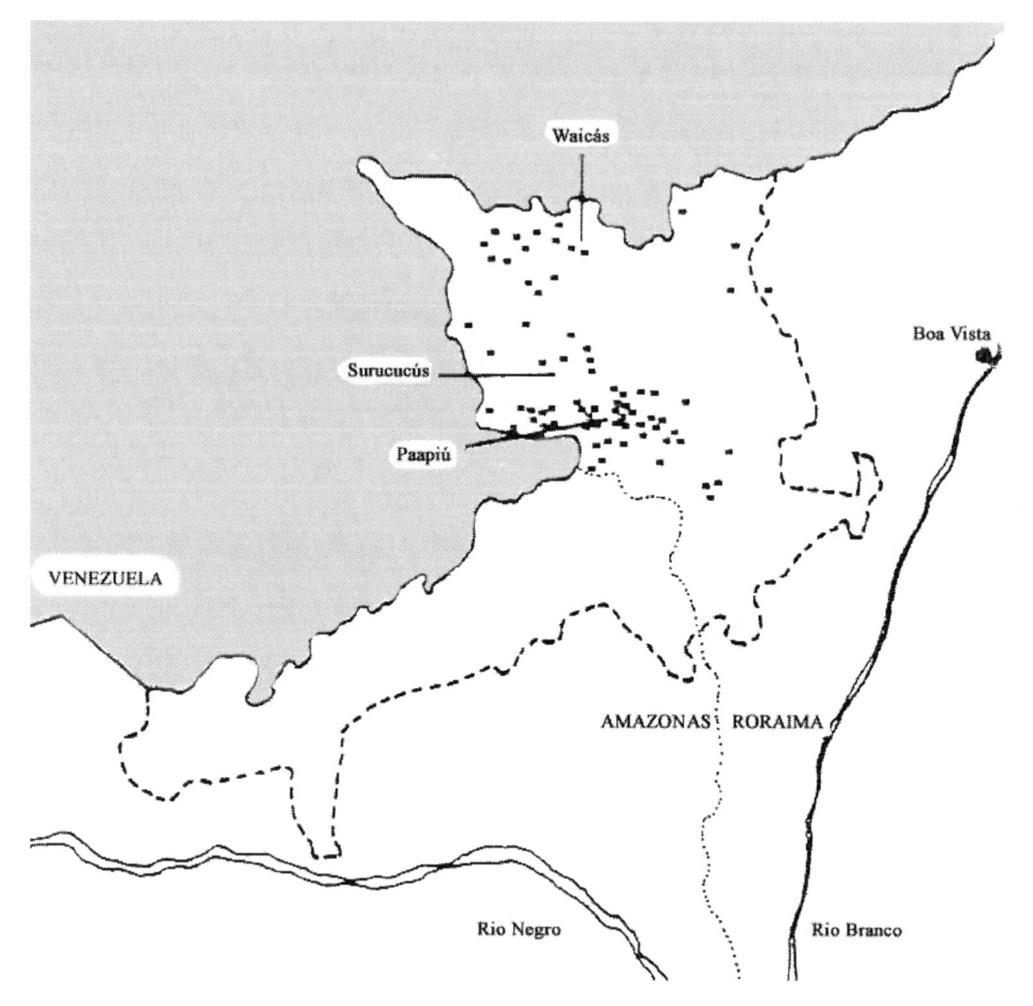

Figure 11.5 *Location of clandestine airstrips until February 1990*

Source: Ação pela Cidadania (1990, p16)

overlap between the points of aerial entry and supply for the gold miners and the indigenous villages. The central location of Paapiú (Figure 11.5) within an area of concentration of airstrips placed its inhabitants in immediate, uncontrolled contact with hundreds of men whose primary goal was to exploit the mineral riches of the forest at any cost, and whose ignorance and greed translated into aggression and violence towards the villagers.

The incidence of malaria among the Yanomami in Brazil quadrupled, with 80 to 90 per cent of the populations of some villages infected. The average mortality rate, based on confirmed deaths, for all Yanomami communities during 1987–1989 was 12.6 per cent, and 14.7 per cent for communities near mining centres such as Paapiú and Surucucús (CCPY, 1990). The actual figures were probably much higher, considering the Yanomami taboo on mentioning the name of a person who has died and the fact that entire villages disappeared as a result of abandonment because of the deaths of their inhabitants.

In 1990, Roraima had the highest number of positive cases of malaria (162.7 per 1000 inhabitants) of the nine states of the Amazon region (Table 11.1). The municipalities in Roraima that had the highest incidences of malaria in 1990 were Boa Vista and Mucajaí (Table 11.2).

At the time that these statistics were calculated, the following explanation was given by the Ministry of Health through documents prepared by the National Health Foundation's Department of Public Health Campaigns in Roraima to account for the high malaria rate in Boa Vista, the state capital:

> *Being the referential center for the whole state, a large number of registered cases converge here. Nevertheless, when grouped according to 'origin of contagion', 48 per cent of the 15,002 cases of malaria for 1990 originated in other municipalities – with 7,290 originating in the municipality of Boa Vista. Of these, 30 per cent are cases registered in the Yanomami area, 17 per cent from rural areas, and 38.4 per cent from urban areas corresponding to 2805 cases whose origins have not been determined. But of these cases it is known that they are patients residing in the city who are involved in prospecting principally in the Yanomami indigenous area and in colonization projects, who return monthly or weekly to the city. The principal locations in the affected rural area are directly related to the entrance and exit from prospecting areas and to a lesser degree colonization projects.* (Ministério de Saúde et al, 1990, p11; English translation by the author)

Table 11.1 *Epidemiological data on malaria in the Amazon region, 1990*

State	*Population of malarial area*	*Slides examined*	*Positive slides*	*Percentage*			
				Falciparum	*ILP*	*IPA*	*IAES*
Acre	422,570	66,545	14,455	39.0	21.7	34.8	15.7
Ampá	269,633	36,773	10,677	40.9	29.0	39.0	13.6
Amazonas	2,146,451	155,197	28,479	26.0	18.4	18.3	7.2
Maranhão	5,353,096	408,264	34,355	46.6	8.6	6.5	7.6
Mato Grosso	2,175,464	351,092	143,853	53.1	41.0	66.1	16.1
Pará	5,369,936	561,623	109,736	51.8	19.5	29.4	10.5
Rondônia	2,188,282	629,611	174,330	36.8	27.7	79.7	29.8
Roraima	**153,245**	**82,473**	**24,937**	**42.2**	**30.2**	**162.7**	**53.8**
Tocantins	1,149,291	94,684	4673	44.1	4.9	4.1	8.2
UB – Total	19,227,968	2,386,262	546,095	44.6	22.9	28.4	12.4

Source: Ministério de Saúde et al, 1990

Note: ILP indicates the percentage of positive cases per slides examined; IPA indicates the number of positive cases per 1000 inhabitants; and IAES indicates the number of slides examined per 1000 inhabitants. Falciparum indicates the percentage of positive cases diagnosed as Plasmodium falciparum.

Table 11.2 *Annual epidemiological evaluation by municipality in Roraima, 1990*

Municipalities	Population	Slides registered in the municipality			Slides distributed by origin		
		Positive	Falcip	Vivax	Positive	Falcip	Vivax
Alto Alegre	12,280	2034	643	1391	2642	789	1853
Boa Vista	190,009	15,002	6926	8076	7290	2964	4326
Mucajaí	**16,379**	**3956**	**1738**	**2218**	**9638**	**4767**	**4871**
Bomfim	15,522	1443	591	852	2540	1006	1534
Caracaraí	12,343	1283	494	790	1417	546	871
Normandia	10,607	361	189	172	568	306	262
São Luis	15,091	751	140	611	751	119	632
S. João da Baliza	13,511	183	62	121	168	48	120
Total	285,742	25,014	10,783	14,231	25,014	10,545	14.469

Source: Ministério de Saúde et al, 1990

The fact that Roraima had the highest number of positive cases of malaria of the entire Brazilian Amazon was clearly linked at the time by state health documents to gold prospecting in the Yanomami area. The high incidence of malaria in the region of Mucajaí was explained as follows:

> *The area of this municipality is considered* the main point of entrance and exit of the Yanomami prospecting area *(the Apiaú and Mucajaí rivers and several landing strips in the rural area of the municipality)* with a large flow of people involved in this activity, *including the very same farmers of the colonization projects who are diverted from their original functions. These same people return to their lodgings carrying malaria (with or without symptoms) adding to the critical mass of infected vectors, located in regions favorable to the growth and spread of anopheles mosquitoes,* where deforestation is intense, transforming it into an area of high malaria transmission. (Ministério de Saúde et al, 1990, p12; English translation and emphasis added by the author)

Because of the international outcry over this grave situation, an emergency health plan for the Yanomami was finally instituted by the Brazilian government in January 1990. Intense domestic and international pressure led Brazil's newly elected president, Fernando Collor de Mello, to order the expulsion of *garimpeiros* from Yanomami lands by federal police and the dynamiting of clandestine airstrips in 'Operation Free Jungle' at a substantial expense to the federal government. Subsequently, anthropologists, missionaries and medical teams, who had been expelled by the government from

Yanomami territory when the gold rush escalated in 1987, were allowed to return to work in the area. They found the Yanomami people, who had been totally deprived of health care during the intervening three-year period, in a dire state. Brazilian anthropologist Alcida Rita Ramos (1995) condemned the military and local and state politicians for their roles in the effort to exploit Yanomami lands without regard for the health and survival of the indigenous inhabitants.

ESTABLISHMENT OF YANOMAMI SANITARY DISTRICT

From the initial emergency health plans, a permanent Yanomami Sanitary District was created in 1991 by FUNASA to reduce the high mortality rate. By 1995 there were 26 health posts within the Yanomami Sanitary District (Figure 11.6), located within five subregions designated by FUNASA in Roraima. These health posts operated under the auspices of seven governmental, non-governmental, and religious organizations. Under this new system, however, the health coverage for the Yanomami was not adequate, comprehensive or consistent throughout the area. Moreover, FUNASA, which coordinated these activities, was plagued by frequent changes of personnel and inadequate funding.

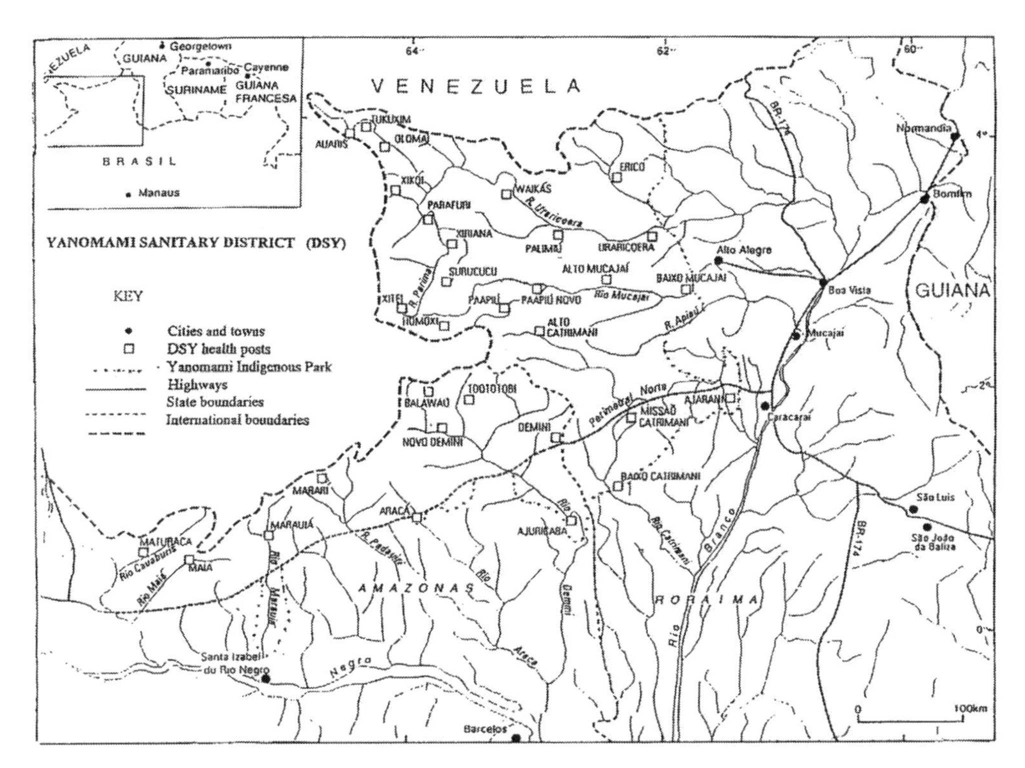

Figure 11.6 *Health posts in the Yanomami Sanitary District, 1994–1995*

Source: FNS-RR (1995)

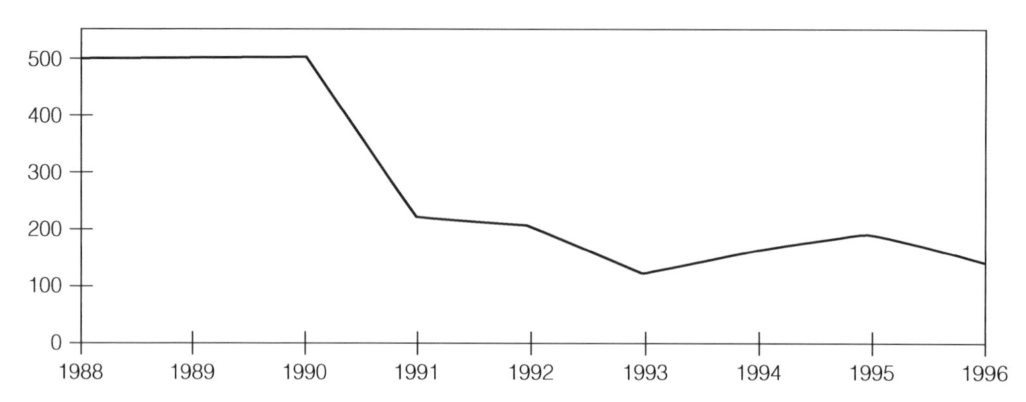

Figure 11.7 *Variation in absolute mortality in Yanomami Indigenous Area, 1988–1996*

Source: DSY/FNS-RR (1997)

The World Bank, which funded a project for malaria control that provided substantial funding for the district until 1996, was reported in a newsletter of the Pro-Yanomami Commission (CCPY, originally the Commission for the Creation of the Yanomami Park) 'to disagree with the federal government's intention to decentralize the activities' of FUNASA, stating that such a move would place the responsibility for Yanomami health on 'local authorities [who] have neither the funds nor the trained staff' to take over its work (CCPY, 1997, pp2–3).

Since 1991, statistics have been maintained, and health care for the Yanomami has been coordinated by the FUNASA for the district. Rates for absolute mortality for the Yanomami Indigenous Area from 1988 to 1996 (Figure 11.7) indicate a dramatic drop

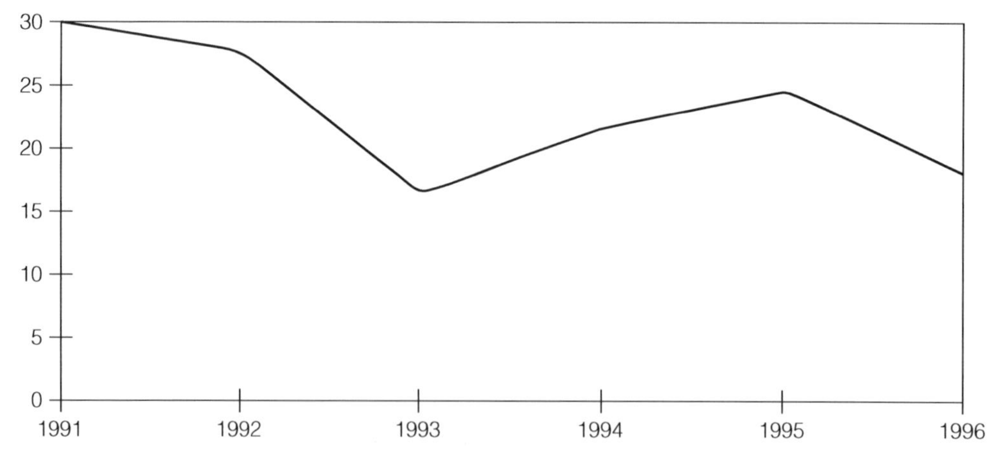

Figure 11.8 *Variation in the coefficient of general mortality, Yanomami Indigenous Area, 1991–1996*

Source: DSY/FNS-RR (1997)

after the arrival of medical teams in 1990. The high mortality rate prior to 1990, during the period of the gold rush, is projected from estimates made by the teams of the Emergency Health Assistance Program based on the disappearance of villages and the deaths encountered in 1990. The estimated number of deaths during the three years 1987–1989 was set at 1500. Because of the lack of an accurate and adequate census of the total Yanomami population in Brazil and given the Yanomami taboo against naming the dead, this number could actually be much higher. The fluctuations in the general mortality rate among the Yanomami in Roraima from 1991 to 1996 (Figure 11.8) mirror actual events that occurred during this period.

YANOMAMI INDIGENOUS RESERVE: CREATION AND RE-INVASIONS, 1990–2000

On 15 November 1991, President Fernando Collor surprised both supporters and opponents of Yanomami land rights by announcing the administrative demarcation of a protected Yanomami territory (CCPY, 1991). Less than six months later, the physical demarcation of 96,000km^2 of Amazon rainforest (an area the size of Washington State in the USA) was completed. Despite strong objections from the military and from mining interests, President Collor signed the decree ratifying the demarcation and legally created the Yanomami Indigenous Reserve on 25 May 1992 (CCPY, 1992). The establishment of this protected reserve on traditional Yanomami lands was the successful culmination of an international campaign begun in the 1970s by founding members of CCPY.

No sooner had the indigenous reserve been established than gold prospectors began re-invading Yanomami territory. By early 1993, an estimated 11,000 *garimpeiros* were once again working illegally within the reserve, threatening the lives of the Yanomami and destroying the environment (Brooke, 1993; CCPY, 1993a). A second 'Operation Free Jungle' was undertaken by the government to remove the gold prospectors (CCPY, 1993a). In July of that same year, 16 Yanomami (including women and children) of the community of Haximu (just across the border in Venezuela) were killed and mutilated in two attacks by illegal Brazilian gold miners (CCPY, 1993b). The massacre threatened to become an international incident when Brazil and Venezuela gave conflicting accounts as to the actual location and, hence, the nationality of the Yanomami victims (CCPY, 1993c).

These events are reflected in variations in the general mortality rate for the Yanomami in Roraima (Figure 11.8). There was a dramatic drop in general mortality after the demarcation of the Yanomami reserve in 1992. A progressive rise in the rate from 1993 until 1995 coincided with the re-invasion of the area by gold miners and the Haximu massacre, as well as difficulties experienced by the Yanomami Sanitary District in providing continued health care to the area because of the administrative and institutional problems of FUNASA, such as a lack of funds and political interference in the indigenous health services (Dias Magalhães et al, 1997). In a 1994 report, two CCPY doctors, Claudio Esteves de Oliveira and Deise Alves Francisco, stated, 'The most important occurrence during this period was the alarming resurgence of malaria. The re-emergence of this disease, since the beginning of the year, was due to the presence of gold miners in Venezuelan territory, bordering Brazil' (Esteves de Oliveira and Alves Francisco, 1994, p5). Half of the total Yanomami population in Brazil suffered from malaria in 1995; this is 15 times the benchmark (30 cases per 1000 inhabitants) used by the WHO to indicate a grave public health situation (DSY/FNS-RR, 1995).

Another epidemic disease threatening the Yanomami is tuberculosis (see Chapter 5, this volume). Its spread was facilitated in the 1990s by an absence of acquired immunity (specific to the Yanomami population) to *Mycobacterium tuberculosis* in combination with a compromised immunological state that accompanied epidemic diseases, especially malaria, and the invasion of their territory by carriers of the disease (gold prospectors and miners). The incidence of tuberculosis among the Brazilian Yanomami in 1995 was 600 per 100,000 inhabitants, compared with 56 per 100,000 for Brazil as a whole in 1992 and 20 per 100,000 for developed countries (DSY/FNS-RR, 1995). The report of a field study on tuberculosis among the Brazilian Yanomami, led by Dr Alexandra O. de Sousa of the Howard Hughes Medical Institute at Albert Einstein College of Medicine in New York, concluded:

> *Discovery of gold in the 1980s attracted 30,000 to 40,000 miners to the center of the indigenous territory with profound consequences for the Yanomami population. During the past three decades of contact with the outsiders, tuberculosis was introduced in the Yanomami population, apparently for the first time. The first reported case among the Indians was dated to 1965, and a small number of cases were noted in the 1970s. By the 1980s the infection spread throughout the area and has become epidemic. According to field reports, the epidemic is characterized by severe disease and high mortality rates.* (Sousa et al, 1997)

Sousa reiterated that the 'level of mortality due to TB is *very* high (100 times higher than in [New York City] during the worst moment of the recent multi-drug-resistant epidemic in 1994)' (Sousa, personal communication, 26 March 1998). Blood tests showed that the Yanomami had higher concentrations of antibodies than Brazilians of European ancestry, but such antibodies are effective against microbes and toxins in the bloodstream, not 'a bacterium like *M. tuberculosis*, which is sequestered in the body's cells' (Fackelmann, 1998, p75). Thus, the Yanomami are less prepared to fight tuberculosis, not having a historically acquired immunity to the disease as do people of European ancestry, including many of the gold miners who introduced the bacterium (see Chapter 12, this volume).

YANOMAMI HEALTH UPDATE: ONGOING STRUGGLE

The gravity of the health situation of the Yanomami and the inability of FUNASA to effectively meet the needs of the Yanomami Sanitary District motivated major changes in indigenous health care in Brazil. Special indigenous sanitary districts were created to improve differentiated health care at the village level for indigenous communities throughout the country. The myriad difficulties involved in actually executing this mandate, however, led FUNASA to restructure its indigenous health system in 2000. This involved decentralizing medical assistance by subcontracting health care in indigenous villages through partnerships with indigenous organizations and NGOs and, in some cases, municipalities. CCPY medical teams already had six years' experience as permanent health-care providers at three health posts funded by FUNASA, so when the Brazilian Ministry of Health invited CCPY 'to extend its health care programs to other regions' (CCPY, 2000, p1), the creation of a new, independent NGO, URIHI-Yanomami Health, was a logical next step for the former medical wing of CCPY. In January 2000, URIHI assumed the responsibility of providing health care for 12 of the 37 regions of the Yanomami Indigenous Reserve (including over half the Brazilian Yanomami population) with a budget of several million dollars (R$6.5 million) for the first 15 months (CCPY, 2000, p1).

The improvement in the health situation of the Yanomami under URIHI was dramatic. A comparison of the general mortality and the infant mortality rates in 1998–1999 and in 2000–2002 among the URIHI-assisted Yanomami population (before and after URIHI assumed responsibility) show reductions of 67 per cent and 64 per cent, respectively. Likewise, from January 2000 until December 2002, there was a 98 per cent reduction in the incidence of malaria in the Yanomami communities that were receiving health care through URIHI (Coordenação Regional de Roraima et al, 2003). This was accomplished initially by making the fight against malaria a priority and by hiring and training medical teams specifically for work in remote village locations (see Chapters 15 and 16, this volume, for similar efforts in Indonesia). Other programmes were instituted by URIHI to combat TB, onchocerciasis (river blindness) and intestinal parasites, and to provide regular childhood immunization as part of a broader campaign focusing on the health needs of Yanomami children. Unfortunately, this effective, village-centred health programme lasted only a few years.

In early 2004, political and economic pressures prevailed when the Brazilian Ministry of Health detailed new guidelines that were intended to strengthen the administrative role of the federal government in indigenous health services by curtailing the partnerships established in the previous reform of 1999. This controversial move forced URIHI to end its partnership with FUNASA and suspend its medical services to the Yanomami, whose health situation has steadily deteriorated. The lack of medications and the precariousness of the services provided at the present time by FUNASA were denounced by Yanomami leaders thoughout 2005, a year for which recent statistics indicate an increase in the incidence of malaria by 164 per cent from the previous year (CCPY, 2006a). In a letter dated 25 April 2006, Yanomami leaders in Amazonas State decried the continued increase in malaria cases in their communities. Furthermore, in May 2006, medical care in the Yanomami Indigenous Reserve was paralysed by a general strike of the employees of the University of Brasilia, the institution contracted by FUNASA to provide health care to the Yanomami. The strike was in response to a delay in salary payments to the university employees. As a result, 70 health workers temporarily left Yanomami territory to protest against the neglect and indifference of FUNASA (CCPY, 2006b).

The absence of personnel at the health posts has wider implications for the health and welfare of the indigenous communities and for the protection of the forest. In May 2006, the

Pro-Yanomami Commission noted that a lack of outside supervision (i.e., at health posts) encouraged the presence of gold prospectors and the intensification of their illegal activities within the Yanomami Reserve. The National Indian Foundation estimated that there were 700–800 gold prospectors in Roraima at that time, most within the protected Yanomami area, and the fact that, even when apprehended, few are punished by the local judicial system leads the *garimpeiros* to believe that they can continue their destructive activities with impunity (CCPY, 2006b). As the presence of these gold miners increases, the incidence of malaria and the general mortality among the Yanomami increase as well. When this situation is accompanied by a decrease or absence of medical services in these indigenous communities, the result is homicidal, as documented during the gold rush of 1987–1990.

The case of the Yanomami in Brazil over the past three decades illustrates the intimate connection between the health and survival of a remote forest people and the state of the standing forest, or conversely between their deteriorating health and forest disturbances. The outside exploitation of natural resources disrupts the traditional subsistence activities as well as the social fabric of indigenous communities not only by polluting the rivers, destroying the land and scaring off game animals, but also by the introduction of exogenous diseases, such as malaria, TB and respiratory infections. The history of European conquest bears witness to the genocidal effects of this deadly combination. Today, so-called economic development and the incessant demand for non-renewable resources continue to threaten the health of the remaining tropical forests and their indigenous inhabitants. Despite a new awareness among scientists of the value of indigenous cultures and traditional knowledge, the importance of highly biodiverse ecosystems is not appreciated by most members of the political and corporate establishment. Unless this attitude changes, the health of forests will be compromised and their inhabitants (human, animal and plant) remain at high risk of extinction.

ACKNOWLEDGEMENTS

I would like to thank Claudia Andujar, former coordinator of the Pro-Yanomami Commission (CCPY) in São Paulo, for her assistance in acquiring information concerning the CCPY Health Program and for her friendship and support over many years; and Dr Deise Alves Francisco, former coordinator of the CCPY Health Program and codirector of URIHI-Yanomami Health, for providing the statistical data for an earlier unpublished manuscript that provided the foundation for this chapter. My thanks also go to two anonymous reviewers for their helpful suggestions and to Dr Hein van der Voort for useful comments and critical proofreading. Finally, I would like to acknowledge the Rhode Island College Faculty Research and Faculty Development Funds for financial support during recent field trips to Brazil.

REFERENCES

Ação pela Cidadania (1990) *Yanomami: A Todos os Povos da Terra,* Commissão pela Criação do Parque Yanomami (CCPY)/Centro Ecumênico de Documentação e Informação (Cedi)/Conselho Indigenista Missionário (Cimi)/Núcleo de Direitros Indígenas (NDI), São Paulo, Brazil

Albert, B. (2006) 'Os Yanomami e sua terra', www.proyanomami.org.br/v0904/index.asp?pag=htm&url=http://www.proyanomami.org.br/base_ini.htm#top (accessed 22 April 2006)
Brooke, J. (1993) 'Brazil is evicting miners in Amazon', *New York Times*, 8 March, pA7
CCPY (1990) *URIHI: Report No. 10 (July)*, São Paulo, Brazil
CCPY (1991) *Update 49 (November)*, São Paulo, Brazil
CCPY (1992) *Update 57 (June)*, São Paulo, Brazil
CCPY (1993a) *Update 63 (January)*, São Paulo, Brazil
CCPY (1993b) *Update 71 (August)*, São Paulo, Brazil
CCPY (1993c) *Update 72 (September)*, São Paulo, Brazil
CCPY (1997) *Update 93/94 (May–July)*, São Paulo, Brazil
CCPY (2000) *Pro-Yanomami Boletim Online 01* (Janeiro), www.proyanomami.org.br/boletimMail/yanoBoletim/html/Bulletin_1.htm, accessed 28 May 2006
CCPY (2006a) *Pro-Yanomami Boletim Online 75* (Janeiro), www.proyanomami.org.br/boletimMail/yanoBoletim/html/Boletim_75.htm, accessed 28 May 2006
CCPY (2006b) *Pro-Yanomami Boletim Online 78* (Maio), www.proyanomami.org.br/boletimMail/yanoBoletim/html/Boletim_78.htm, accessed 28 May 2006
Committee for the Creation of the Yanomami Park (1979) 'Yanomami Indian Park: Proposal and justification', in A.R. Ramos and K.I. Taylor, *The Yanoama in Brazil 1979*, Anthropology Resource Center (ARC)/International Work Group for Indigenous Affairs (IWGIA)/Survival International (SI) Document 37, Copenhagen, Denmark, pp99–170
Coordenação Regional de Roraima, Distrito Sanitário Especial Indígena Yanomami, URIHI-Saúde Yanomami (2003) *Plano Distrital de Saúde*, Fundação Nacional de Saúde (FUNASA)/Ministério de Saúde (MS)/Departmento de Saúde Indígena, Boa Vista, Roraima, Brazil
Dias Magalhães, E., Batista de Azevedo, M., Soto Venegas, S., Gomes Cavalcanti, L., Camaroti, L. and Almeida Pereira, F. (1997) *Mortalidade na Área Indígena Yanomami no Período de 1991 a 1996*, Fundação Osvaldo Cruz (FIOCRUZ/CEPIRR), Boa Vista, Roraima, Brazil
DSY/FNS-RR (Distrito Sanitário Yanomami/Fundação Nacional de Saúde-Roraima) (1995) Saúde Yanomami – *1995*, internal document, Ministério de Saúde (MS)/DSY/FNS-RR, Boa Vista, Roraima, Brazil
DSY/FNS-RR (1997) internal document, DSY/FNS-RR, Ministério de Saúde, Boa Vista, Roraima, Brazil
Esteves de Oliveira, C. and Alves Francisco, D. (1994) *Report on the Health Activities in the Yanomami Area*, CCPY, São Paulo, Brazil
Fackelmann, K. (1998) 'Tuberculosis outbreak: An ancient killer strikes a new population', *Science News* 153: 73–75
FNS-RR (Fundação Nacional de Saúde-Roraima) (1991a) 'Gráfico demonstrativo dos casos de malária: Lâminas examinadas e positivas, Estado de Roraima, Período – 1962/90', internal document, Ministério de Saúde, Boa Vista, Roraima, Brazil
FNS-RR (1991b) 'Grafico demonstrativo populacional do Est. [Estado] de Roraima, Período - 1962/90', internal document, Ministério de Saúde, Boa Vista, Roraima, Brazil
FNS-RR (1995) internal document, Ministério de Saúde, Boa Vista, Roraima, Brazil
Gomes, S. (1986) 'Legal Project No. 379/85', *Commission for the Creation of the Yanomami Park (CCPY)*, URIHI Report No. 2, CCPY, São Paulo, Brazil, pp1–25
Levy-Bruhl, D. (1985) internal document, Medicos do Mundo/Comissão pela Criação do Parque Yanomami-CCPY, São Paulo, Brazil
Milliken, W. and Albert, B. with Goodwin Gómez, G. (1999) *Yanomami: A Forest People*, Royal Botanic Gardens, Kew, UK
Ministério de Saúde (MS)/Fundação Nacional de Saúde (FNS)/SUCAM-RR [Public Health Supervisory Office in Roraima] (1990) 'Reunião com os 30 municípios com maior registro de malária em 1990', internal document

Ramos, A.R. (1979) 'Yanoama Indians in northern Brazil threatened by highway', in A.R. Ramos and K.I. Taylor (eds) *The Yanoama in Brazil 1979,* Anthropology Resource Center (ARC)/International Work Group for Indigenous Affairs (IWGIA)/Survival International (SI) Document 37, Copenhagen, Denmark, pp1–42

Ramos, A.R. (1995) 'O papel político das epidemias o caso Yanomami', in M. Bartolomé (ed) *Ya no hay lugar para cazadores: procesos de extinción y transfiguración en America,* ABYA-YALA, Quito, Ecuador, pp55–89

Sousa, A., J. Salem, Lee, F., Verçosa, M., Cruaud, P., Bloom, B., Lagrange, P. and David, H. (1997) 'An epidemic of tuberculosis with a high rate of tuberculin anergy among a population previously unexposed to tuberculosis, the Yanomami Indians of the Brazilian Amazon', *Proceedings of the National Academy of Sciences,* USA, 94: pp13227–13232

Sponsel, L. (1997) 'The master thief: Gold mining and mercury contamination in the Amazon', in B.R. Johnston (ed) *Life and Death Matters: Human Rights and the Environment at the End of the Millennium,* Altamira Press, Walnut Creek, CA, pp99–127

Taylor, K.I. (1979) 'Development against the Yanoama: The case of mining and agriculture', in A.R. Ramos and K.I. Taylor (eds) *The Yanoama in Brazil 1979,* Anthropology Resource Center (ARC)/International Work Group for Indigenous Affairs (IWGIA)/Survival International (SI) Document 37, Copenhagen, Denmark, pp43–98

12

Biodiversity, Environment and Health among Rainforest-Dwellers: An Evolutionary Perspective

Alain Froment

Rainforest ecosystems are the icon and the symbol of concern about biodiversity and planet 'health', and indeed, they are critically important in global ecology. The total surface covered by rainforests is 11 million square kilometres, the equivalent of South America (or 8 per cent of the global land mass), and 200 million inhabitants (or 3.3 per cent of the world's population) live in such areas. However, only 12 million people, or 0.2 per cent of humanity, directly live from tropical forest products, a very low proportion compared with, for instance, the world's savannah areas. In some human societies, forests are considered dangerous places full of evil. The Hindi word *jangal* ('jungle'), for instance, means 'uninhabited world'. In Latin, the words for 'forest' (*silva*) and 'savage' have the same root, and *foris*, which gave us 'forest', means 'outside, remote' (and *forasticus*, 'wild'). Referring to Africa, Zerner (2003) invented the concept of 'viral forests', full of new threats. Yet paradise is often imagined as a luxuriant, moist and warm landscape, and some great civilizations, like the Maya in Mexico, the Khmer in Cambodia and the people of Borobudur in Java, built their achievements in rainforests (Diamond, 2005). The tropical forest environment thus carries an ambivalent image, and this chapter considers the health issues related to life in such an environment.

It can be hypothesized that medical assessment allows us to measure the adequacy of fit between a society and its environment: the better adapted are the inhabitants, the healthier they will be. In the following discussion we consider some epidemiological problems related to nutritional status, infectious diseases and chronic diseases. A new conceptual approach is needed to encompass such a wide range of pathologies; for doing so, Grmek (1969) proposed the concept of pathocenosis (cf. biocenosis, the universe of pathogens) to refer to the complex interdependence among all the diseases observed within a certain population at a certain time, It is defined by three propositions:

1 it is a *system*, with specific structural properties;
2 the distribution of each disease is influenced by *all* the others; and
3 pathocenosis tends towards *equilibrium* in a stable ecological situation.

Some of the consequences discussed in this chapter are based on the idea that the immune system response directed against one pathogen may affect the expression of others (for instance, malaria and intestinal worms). At this point, truly testing the hypothesis would require an exhaustive counting of all diseases met in moist or dry tropical environments. Such comprehensive data are not yet available, but this is a fruitful research direction for those of us concerned about public health.

The ultimate goal of this chapter is to briefly review the health problems of rainforest-dwellers, with a focus on some biological adaptations as they relate to the constraints of the milieu (nutrition and disease) and the pressures of new social needs (sedentarization) (see Chapter 13, this volume).

HIGH LEVELS OF BIODIVERSITY, BOTH 'GOOD' AND 'BAD'

A major characteristic of tropical rainforests is their phenomenal biodiversity. They display an enormous vegetal biomass and about one ton of mammals per square kilometre (Thackeray, 1995). However, it would be a mistake to believe that such diversity represents a pristine wilderness. Recent archaeological investigations have demonstrated a long history of human occupation and management in tropical forests, dating back several millennia (Mercader, 2002), which have radically modified forested landscapes. Tropical forests must realistically be considered anthropo-systems, if not gardens (Balée, 1989), where encouraged species, like red *Pandanus* in New Guinea or oil palm (*Elaeis guineensis*) in Central Africa, are markers of ancient human occupation.

Despite this impressive biodiversity, wild edible foods may be limited for humans. Bailey et al (1989) even postulated that no human settlement could survive in equatorial forests without cultivating crops, which means that Pygmy hunters could not live independently of farmers. Though this view is exaggerated (Bahuchet et al, 1991), forest peoples may experience some seasonal hunger: that is, although their caloric requirements are covered, a lack of favourite foods (such as meat or fish) induces observable psychological stress (Pagezy, 1982). Close to the equator, more than 50 per cent of the total caloric intake is derived from wild game (Cordain et al, 2000), and the threats to wildlife by logging activities and deforestation are seriously jeopardizing this resource. Nevertheless, if biodiversity can be celebrated as beneficial for human health (Chivian, 2002), one neglected but important aspect can be very harmful: microbial diversity.

The inverse correlation between biodiversity and latitude is known as Rapoport's rule. This debated rule has recently been shown to apply to microbes: there are up to 40 times more microbial species in the tropics than in cold areas (Guernier et al, 2004) because humid and warm environments clearly act as 'incubators' for germs. Viruses and bacteria are difficult to count exhaustively, especially when the diseases they induce result in non-specific symptoms, like fever or headache and body ache; they are then often labelled as flu or malaria. Conversely, parasites, in part because of their bigger size, are easier to identify; among humans there are about 270 parasite species, 16 per cent of which are strictly dependent on humans. Study of these latter parasites allows assessment of the relationships between a contaminated environment (through soil, food, insects or other means of transmission) and people. It has long been known that pathogenic diversity is related to the physical characteristics of the environment. Dunn (1977) noted that inhabitants of desert areas had a small number of parasites – only one, for instance, in the desert of Central Australia, and three among the Kalahari San ('bushmen') of southern Africa – but in humid rainforests the picture reverses: 20 species among the Pygmies of Central Africa, and 22 among the Semang of Malaysia.

In Cameroon, a country stretching southward from the Saharan landscapes around Lake Chad to the deep forest close to the equator, Ratard et al (1991) demonstrated a clear correlation between climatic factors and intestinal parasites. For roundworm (*Ascaris lumbricoides*), the percentage of infected children varied from 1 per cent in the dry north to 84 per cent in the humid south; and for whipworm (*Trichuris trichiura*), a good marker of faecal hygiene, the figures varied from 2 per cent to 98 per cent (Figure 12.1). In a 2001 survey we conducted in the Ntem valley, which borders Equatorial Guinea, 2° north of the equator, 14 per cent of the whole population and 28 per cent of the children under five suffered from diarrhoea at the time of our visit. It is important to remember that diarrhoeal diseases are the first cause of malnutrition and death in tropical countries.

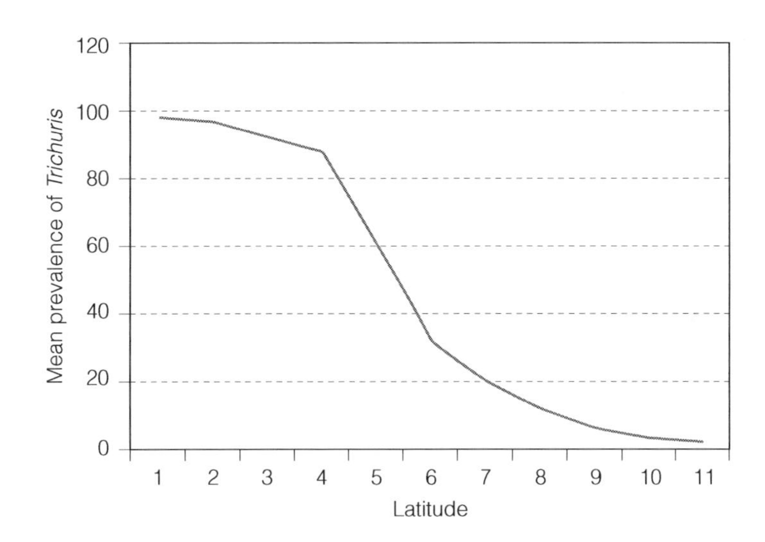

Figure 12.1 *The frequency of whipworm* (Trichuris) *infection with latitude, between the southern (2°N) and northern (12°N) borders of Cameroon*

WAYS OF LIFE AND POPULATION DENSITY

Mobility is the main characteristic of the hunting-gathering way of life. In forests, even farmers move their fields every two to three years because low inherent soil fertility leads to swidden (rotating) as opposed permanent agriculture. It is rare that hunters' demographic density goes beyond one person per square kilometre. Partly because carrying children is such a burden, family size is usually small, and long birth spacing is obtained by prolonged breastfeeding and low body fat in women (Pennington, 2001; also see Table 12.1). Among African agriculturalists the mean population density is around eight inhabitants per square kilometre. Hunters typically live in small bands of 10 to 40 persons covering a territory of about 100km^2, while the average density among swidden farmers is 230 to 330 persons for a territory of 170 to 300km^2 (Bahuchet, 2000).

According to the WHO, the average lifespan in tropical countries is around 47 years; a majority of people in the world live in urbanized areas where access to hospitals is easier and the environment healthier than in remote places. Our own unpublished surveys, conducted during the impact studies of the Chad–Cameroon pipeline between 1999 and 2004, showed that the lifespan in rural zones was astonishingly low, and even shorter in dry areas than in forested areas: the mean age at death was 19 years in northern Cameroon, 33 in the south. The low life expectancy in the north may be related to the lesser availability of health care and also the dependence of food supplies on the low rainfall regime of the Sahelian areas. Father Dhelemmes, a missionary who lived among Cameroon's Pygmies between 1950 and 1980 and conducted a very detailed (unpublished) census, concluded that the lifespan of the Baka Pygmies was around 23 years. Here also, the people have less access to medical facilities than do villagers, as illustrated by the disease 'yaws'. This infection, caused by treponemal bacteria and akin to venereal syphilis but transmitted by direct contact with infected skin or by flies, was widespread in all rainforests of the world a few decades ago. Because it is very sensitive to a single dose of penicillin, it was eradicated from villages during colonial times, but it is still frequent among Pygmies. However, in a context of scarce natural resources, the theory of evolution suggests that a shorter lifespan may maintain a good balance between population and food. Even when food is abundant, life can be dramatically shortened by the heavy burden of diseases in moist areas.

Mobility and small group size have advantages for preventing epidemics. For instance, nomadic societies avoid accumulating waste by moving their camps when the pressure of ecto-parasites increases, preventing related faecal pollution. Conversely, old villages built on the accumulation of trash harbour rodents, which are vectors of many infectious diseases, from plague to hantaviruses. Also, small, isolated human groups suffer no crowding diseases, like measles or smallpox, because these diseases induce a lifelong immunity to the survivors; the virus then requires a large number of individuals – perhaps half a million for measles – to complete its cycle among naïve subjects.

Knapen (1998) notices that the practice of silent barter, used by Borneo's Punan hunters to trade with their neighbouring farmers, allowed the circulation of goods while avoiding direct contacts. The opening of roads, however, provided an excellent opportunity for microbes or parasites like *Trypanosoma*, carried by the tsetse fly, to spread. Small bands of hunters are less a target for biting arthropods than crowded villages; moreover, the most dangerous disease, malaria, is transmitted by a mosquito that is rare in dense forest, except where clearings have been created for agriculture. All the above factors made the traditional hunting and gathering way of life healthier than others. But nowadays, it is almost impossible to meet such traditional people; most have been sedentarized either for political reasons, forced by the local governments that seek to control them better, or by volunteer settlement along the roads, to have better access to trade, schools and hospitals. Box 12.1 describes a dangerous aspect of hunting practices in Central Africa today.

BOX 12.1 MALEVOLENT WIND OR SORCERER'S REVENGE? EBOLA HAEMORRHAGIC FEVER, HUMANS AND WILDLIFE IN CENTRAL AFRICA

Sally A. Lahm

Among the most recognized and exotic medical mysteries, Ebola haemorrhagic fever and the closely related Marburg filovirus evoke fearful images of certain and horrific death in impenetrable tropical jungles. Since the scientific discovery of Ebola in the Democratic Republic of Congo, formerly Zaïre, in 1976, four distinct subtypes of the virus have been described: Zaïre, Sudan, Reston and Côte d'Ivoire. Epidemics of Ebola Sudan have occurred in forest-savannah mosaic habitats in southern Sudan and northwestern Uganda; the other three subtypes appear to be exclusively tropical forest viruses of Africa and Asia (Peters et al, 1996). No wild animals were known to be associated with any of the four Ebola outbreaks in Africa prior to November 1994, when a new virus, Ebola Côte d'Ivoire, was discovered in a wild chimpanzee community in Taï National Park (Le Guenno et al, 1995). A concurrent outbreak occurred in humans in several gold-panning communities of Gabon, Central Africa, during which some residents reported finding dead gorillas, although no tissue samples from the animals were obtained for analyses. This epidemic was originally attributed to yellow fever (Georges-Courbot et al, 1997). A second epidemic in the same region of Gabon 14 months later further confirmed the vulnerability of apes to Ebola infection, since the primary human victims had contracted the virus from an infected chimpanzee carcass they found and butchered for consumption. A third epidemic in a logging camp 250km west of this site in July 1996 was not officially connected with animal deaths, although an Ebola-infected chimpanzee carcass was discovered 85km southwest of the camp two months later (Georges et al, 1999).

The association of large numbers of animals, particularly apes, with Ebola was documented during a series of sequential outbreaks in the remote Mékambo region of northeastern Gabon and in adjacent northwestern Congo Republic in 2001–2002. Mammal species implicated included chimpanzees, gorillas, antelopes, monkeys and porcupines, although infection was confirmed only in apes and one species of duiker antelope. Most of the primary human cases in these outbreaks were linked to physical contact with animal carcasses (Rouquet et al, 2005). A further five outbreaks in Congolese villages followed, the most recent in April 2005. It is now believed that many gorilla and chimpanzee populations in Gabon and Congo Republic suffered high mortality from Ebola (Walsh et al, 2003). This has been best documented in Congo Republic: 95 per cent of habituated gorillas died or disappeared and were presumed dead from Ebola infection in the Lossi Sanctuary (Leroy et al, 2004), and an estimated 97 and 77 per cent of a total of 377 group-living and solitary gorillas, respectively, suffered a similar fate at a research site in Odzala-Kokoua National Park (Caillaud et al, 2006).

The propagation methods of Ebola remain unknown, although three species of bats were recently identified as potential reservoir hosts (Leroy et al, 2005; see Chapter 8, this volume). The ranges of the three bat species encompass humid forest areas of Central and coastal West Africa, so they may be important links in the transmission chain of Ebola Côte d'Ivoire as well. Moreover, results from long-term research in Gabon and Congo Republic suggest that there have been numerous additional multispecies outbreaks affecting only wildlife, as well as recurring human and wild animal exposure to Ebola, in the known epidemic zone and in parts of these countries where it has not been recorded (Lahm et al, 2007).

The question remains whether Ebola is an ancient virus of humid tropical forests or a recently emerged pathogen expanding its range because of more favourable conditions (Walsh et al, 2005). Bantu villagers and Pygmies in epidemic zones could not recall disease outbreaks of this virulent nature, but asymptomatic and mild cases of Ebola are known to occur (Leroy et al, 2000), many tropical diseases have similar symptoms, and elders have recalled villages and people abandoned because of contagious illness, flight from sorcery and fear of punishment by ancestors or spirits. In fact, many victims and their close contacts have attributed Ebola epidemics in Gabon to retaliation of forest spirits, sorcerers or powerful individuals for improper behaviour, such as a malevolent, disease-infected wind sent by forest guardian spirits, and in a more modern context, the deliberate contamination of wildlife with Ebola virus during experimental medical field procedures (Lahm et al, 2007).

Scavenging animal carcasses, an ancient foraging strategy of humans and their ancestors (Lewis, 1997), is still widely practised in remote areas of Central Africa (Bahuchet, 1985; Lahm, 1993). Scavengers are at great risk of contracting Ebola and transmitting it to their families and communities, as are hunters who may inadvertently kill infected animals. Extensive and prolonged public education campaigns are urgently needed to discourage scavenging and promote safe hunting and trapping techniques to procure wild meat. Gabonese utilize and consume a wide range of animals, from birds, reptiles and rodents to primates and elephants, including many species confirmed or suspected to have died from Ebola (Lahm, 1993; Lahm et al, 2007). By collaborating with rural community residents in the continuous monitoring of wildlife mortality (Rouquet et al, 2005) and applied field research, we will comprehend more fully the natural history, etiology and distribution of Ebola hemorrhagic fever and be better able to safeguard public health.

PHYSICAL FITNESS AND DIET

During the whole of human evolution, germs have certainly been the most dangerous predators of humankind. Though there have been both cultural and genetic adaptations to infection, as in the well-documented case of malaria (Etkin, 2003), some 1 million to 2 million deaths annually are due to this parasite, and infectious diseases in general still rank as the first cause of death in developing countries (see Chapter 9, this volume). According to WHO statistics for 2002 (www3.who.int/whosis/), of 156,000 daily deaths in the world, 41,000 are due to infections, compared with 45,000 due to cardiovascular causes and 20,000 to cancers. In Africa, transmissible diseases account for 64 per cent of overall mortality, compared with 11 per cent in Asia and 5 per cent in North America. The heavy mortality attributable to infections means that instead of the survival of the fittest, people who reach adulthood can be considered to be the luckiest – those least exposed to infectious disease. Thus the survivors are not necessarily gifted with an outstanding physical superiority or resistance.

Our closest cousins, the great apes, are adapted to forests, and it seems that the development of humankind was triggered by the conquest of more open landscapes. Toumai, the oldest known hominid, dating back 7 million years, was discovered in Chad in a dry environment (Brunet et al, 2005), and stable isotopes analyses tend to indicate that Australopithecines were eating savannah food (Sponheimer and Lee-Thorp, 1999).

The question is then whether humankind is physically adapted to life in the forest. The most noticeable morphological change of mammals living in rainforests is a reduction in size (cf antelopes, elephants), and the Pygmy morphotype, which is genetic, could be such an adaptation. But many forest populations, like the Bantus in Central Africa, display a statistically significant higher stature (166cm for Fang men from Cameroon, compared with 158cm for the neighbouring Bakola Pygmies and 144cm for the Mbuti of the Ituri forest; Froment, 1993), without showing any sign of biological maladaptation. It is possible that a short stature allows better mobility and hunting success in forests.

After hunter-gatherers survive infancy, they usually display excellent fitness and resistance to stress. Their adiposity (or fat) is lower than farmers', and as seen in Table 12.1, body mass index (BMI) is always lower among foragers than among agriculturalists, which proves that despite a convenient diet, the hunter's nutritional status is fragile, with low reserves. For unclear reasons, present-day foragers all have a shorter stature than nearby farmers (even though the farmer groups all descend from hunter ancestors). Although some genetic reasons for short stature have been demonstrated for African Pygmies (Froment, 2001), in other rainforest populations in very different ecological systems, short stature has also been observed (Holmes, 1993). Such differences may well be due to a combination of genetics and chronic malnutrition.

Table 12.1 *Body mass index of former hunter-gatherers versus farming neighbours*

Population (country)	*Men*	*Women*	*Source*
Africa			
Coastal Kola Pygmies (Cameroon)	20.2	19.7	Koppert et al (1993)
Mvae farmers (coastal Kola neighbours)	22.0	22.5	
Continental Kola Pygmies (Cameroon)	20.0	19.8	Kesteloot et al (1996)
Bulu and Ngumba farmers (continental Kola neighbours)	20.7	21.0	
Baka Pygmies (eastern Cameroon)	20.7	20.2	Froment (2006, unpublished survey)
Bangandou (Baka neighbours)	21.1	20.9	
Zime (Baka neighbours)	21.9	21.1	
Efe Pygmies (Democratic Republic of Congo)	20.2	20.2	Bailey et al (1993)
Lese farmers (Efe neighbours)	21.6	21.7	
Southeast Asia			
Tubu Punan (Borneo, remote)	20.6	19.9	Dounias et al (2004), Strickland and Duffield (1998)
Tubu Punan (Borneo, suburban)	19.9	19.6	
Iban farmers (Borneo)	20.9	22.2	

Forty years ago, Mann et al (1962) noticed that cardiovascular diseases, hypertension and diabetes were less frequent among Pygmies than among villagers. This situation may be related to the 'thrifty genotype' hypothesis (Neel et al, 1998), which posits that very long periods of food uncertainty modulated the ability of the human body to store reserves; this strategy was adaptive during the eons before humans took up agriculture and animal husbandry but turns out to be detrimental in an affluent context. Some nutritionists have celebrated the 'Palaeolithic' diet (Eaton and Eaton, 2000), rich in protein and in fibre, low in salt and sugar, and no milk, combined with a healthy way of life, with no smoking or drinking, and more exercise and less stress than nowadays. Kesteloot et al (1996) showed that the difference in hypertension figures between Cameroonian Baka and Bakola Pygmies, on the one hand, and Bantu farmers, on the other, no longer exists. Yamauchi et al (2000) also found that a small percentage of people (5 of 147) among the Baka can now be considered overweight (BMI>25). On the other hand, Benyshek and Watson (2006), in a worldwide comparison of 94 societies, are questioning the thrifty genotype hypothesis on the basis that regular and severe food shortages are no more frequent, at least nowadays, among foragers than among agriculturalists.

Although meat consumption is extremely high among Bakola Pygmies – 200g daily on average, 285g for a male adult (Koppert et al, 1993) – this iron- and protein-rich diet does not prevent them and other forest groups from suffering from anaemia (Table 12.2), compared with savannah people, who are almost vegetarians. Malaria, which is hyperendemic during the entire year in the forest, and the burden of intestinal parasites explain this difference and contribute to the growth impairment (stunting) observed among forest peoples: 45 per cent of south Cameroon's Bantu children suffer from growth retardation, defined as a height for age inferior to the third percentile of the reference curve; the figure for savannah children is 15 per cent. Pygmies also have a higher immunoglobulin rate, which means a higher exposure to infections (Table 12.2).

Table 12.2 *Blood levels of haemoglobin, albumin and immunoglobulins G (IgG or gamma-globulins) in Cameroonian adults, forest versus savannah*

Population	*Men*				*Women*			
	n	*Hb g%*	*Albumin g/l*	*IgG g/l*	*n*	*Hb g%*	*Albumin g/l*	*IgG g/l*
Mvae (forest)	54	12.0 ± 1.9	31.6 ± 4.7	28.4 ± 6.9	62	11.4 ± 1.6	32.0 ± 6.5	27.4 ± 7.2
Bakola (Pygmy) (forest)	29	12.4 ± 1.7	32.4 ± 5.1	34.9 ± 8.4	39	11.5 ± 1.5	31.5 ± 6.0	34.1 ± 7.6
Duupa (savannah)	36	14.5 ± 1.4	26.3 ± 4.4	37.5 ± 7.9	35	12.7 ± 1.7	25.1 ± 4.6	39.0 ± 8.7
Koma (savannah)	54	14.7 ± 1.3	25.1 ± 7.6	28.5 ± 8.6	18	13.1 ± 1.2	24.0 ± 3.7	31.8 ± 8.5

Source: Froment and Koppert, 1994

ECO-BIOCULTURAL TRANSITION: FUTURE RISKS AND BENEFITS

Before the introduction of agriculture, the characteristics of rainforest hunter-gatherers' past health and environment interactions can be summarized as follows (Froment, 2001):

- high biodiversity in the environment, but possibly limited starchy food;
- high biodiversity of microbial pathogens;
- high mobility, with light burdens and few children;
- low population density;
- good cardiovascular fitness and low adiposity;
- diet rich in proteins and a healthy way of life;
- possibly a 'thrifty genotype';
- no crowding diseases (measles, smallpox);
- low level of airborne and food-borne parasites;
- life hazards (hunting accidents, venomous animals);
- exposure to wild zoonoses;
- some domestic pollution (fireplaces in huts; see Chapter 5, this volume); and
- high overall mortality and short life expectancy.

Gage (2005) recently questioned the assumption that agriculture and modernization have been detrimental to human health, and showed not only that life expectancy has increased and transmissible diseases declined (see Chapter 2, this volume), but also that degenerative diseases are now declining in modern societies. Lawrence et al (1980) observed in Amazonia that the level of intestinal parasites was not lower in the newly contacted villages than in more acculturated ones. Most of the present-day foragers in Asian, African or South American tropical forests suffer a heavy load of parasites (Table 12.3).

What really differs between hunters and farmers is more a philosophy of life than technical differences. For instance, in Africa, the Pygmies and Bantus have shared a complementary economy for centuries, a history attested by the fact that in most cases the Pygmies adopted the language of their neighbours and often work in Bantu fields to earn some money or food, an economically rewarding strategy (Leclerc, 2001). But despite this proximity, the hunters did not turn to food production until recently. One of the reasons is that a foraging lifestyle allows for greater leisure, whereas agriculture – as made clear in Genesis, when God expels Adam and Eve from the Garden of Eden – demands more consistent hard labour. Sahlins (1972) clearly showed that hunting and gathering is less time-consuming – three to five hours of labour per day – and certainly more fun than cultivating. That is why foragers have resisted forced sedentarization, and now that such resistance has ended, foragers are tempted to replace the resources provided by the forest with those supplied by NGOs or governments (e.g., Amerindians in French Guyana, who all receive monthly pensions). What they appreciate most are two luxury goods brought by the Occidental model, alcohol and tobacco; unfortunately, these two addictions lead to many health hazards, one of the most preoccupying being tuberculosis.

Table 12.3 *Burden of parasites in sedentarized hunter-gatherers*

Population	*Hookworms* (Ankylostoma *and* Necator *spp)*	*Whipworms* (Trichuris *spp)*	*Roundworms* (Ascaris *spp)*	*Amoebas*	
				Pathogenic	*Non-pathogenic*
SOUTHEAST ASIANS					
Tubu Punan (Indonesia)	35	9	60	5	6
Semang (Malaysia)	93	56	12	9	30
Temiar (Malaysia)	78	23	2	3	18
Jahut (Malaysia)	52	29	20	8	28
Semai (Malaysia)	74	12	13	10	39
Jakun (Malaysia)	64	62	65	3	31
Semelai (Malaysia)	70	72	71	6	17
Temuan (Malaysia)	79	91	59	12	37
AFRICAN PYGMIES					
Mbuti (DRC)	85	70	57	36	–
Aka (CAR)	71	–		–	–
Kola (Cameroon)	–	85	51	–	–
Medjan (Cameroon)	–	83	90	–	–
AMERINDIANS					
Yanomami (Brazil)	59	80	86	49	85
Ticuna (Colombia)	83	77	76	69	55
Palikur (French Guyana)	90	19	76	31	16
Campa (Peru)	45	20	28	21	37
Xingu (Brazil)	81	–	18	61	87

– = no data.

Sources: Punan: Dounias and Froment, 2006; Semang, Temiar, Jahut, Semai, Jakun, Semelai and Temuan: Dunn, 1972; Mbuti: Mann et al, 1962; Aka: Pagezy, 1985; Kola and Medjan: Froment, 2001; Yanomami: Lawrence et al, 1980; Holmes, 1984; Ticuna: Restrepo, 1962; Palikur: Bruno, 1978; Campa: Eichenberger, 1966; Xingu: Baruzzi, 1970

Whatever their preferences, the traditional foraging way of life can no longer be sustained. An acute problem is related to the present degradation of forest ecosystems. One important question is: did deforestation create new opportunities for some diseases or vectors to thrive? The recent spread of frightening diseases like filoviruses (such as Ebola and Marburg), retroviruses – in particular, human immunodeficiency viruses (HIVs) and human T-lymphocyte viruses (HTLVs) – and arthropod-borne viruses (or arboviruses, such as Dengue fever) have created worldwide fear (see Chapters 8 and 10, this volume). However, many emergent diseases are observed in Africa (Ashford, 1991), despite the fact that major ecosystem attacks are worse in Indonesia and Amazonia. The reason is, in part, that Africa is the place where most other primates co-evolved with primitive humans, and many infectious diseases can be shared between the two groups (Wolfe et al, 2004). But it is only recent improvements in global transport that have allowed rare diseases restricted to remote areas to circulate at a worldwide level. For example, it is now thought that AIDS appeared nearly a century ago (Rambaut et al, 2001) but spread only 30 years ago, when the development of infrastructure and towns acted as a booster.

The second question relates to the shrinkage of forest resources for their inhabitants. A major concern is the disappearance of wild game, which supplies a fair amount of proteins. Some authors also draw attention to the disappearance of medicinal plants, though their efficacy is often still to be proved (see Chapters 3 and 10, this volume). In this epidemiological transition (Wirsing, 1985) for forest-dwellers, new pathologies, like alcoholism, violence and poverty, are emerging related to psychosocial distress. One consequence of sedentarization along the roads is the exposure to infectious diseases, among which measles is especially lethal for children. Sexually transmitted infections are also frequent, and unfortunately, significant numbers of cases of AIDS are now found among marginal societies all over the world (see boxes in Chapter 10, this volume). Conversely, better access to medical care explains the demographic expansion of hunter-gatherers, and the inescapable trend towards settlement and agriculture. Many hunter-gatherer families wish their children to go to school, even though they know it can impair the transmission of traditional knowledge. Facing such difficult decisions, rainforest-dwellers, who have demonstrated their ability to adapt, must be free to choose their own direction, and local governments must recognize their rights and work to secure their future, including full access to land and, perhaps even more important, freedom of choice.

REFERENCES

Ashford, R.W. (1991) 'The human parasite fauna: Towards an analysis and interpretation', *Annals of Tropical Medicine and Parasitology* 85: 189–198

Bahuchet, S. (1985) *Les pygmées Aka et la forêt centrafricaine ethnologie écologique*, Ethnosciences 1, SELAF, Paris

Bahuchet, S. (ed) (2000) *Les peuples des forêts tropicales aujourd'hui*, Rapport de Synthèse, Programme Avenir des Peuples des Forêts Tropicales, European Union, Folon s.a., Brussels, Belgium

Bahuchet, S., Mac Key, D. and de Garine, I. (1991) 'Wild yams revisited: Is independence from agriculture possible for rain-forest hunters-gatherers?' *Human Ecology*, 19: 213–243

Bailey, R.C., Head, G., Jenike, M., Owen, B., Rechtman, R. and Zechenter, E. (1989) 'Hunting and gathering in tropical rain forest: Is it possible?' *American Anthropologist* 91: 59–82

Bailey, R.C., Jenike, M.R., Ellison, P.T., Bentley, G.R., Harrigan, A.M. and Peacock, N.R. (1993) 'Seasonality of food production, nutritional status, ovarian function and fertility in Central Africa', in C.M. Hladik, A. Hladik, O.F. Linares, H. Pagezy, A. Semple and M. Hadley (eds) *Tropical Forests, People and Food: Biocultural Interactions and Applications to Development*, United

Nations Educational, Scientific and Cultural Organization (UNESCO), Paris, France, pp387–402

Balée, W. (1989) 'The culture of the Amazonian forest', *Advances in Economic Botany* 7: 1–21

Baruzzi, R.G. (1970) 'Contribution to the study of the toxoplasmosis epidemiology: Serologic survey among the Indians of the Upper Xingu river, Central Brazil', *Revue de l'Institut de Médecine Tropicale* 13: 356–362

Benyshek, D.C. and Watson, J.T. (2006) 'Exploring the thrifty genotype's food-shortage assumptions: A cross-cultural comparison of ethnographic accounts of food security among foraging and agricultural societies', *American Journal of Physical Anthropology* 131: 120–126

Brunet, M., Guy, F., Pilbeam, D., Lieberman, D.E., Likius, A., Mackaye, H.T., Ponce de Leon, M.S., Zollikofer, C.P. and Vignaud, P. (2005) 'New material of the earliest hominid from the Upper Miocene of Chad', *Nature* 434: 752–755

Bruno, A.A.G. (1978) 'Condições sanitárias de escolares em zonas rurais do território federal do Amapá', MSc dissertation, Universidade Estadua de Campinas, Campinas, Brazil

Caillaud, D., Levréro, F., Cristesçu, R., Gatti, S., Maeva, D., Douadi, M., Gautier-Hion, A., Raymond, M. and Menard, N. (2006) 'Gorilla susceptibility to Ebola virus: The cost of sociality', *Current Biology* 16(13): 489–491

Chivian, E. (ed) (2002) *Biodiversity: Its Importance to Human Health,* Center for Health and the Global Environment, Harvard Medical School, Boston, MA, available at www.med.harvard.edu/chge/resources.html

Cordain, L., Miller, J.B., Eaton, S.B., Mann, N., Holt, S.H. and Speth, J.D. (2000) 'Plant–animal subsistence ratios and macronutrient energy estimations in worldwide hunter-gatherer diets', *American Journal of Clinical Nutrition* 71: 682–692

Diamond, J. (2005) *Collapse: How Societies Choose to Fail or Succeed,* Viking Penguin Books, London

Dounias, E., Kishi, M., Selzner, A., Kurniawan, I. and Levang, P. (2004) 'No longer nomadic: Changing Punan Tubu lifestyle requires new health strategies', *Cultural Survival Quarterly* 28(2): 15–20, available at http://209.200.101.189/publications/csq/csq-article.cfm?id=1761

Dounias, E. and Froment, A. (2006) 'When forest-based hunter-gatherers become sedentary: Consequences for diet and health', *Unasylva* 57: 26–33

Dunn, F. (1972) 'Intestinal parasitism in Malayan aborigines', *Bulletin of the World Health Organization* 46: 99–113

Dunn, F.L. (1977) 'Health and disease in hunter-gatherers: Epidemiological factors', in D. Landy (ed) *Culture, Disease, and Healing,* Macmillan, New York, NY, pp99–113

Eaton, S.B. and Eaton, S.B. (2000) 'Paleolithic vs. modern diets: Selected pathophysiological implications', *European Journal of Nutrition* 39: 67–70

Eichenberger, R.W. (1966) 'Una filosofía de salud pública para las tribus indígenas amazónicas', *America Indigena* 26: 119–141

Etkin, N.L. (2003) 'The co-evolution of people, plants, and parasites: Biological and cultural adaptations to malaria', *Proceedings of the Nutrition Society* 62: 311–317

Froment, A. (1993) 'Adaptation biologique et variation dans l'espèce humaine: Le cas des Pygmées d'Afrique', *Bulletins et Mémoires de la Société d'Anthropologie de Paris* 5: 417–448

Froment, A. (2001) 'Evolutionary biology and health of hunter-gatherer populations', in C. Panter-Brick, R. Layton and P. Rowley-Conwy (eds) *Hunters-Gatherers: An Interdisciplinary Perspective,* Cambridge University Press, Cambridge, pp239–266

Froment, A. and Koppert, G. (1994) 'Comparative food practices in African forest and savanna populations and their biological consequences', in B. Thierry, J.R. Anderson, J.J. Roeder and N. Herrenschmidt (eds) *Current Primatology, Vol. I, Ecology and Evolution,* University of Louis Pasteur, Strasbourg, France, pp161–174

Gage, T.B. (2005) 'Are modern environments really bad for us? Revisiting the demographic and epidemiologic transitions', *American Journal of Physical Anthropology* 41: 96–117

Georges, A.J., Leroy, E.M., Renaut, A.A., Benissan, C.T., Nabias, R.J., Ngoc, M.T., Obiang, P.I., Lepage, J.P.M., Bertherat, E.J., Bénoni, D.D., Wickings, E.J., Amblard, J.P., Lansoud-Soukate, J.M., Milleliri, J.M., Baize, S. and Georges-Courbot, M.-C. (1999) 'Ebola haemorrhagic fever outbreaks in Gabon, 1994–1997: Epidemiologic and health control issues', *Journal of Infectious Diseases* 179: S65–75

Georges-Courbot, M.-C., Sanchez, A., Lu, C-Y., Baize, S., Leroy, E., Lansoud-Soukate, J., Tévi-Bénissan, Georges, A.J., Trappier, S.G., Zaki, S.R., Swanepoel, R., Leman, P.A., Rollin, P.E., Peters, C.J., Nichol, S.T. and Ksiazek, T.G. (1997) 'Isolation and phylogenetic characterization of Ebola viruses causing different outbreaks in Gabon', *Emerging Infectious Diseases* 3: 59–62

Grmek, M. (1969) 'Préliminaires d'une étude historique des maladies', *Annales ESC* 24: 1437–1483

Guernier, V., Hochberg, M.E. and Guégan J.F. (2004) 'Ecology drives the worldwide distribution of human infectious diseases', *PLoS Biology* 2: 740–746

Holmes. R. (1984) 'Non-dietary modifiers of nutritional status in tropical forest populations of Venezuela', *Interciencia* 9: 386–391

Holmes, R. (1993) 'Nutritional anthropometry of South American indigenes: Growth deficits in biocultural and development perspective', in C.M. Hladik, A. Hladik, O. Linares, H. Pagezy, A. Semple and M. Hadley (eds) *Tropical Forests: People and Food*, MAB Series Vol.13, Parthenon-UNESCO, London, pp349–356

Kesteloot, H., Ndam, N., Sasaki, S., Kowo, M. and Seghers, V. (1996) 'A survey of blood pressure distribution in Pygmy and Bantu populations in Cameroon', *Hypertension* 27: 108–113, available at http://hyper.ahajournals.org/cgi/content/full/27/1/108

Knapen, H. (1998) 'Lethal diseases in the history of Borneo mortality and the interplay between disease environment and human geography', in V.T. King (ed) *Environmental Challenges in South-East Asia*, Curzon Press, Richmond, Surrey, UK, pp69–94

Koppert, G., Dounias, E., Froment, A. and Pasquet, P. (1993) 'Food consumption in three forest populations of the southern coastal area of Cameroon', in C.M. Hladik, A. Hladik, O. Linares, H. Pagezy, A. Semple and M. Hadley (eds) *Tropical Forests: People and Food*, MAB Series Vol.13, Parthenon-UNESCO, London, pp295–311

Lahm, S.A. (1993) 'Ecology and economics of human/wildlife interaction in northeastern Gabon', PhD dissertation, New York University, NY

Lahm, S.A., Kombila, M., Swanepoel, R. and Barnes, R.F.W. (2007) 'Morbidity and mortality of wild animals in relation to Ebola haemorrhagic fever outbreaks in Gabon and Congo Republic, 1994–2003', *Transactions of the Royal Society of Tropical Medicine and Hygiene*, 101, 64–78

Lawrence, D.N., Neel, J.V., Abadie, S.H., Moore, L.L., Adams, L.J., Healy, G.R. and Kagan, I.G. (1980) 'Epidemiologic studies among Amerindian populations of Amazonia. III. Intestinal parasitoses in newly contacted and acculturating villages', *American Journal of Tropical Medicine and Hygiene* 29: 530–537

Leclerc, C. (2001) 'En bordure de route: Espace social, dynamisme et relation à l'environnement chez les Pygmées Baka du sud-est Cameroun', PhD thesis, Anthropology, University of Paris X-Nanterre, France

Le Guenno, B., Formenty, P., Wyers, M., Gounon, P., Walker, F. and Boesch, C. (1995) 'Isolation and partial characterization of a new strain of Ebola virus', *Lancet* 345: 1271–1274

Leroy, E.M., Baize, S., Volchkov, V.E., Fisher-Hoch, S.P., Georges-Courbot, M.C., Lansoud-Soukate, J., Capron, M., Debré, P., McCormick, J.B. and Georges, A.J. (2000) 'Human asymptomatic Ebola infection and strong inflammatory response', *Lancet* 355: 2210–2215

Leroy, E.M., Rouquet, P., Formenty, P., Souquière, S., Kilbourne, A., Froment, J.-M., Bermejo, M., Smit, S., Karesh, W., Swanepoel, R., Zaki, S.R. and Rollin, P.E. (2004) 'Multiple Ebola virus transmission events and rapid decline of Central African wildlife', *Science* 303: 387–390

Leroy, E.M., Kumulungui, B., Pourrut, X., Rouquet, P., Hassanin, A., Yaba, P., Delicat, A., Paweska, J., Gonzalez, J.-P. and Swanepoel, R. (2005) 'Fruit bats as reservoirs of Ebola virus', *Nature* 438: 575–576

Lewis, E. (1997) 'Carnivoran paleoguilds of Africa: Implications for hominid food procurement', *Journal of Human Evolution* 32: 257–288

Mann, G.V., Roels, A., Price, D.L. and Merrill, J.M. (1962) 'Cardiovascular disease in African Pygmies: A survey of the health status, serum, lipids, and diet of Pygmies in Congo', *Journal of Chronic Diseases* 15: 341–371

Mercader, J. (ed) (2002) *Under the Canopy: the Archaeology of Tropical Rain Forests*, Rutgers University Press, Rutgers, NJ

Neel, J.V., Weder, A.B. and Julius, S. (1998) 'Type II diabetes, essential hypertension, and obesity as "syndromes of impaired genetic homeostasis": The "thrifty genotype" hypothesis enters the 21st century', *Perspectives in Biology and Medicine* 42: 44–74

Pagezy, H. (1982) 'Seasonal hunger as experienced by the Oto and Twa of a Ntomba village in the equatorial forest (Lake Tumba, Zaire)', *Ecology of Food and Nutrition* 12: 139–153

Pagezy, H. (1985) *Etat de nutrition de la mère et de l'enfant de 0 à 4 ans en liaison avec les facteurs de l'environnement biologique et culturel*, contrat MRT N°82L1294, Ministère de l'Industrie et de la Recherche, Paris

Pennington, R. (2001) 'Hunter-gatherer demography', in C. Panter-Brick, R. Layton and P. Rowley-Conwy (eds), *Hunters-Gatherers: An Interdisciplinary Perspective*, Cambridge University Press, Cambridge, pp170–204

Peters, C.J., Sanchez, A., Rollin, P.E., Ksiazek, T.G. and Murphy, F.A. (1996) 'Filoviridae: Marburg and Ebola viruses', in B.N. Fields, D.M. Knipe and P.M. Howley (eds) *Fields Virology*, 3rd edn, Lippincott-Raven, Philadelphia, PA, pp1161–1176

Rambaut, A., Robertson, D.L., Pybus, O.G., Peeters, M. and Holmes, E.C. (2001) 'Human immunodeficiency virus: Phylogeny and the origin of HIV-1', *Nature* 410:1047–1048

Ratard, R.C., Kouemeni, L.E., Ekani Bessala, M.M., Ndamkou, C.N., Sama, M.T. and Cline, B.L. (1991) 'Ascariasis and trichuriasis in Cameroon', *Transactions of the Royal Society of Tropical Medicine and Hygiene* 85: 84–88

Restrepo, M. (1962) 'Estudio parasitológico de una región del Amazonas colombiano', *Antioquia Medica* 12: 462–484

Rouquet, P., Froment, J.-M., Bermejo, M., Kilbourne, A., Karesh, W., Reed, P., Kumulungui, G., Yaba, P., Délicat, A., Rollin, P.E. and Leroy, E.M. (2005) 'Wild animal mortality monitoring and human Ebola outbreaks in Gabon and Republic of Congo 2001–2003', *Emerging Infectious Diseases* 11: 283–290

Sahlins, M. (1972) *Stone Age Economics*, Aldine, Chicago, IL

Sponheimer, M. and Lee-Thorp, J.A. (1999) 'Isotopic evidence for the diet of an early hominid, *Australopithecus africanus*', *Science* 283: 368–370

Strickland, S.S. and Duffield, A.E. (1998) 'Nutrition and ecosystems in Sarawak: The role of the areca nut', *Asia Pacific Journal of Clinical Nutrition*, 7(3/4): 300–306, available at www.healthyeatingclub.com/APJCN/Volume7/vol7.34/Strickland.pdf

Thackeray, J.F. (1995) 'Exploring ungulate diversity, biomass, and climate in modern and past environments', in E.S. Vrba, G.H. Denton, T.C. Partridge and L.H. Burckle (eds) *Paleoclimate and Evolution, with Emphasis on Human Origins*, Yale University Press, New Haven, CT, pp479–482

Walsh, P.D., Abernethy, K.A., Bermejo, M., Beyers, R., DeWachter, P., Ella Akou, M., Huijbregts, B., Idiata Mambounga, D., Kamdem Toham, A., Kilbourn, M., Lahm, S.A., Latour, S., Maisels, F., Mbina, C., Mihindou, Y., Ndong Obiang, S., Ntsame Effa, E., Starkey, M.P., Telfer, P., Thibault, M., Tutin, C.E.G., White, L.J.T. and Wilkie, D.S. (2003) 'Catastrophic ape decline in western equatorial Africa', *Nature* 422: 611–614

Walsh, P.D., Biek, R. and Real, L.A. (2005) 'Wave-like spread of Ebola Zaire', *PLoS (Public Library of Science) Biology* 3(11): 1946–1953

Wirsing, R.L. (1985) 'The health of traditional societies and the effects of acculturation', *Current Anthropology* 26: 303–323

Wolfe, N.D., Switzer, W.M., Carr, J.K., Bhullar, V.B., Shanmugam, V., Tamoufe, U., Prosser, A.T., Torimiro, J.N., Wright, A., Mpoudi-Ngole, E., McCutchan, F.E., Birx, D.L., Folks, T.M., Burke, D.S. and Heneine, W. (2004) 'Naturally acquired simian retrovirus infections in central African hunters', *Lancet* 363: 932–937

Yamauchi, T., Sato, H. and Kawamura, K. (2000) 'Nutritional status, activity patterns, and dietary intakes among the Baka hunter-gatherers in the village camps in Cameroon', *African Study Monographs* 21: 67–82

Zerner, C. (2004) 'The viral forest in motion: Ebola, African forests, and emerging cartographies of environmental danger', in Slater, C. (ed) *In Search of the Rain Forest,* Duke University Press, Durham, NC, pp246–284

13

Sociocultural Dimensions of Diet and Health in Forest-Dwellers' Systems

Edmond Dounias with Carol J. Pierce Colfer

Human beings are primates, endowed with symbolic thought and culture. They not only try to satisfy biological needs but are also driven by cultural wants in both the material and non-material worlds. These wants generate behaviours that are not meant to contribute to human biological adaptation but rather are aimed at providing psycho-cultural rewards in the symbolic field, which may in turn have some influence on material wellbeing.

In this chapter we highlight some of the cultural aspects of human diet and health (see Chapter 4, this volume). In trying to convey the significance of cultural difference, we have struggled with a problem that is common for anthropologists and ethno-biologists wanting to share their analyses: we tend to see things in a very holistic manner, and pulling the health-related aspects of human behaviour out of their cultural context is somehow alien. To avoid misrepresentation, we must explain some points in detail to show how local people's symbolic understandings affect their understanding of health and their practices that lead to (or diminish) health.

Rather than covering the whole spectrum of cultural factors that influence the diet and health of human beings, we have selected a short list of diverse topics – cultural adaptation, food choices, biological optimization, cultural forms of hunger, treatment of high-risk groups, traditional medicines, and social change – that clearly highlight health-related aspects of culture. We illustrate these topics with experiences in Central Africa, Indonesia and Amazonia among forest groups whose cultures are intimately connected with forests (Figure 13.1). By briefly exploring these topics, we hope to bring out the tremendous complexity of the cultural dimension of human–forest interactions. Box 13.1 highlights the methods and approaches we implemented during our research. Box 13.2 provides a sense of the holistic nature of culture's health-related aspects through the particular case of the Fang of southern Cameroon, by detailing some crucial elements of their health-related belief and symbolic system.

Food, health and cultural adaptations

Since the concept of adaptation – defined as a continuing process of all organisms to achieve a better environment–organism fit in responding to changes in their environment (Appell, 1986, p44) – was first introduced, it has been used in many ways. Of these

Figure 13.1 *Ethnic groups mentioned as examples in this chapter*

BOX 13.1 COMPARISON AS A FUNDAMENTAL TOOL IN DIFFERENTIATING SOCIOCULTURAL CHOICES AND BIOPHYSICAL CONSTRAINTS

In southern and coastal Cameroon (Koppert et al, 1993; Froment et al, 1996) as well as in East Kalimantan (CIFOR, 2006), we have been carrying out studies among rural ethnic groups that are no longer in a situation of pristine isolation and have become involved in a broader cash and market economy during the past half-century. The analytic framework adopted to undertake the anthropology of food among these groups consists of a blending of three complementary grids of comparison (Figure 13.2).

- Cultural. We synchronically studied distinct ethnic groups sharing the same ecosystem. This 'synecological' referential offers the opportunity to analyse the discrete cultural choices elaborated by societies confronted with the same ecological constraints. Cameroon: Mvae trappers, Yasa fishers, Kola Pygmies. Kalimantan: Tubu Punan, Belaka Punan.
- Ecosystemic. We synchronically studied the same ethnic group (the Mvae trappers and farmers) living in two contrasted forest ecosystems. This 'auto-ecological' referential provides a choice field to analyse how ecological constraints affect the strategies of distinct communities with the same origin. Cameroon: coastal Mvae trappers in the evergreen coastal forest, inland Mvae trappers in the semi-deciduous continental forest. Kalimantan: remote Tubu Punan, peri-urban Tubu Punan.
- Diachronic. It is practically a truism to say that culture is mutable, and that people change. Nevertheless, studies that have explored the dynamics of traditional societies over a fairly

long time are few and far between. We also studied the same population living in the same ecosystem but at two distinct periods. Cameroon: coastal Mvae trappers studied by Günter Tessmann in the 1890s, coastal Mvae trappers studied by Edmond Dounias in the 1990s. Kalimantan: medical archives on nomadic Punan in the early 20th century, Tubu Punan studied by CIFOR in the 2000s.

The main results described here concern the Kalimantan case. The inter-ethnic referential clearly reveals, among other things, the diverging diet strategies, the distinct traditional healing practices, and the contrasted economic choices between the Belaka and the Tubu (two groups of Punan who know each other, both having lived in the same forest environments and both formerly nomadic hunter-gatherers). Use (or abuse) of the generic term 'Punan' is dangerous, because it overshadows fundamental differences between communities whose cultures do not respond in the same ways to the similar environmental constraints conditioning their diets and health status.

The ecosystemic referential offered the opportunity to compare the needs and wants of two Tubu Punan communities that constantly interact and share the same culture, language and origins, but confront totally different environments. Results reveal interesting trade-offs between the two communities. The peri-urban Punan, who agreed to leave the Tubu watershed 30 years ago to settle near the city of Malinau, enjoy easy access to small-town facilities (schools, dispensaries, markets, job opportunities) and appear economically richer than their parents who chose to stay in the forest. But the former suffer from many social ailments (insecurity, depression, stress, drug addiction, violence) caused by marginalization, social injustice and individualistic attitudes (see Box 10.2 and Chapter 15, this volume). Those who

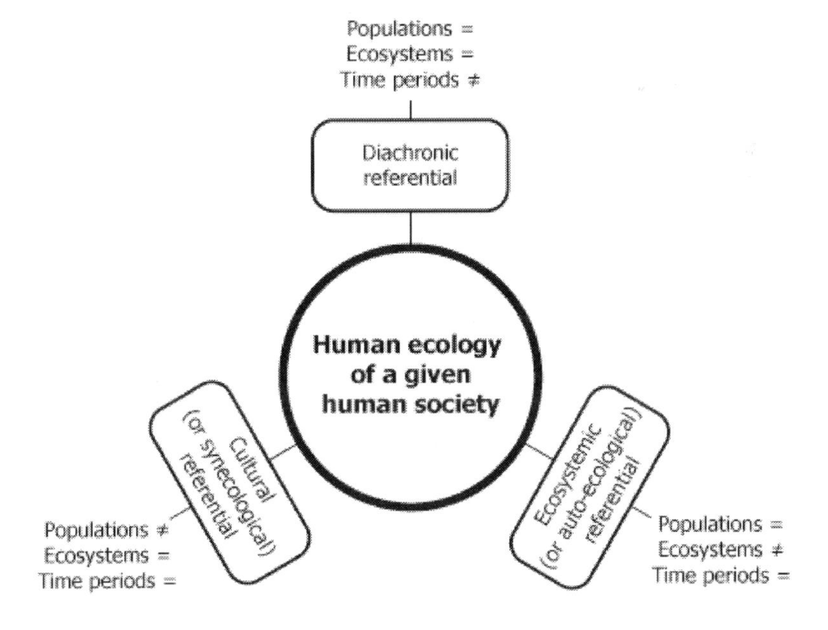

Figure 13.2 *The three complementary 'referentials' used for the study of adaptive human ecology*

Source: adapted from Dounias (1996)

live in the forest envy the material comfort of their peri-urban relatives, but the forest-dwellers are much better off in terms of diet and physical fitness, and they profit from the absence of the land-tenure conflicts that trouble the more urban Punan. The rural Punan appreciate free access to all kinds of forest resources but pay a heavy tribute in infant mortality because of epidemic outbreaks that erratically occur upstream.

The diachronic referential reveals clearly that the remote Tubu Punan, who renounced a nomadic lifestyle and now live in permanent villages, are much more exposed to transmissible and contagious diseases than their grandparents. Their pharmacopoeia is insufficient to treat diseases like malaria, measles and smallpox, which rarely afflicted their ancestors while they were still forest nomads. A sedentary lifestyle in a pacified environment has reduced warfare and head-hunting, but life in permanent villages has considerably increased the risk of contracting vector-borne, zoonotic and crowding diseases (see Chapter 12, this volume), and the remoteness of the settlements along a poorly navigable watershed excludes these communities from access to dispensaries (see Chapter 15, this volume).

multiple possibilities, it is necessary to make a fundamental distinction between biological and psycho-cultural adaptation. Adaptation always involves the sociocultural world of which the individual is a member, and therefore it is necessary to view the psycho-cultural dimension of adaptation as a fundamental characteristic of any human society. Nevertheless, until recently the dominant tendency among biological anthropologists has been to look upon people's feeding behaviour and health status through the lens of biological adaptation alone. Interest in the cultural dimension of adaptive responses was strictly limited to practical and material matters, and culture was narrowly reduced to food habits, preferences, choices and overall dietary strategies that contributed to academic discussion about ecology, nutrition and overall subsistence within an evolutionary perspective.

Food is certainly the most appropriate and integrative topic within which to investigate the contributions of culture to biological adaptation and nutritional success (Garine, 1972; see Figure 13.3). However, many cultural behaviours that come into play in the field of food choices incorporate traits – like hedonism – that can hardly be seen just as adaptive responses (Durham, 1976).

Wild yam gathering by the Baka hunter-gatherers of eastern Cameroon is an example that highlights the need to investigate jointly the ecological and psycho-cultural dimension of food. Dounias (2001) has demonstrated that the Baka have a detailed knowledge of the ecology and biology of yam vines, which are extremely difficult to observe in the forest undergrowth. The Baka also have a dynamic understanding of the plant's growth cycle, and they know how to take advantage of the yam's double capacity to reproduce both sexually and vegetatively. Ethno-linguistic data have revealed the unique status of yams in Baka plant nomenclature. Baka terms applied to yam morphology are specifically drawn from terms referring to human anatomy, and the Baka also have a wide range of names for the kinds and parts of tubers as well as their consistency and taste. But the harvesting procedures are certainly the key aspect of yam expertise. Dounias uses the term 'paracultivation' to define perennial harvesting practices

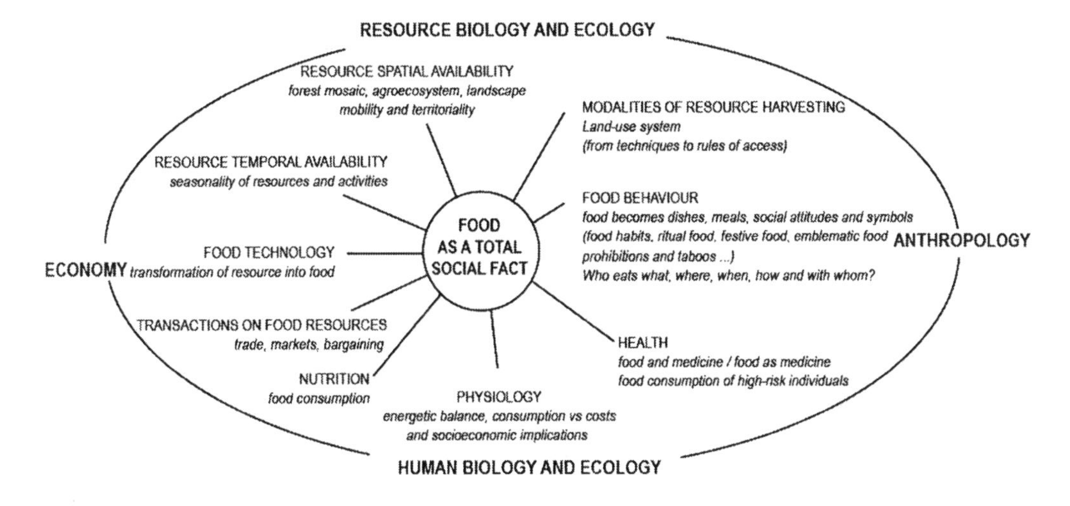

Figure 13.3 *Conceptual framework for the study of the anthropology of food*

aimed at managing yam production while keeping the tuber plants in their original forest environment; these procedures are accompanied by social rules protecting the rights of ownership over a supposedly 'wild' resource (the plant is cared for, protected, owned, managed and eventually inherited as a private possession); it has even determined the technical design of a digging tool – an auger – that is perfectly adapted to the constraints of a nomadic lifestyle: it is efficient, easy to make and, most important, ephemeral. But beyond their primary function as food, wild yams occupy the full status of a cultural good. They appear in matrimonial payments, prestige dishes and the pharmacopoeia, and they play a central role as a ritual object in the complex interactions among the Baka, the elephant and the mighty spirit of the forest. Yams provide edible tubers that are rich in carbohydrates and proteins. But yams are not only good to eat, they are also good to think about. They not only contribute to the nutritional integrity of the Baka, they also mediate their necessary relationships with the invisible world and the spirits who are seen to exert control over forest resources. This not-atypical example clearly illustrates how ecological perspectives should not be separated from cultural aspects, or people's perceptions of their resources.

FOOD CHOICES AND WELLBEING

The multiplicity of cultural factors influencing food choices and inducing what Igor de Garine describes as the 'biological arbitrariness of food habits' (Garine, 1993) are specific to humankind and have contributed to its originality. To understand the dynamics of maintaining a food system and the rationality that is built on forest resources, it is essential to explore the 'emic' dimension of food. 'Emic', a linguistic and cultural anthropological term, 'distinguishes the understanding of cultural representations from the point of view of a native of the culture' from the opposing term, 'etic', meaning 'from the point of view of an outside observer of the culture' (Barfield, 1997, p148).

Among the Aka Pygmies of the Central African Republic, for instance, the feeling of being replete – required for wellbeing – can be obtained only from an essentially meat-and-honey diet eaten to satiety (Bahuchet, 1985; Thomas, 1987). The high value attributed to both these foodstuffs is also noticeable in their pre-eminence in the prominent rituals of life. For the Aka, 'health' is inseparable from 'life': the same word is used for both concepts, and both can find full expression only in conditions of equilibrium and harmony.

What is bad for a particular cultural community is not necessarily bad for others. The perceived quality of food is obviously related to the gustatory preferences of the consumer. However, the cultural dimension has such a large impact on food choice that separating the psychological from the physiological aspects of food perception (or taste) has become a challenging task. From an adaptive, biological perspective, the unpleasant bitter taste of many plants, due to substances like alkaloids, expresses the fact that the plant is poisonous – a result of its adaptation to potential plant eaters – and our adaptation to the potential danger. In that respect, bitter cassava cultivars are potentially more toxic because of their cyanide and cyanogenic glucosides content. Nevertheless, many farming societies throughout the world express a clear preference for bitter cassava cultivars.

This preference for bitter over sweet tastes may appear to contradict the general trend that humans select the less toxic cultivars of a given food plant (Johns, 1990). Westerners typically consider human exposure to naturally occurring cyanides to be wholly undesirable and detrimental to health. But studies have shown that sub-acute exposure may sometimes provide a range of metabolically positive effects for certain consumers. Dietary cyanogenic glycosides may subtly modify the oral biology, for instance by enhancing salivary thiocyanate concentrations that are known to kill oral bacteria and reduce caries and tissue ulceration rates (Jackson, 1994). Among several arguments advanced to justify cultural preferences for bitter cassava (Dufour, 1993), a culinary explanation is that the quality of starch in most bitter cultivars is more suitable for certain foods, responding to locally desired sensory properties, such as touch, smell and taste (Lathrap, 1973). A similar argument was given by the Kubu hunter-gatherers of Sumatra, who used to maintain groves of wild yams, hidden in the undergrowth of natural forest, consistently choosing toxic species for these groves (Dounias, 2005).

BIOLOGICALLY SUB-OPTIMAL NUTRITION

By trying to avoid psychological unrest through satisfying symbolic demands, sometimes at the expense of optimal nutritional options, traditional societies are inclined to adopt what could be interpreted as a 'philosophical approach' to biological fitness: they make cultural choices that are not necessarily meant to respond to the principle of optimality (Foley, 1985). The particularity of human society may even lie in a recurrent inability to systematically achieve ideal biological fitness (Garine, 1991). A good illustration is the consumption of a particular type of 'food' that can sometimes be considered medicine and that may be classed as a drug in the West. Substances like qat, coca, tobacco and alcohol have been vehicles for different forms of social interactions. Their use has played a role in the expression and maintenance of a particular social order, and their exchange was used to draw attention to related verbal exchanges. Among the Tukanoan- and Witotoan-speaking groups of northern Amazonia, the exchange of such products was used as a frame and pretext for heightened social interactions (Hugh-Jones, 1999). When the social frame is altered by trends – like commoditization – that encourage increasingly individualistic behaviours, the consumption of these substances turns into real drug addiction, with all its deleterious effects and social ailments.

CULTURAL FORMS OF HUNGER

Even though forests in equatorial latitudes provide a constant supply of various potential foods that reduce the risk of food shortage (Garine and Harrison, 1988), forest-dwellers are known to undergo what they perceive as 'lean periods' in food supply. One may refer here to psychological stress (Harrison, 1982), which has a cultural component exhibiting tangible effects and possibly negative health consequences.

Some authors have attempted to demonstrate how the lack of culturally valued foods, such as meat, sometimes causes psychological unrest that may in turn have negative biological consequences, such as cardiovascular ailments, vulnerability to infection or low growth rates among children (Bahuchet, 1985; Pagezy, 1988). Among the Aka Pygmies of Central Africa, for instance, meat hunger – even in the absence of measured dietary deficiencies – is said to bring about tiredness, loss of vital strength and, in turn, illness resulting from effects expressed at very diverse levels with a blending of material and symbolic aspects (Thomas, 1987). Meat is perceived as essential for health because it is a sign of the hunter's condition as well. Hunting requires all one's mental and physical abilities, and thus abundance of meat reflects a healthy hunter (and through him, the vitality of the whole community), with which meat is unconsciously and symbolically associated (Motte-Florac et al, 1993). Similar cravings also exist for fish and other seasonal water-related resources. They are likewise expressed among farming societies, especially when cultural 'superfoods' (Jelliffe, 1967) are dwindling; these are carbohydrate staples like rice and sago throughout Southeast Asia (Ellen, 1979; Conklin, 1980) and cassava among Amerindians (Lancaster et al, 1982).

FOOD AND NUTRITION AMONG 'HIGH-RISK' GROUPS

High-risk groups, as commonly defined by epidemiologists, are members of the community who are more likely than others to face health problems because of environmental constraints. These groups include pregnant and lactating women, infants with increased nutritional requirements and elderly people facing other types of nutritional risks (Pagezy and Garine, 1990). A luminous example of the social management of the diet and health of high-risk individuals is the case of the primiparous mother among the Ntomba of the Democratic Republic of the Congo (Pagezy, 1983; see Chapter 7, this volume, for more gender-related health-care implications). A few days after childbirth, the first-time mother and her newborn are escorted by relatives back to her mother's house, usually in another village. The young mother and child stay there for two to four years and benefit from very special care. She stays in a seclusion hut, respects numerous food and sexual taboos, uses tools that no one else is authorized to touch, and avoids any physical activity related to food production and preparation. She receives abundant food and daily exposes her fat body (considered healthy by the Ntomba), elaborately attired and made up to focus everyone's attention on her (Figure 13.4). During this long period of seclusion, over-feeding and intensive care have beneficial consequences for the health status of both the mother and her firstborn, who in other contexts would suffer from low birth weight and weakness. Beyond the obvious adaptive implications of such carefully and benignly scripted primiparity, the primiparous mother incarnates the 'true mother' and radiates a symbolic image of purity and good health that is a source of pride and psycho-cultural wellbeing for the whole community.

Figure 13.4 *Chantal Wale, an Ekonda primiparous mother, poses with her daughter at the end of several years of seclusion, in front of a local painting glorifying primiparity*

Source: Hélène Pagezy, 1991

FOOD AND TRADITIONAL MEDICINE

The interactions between medical and nutritional systems have gained a broader audience because of the works of Nina Etkin and collaborators (see Etkin, 2001). In the forest regions of Africa, food and medicine are areas in which people have knowledge – as these pertain to themselves and to their environment. Fieldwork carried out in southern coastal Cameroon shows how beliefs and practices regarding food and health must be approached as a whole (Hladik et al, 1990; 1993). Such interactions can be assessed through local representations of body and health in relation to nutrition but also through medical practices where food is used in the treatment of illness. Some practices of traditional medicine use nourishment as an integral part of the treatment. The therapeutic dish (like chicken soup among US Jews) is a special kind of culinary preparation that in parts of Africa is used to cure victims of witchcraft, food taboo transgressors, and the possessed. The medicinal plants added to the dish constitute a magical protection for patients and help them recover. In the representations of the etiology of diseases caused by sorcery, food is used metaphorically – as in the case of *èvú*, in which the sorcerer symbolically devours his victim (Box 13.2). Among the Yasa fishers in southern coastal Cameroon, where both the biological reality and the symbolic aspects of health are simultaneously considered, food is used in various ways during rituals dedicated to the treatment of possession. During exorcism rites, for instance, a special preparation is distributed to all the participants in the ritual: patients, drummers and the public (Figure 13.5). It is this protection that makes the consumers invisible to their

Figure 13.5 *Exorcism performance by Anasthasie Njuke, a renowned Yasa healer in southern Cameroon*

Source: Edmond Dounias, 1985

enemies. Medical treatments in this area concern not only the sick individual but the entire community (Garine, 1988; 1990).

Most of the traditional medicines shared as food during these public rituals are forest products, and many are lost with the degradation or change of forest environments (Shanley and Luz, 2003; see also Chapter 3, this volume). Often the environment in which these plants grow is considered essential for the plants to elaborate medicinal powers (Falconer, 1990). In a context of drastic social change, traditional healing may seem inefficient in dealing with dangerous new emerging diseases (see Chapters 12 and 16, this volume). This is, for instance, the case for the Punan, former hunter-gatherers of East Kalimantan, who are confronted today with diseases that they did not meet when they were nomadic (Voeks and Sercombe, 2000). Traditional healers and wise elders are consequently losing their political influence, and social conflict has become recurrent between generations. A lack of social controls results in dramatic misuse of manufactured pills. Self-medication and related addictions degrade immunity and have become a major health problem for the Punan (Dounias and Froment, 2006).

BOX 13.2 COURTYARD AND BACKYARD: MANAGING EXPOSURE TO DISEASE

The 'Fang' of Central Africa comprise diverse ethnic groups – Fang, Mvae, Ntumu, Nzaman, Okak – who speak different dialects of the Fang language subgroup. They understand each other's languages and share common cultural traits. Courtyard and backyard are fundamental features of the habitat of the Fang in southern Cameroon. These land uses represent physical as well as cultural poles and serve opposed functions that mark out the life history of the Fang, social relationships and symbolically rich settings for everyday life and rituals.

A village has a common courtyard, which is indisputably a public space. It is expected to be a pleasing place that must appear friendly and warm to visitors. People commonly speak loudly in the courtyard, and communication tends to be effusive. Festive events and related dance, music, food and recreational performances always take place in the courtyard. The courtyard is also a privileged floor for communication. This is where the Fang organize the tribunal forum (an integral part of their indigenous legal system). The calling drum sits enthroned there on a special tripod on a platform, or it is sometimes hung in a tree so that its sound is amplified and its message is carried farther. The courtyard is the floor dedicated to leisure and resting. It is also characterized by bare soil and little vegetation, and thus it prevents the proliferation of undesirable animals (parasites, ticks, sand flies, tabanid flies, herbivores, snakes, weaver birds). Having a wide deforested area also regulates the resprouting of spontaneous seedlings, since the many birds living in the forest undergrowth and other seed dispersers are discouraged from flying over or running through such large open spaces. (Colfer noted a similar pattern among her Javanese transmigrant neighbours in Sumatra, who rather compulsively cleaned their houseyards, to keep snakes and other vermin away and to demonstrate to neighbours their industry and tidiness.) Maintaining the soil surface flat and free of vegetation and other debris also prevents puddles and thus limits the proliferation of mosquitoes. The Fang devote a significant amount of time to cleaning, weeding and removing useless roots and stumps from their courtyard. Sweeping the courtyard not only removes garbage and animal leavings and prevents vermin, but also symbolically removes importunate spirits. In front of houses – directly under the roof or in shade trees planted there – the Fang also hang containers to gather clean rainwater. Some courtyard trees with an appropriate architecture are even specifically planted to serve as receptacles for rainwater containers. These trees are also used for hanging washed clothes out to dry.

Within the village, each main compound has its own backyard – a private and even, to some extent, unfriendly space. Access to backyards is not really forbidden but it is implicitly restricted or, better said, discouraged. The backyard is known to be full of magical components from vegetal, animal and mineral sources that are hidden in dense undergrowth mimicking the natural forest. By contrast to the care provided to each of the plants growing in the courtyard, the backyard agroforest looks more like a vegetative mess. This space receives the cooking garbage and all the detritus that is carefully swept away from the courtyard. Backyards also benefit from natural fertilization by fallen leaves and the leavings of wildlife, livestock and humans.

The backyard is a woman's domain. Major access to the backyard is through the backdoor of the kitchen, and the old role of the backyard as a rescue pathway to evade intruders in wartime gave priority to women and their children. Women are in charge of disposing of domestic detritus in the backyard and gathering firewood. They also take care of the crops in the kitchen garden and all non-timber forest products that grow in the undergrowth of the backyard. More importantly, the women take care of countless medicinal plants that are hidden in the backyard and frequently used to treat children's diseases and fecundity-related disorders.

Until recent times, Fang women used to give birth in the backyard, under the protection of magic that discouraged aggressive sorcery. After delivery, the mother buried the placenta in a secret location in the backyard, beyond the reach of witches. Labour and childbirth were carried out in a crouching position. This position has meaning among the Fang, who recurrently use the vagina to symbolize the foundation, the roots and thus the irremovable source of things, as opposed to the head, which symbolizes the growing or extendable extremity sometimes guiding people far from their origins. This symbolic value given to the vagina and the related crouching position that is adopted during delivery magnify an opposition force that is attributed to women in these fundamentally patrilineal, virilocal and usually patrilocal societies. Today, even if Fang women no longer give birth in the backyard, they still respect a period of seclusion after delivery, when the mother and her child are weak and thus particularly susceptible to witchcraft aggression. The seclusion area is always set up along the back wall of the kitchen, which is considered the most secure place within the house.

The ritual life of the Fang is extremely rich and marked by several initiation procedures that are a prerequisite to entering various secret societies. Fang pagan cults still endure even though they have been weakened today by mission activity and government prohibition and are now confined to the discreet environment of the backyard.

Witchcraft and sorcery are condemned by the church and denounced during the public judgement of persons who are suspected of devoting themselves to such forbidden practices. The tribunal always officiates in the courtyard. Despite the highly visible sentencing, witchcraft dictates the daily life of the Fang and profoundly conditions their social relations. The origin of Fang witchcraft beliefs lies in the obvious difference in the destinies of individuals. The Fang have concluded that there are several sorts of people according to whether they do or do not possess the witchcraft being, *èvú,* that predestines one's fortune in life. The Fang then couch their discussion of witchcraft in terms of personal ambition and the search for wealth and glory. Those who have a strong *èvú* are true witches. The *èvú*, for the selfish purposes of its possessor, is accustomed to making spells. It lives on the blood and flesh of humankind and is thus a kind of cannibal.

In ancient times, several secret societies actively counteracted witchcraft aggression. Someone suspected of sorcery was forced to drink poison in public – in the courtyard – as an ordeal to prove his or her innocence. If the person died, the body was carried to the backyard and was autopsied to seek further proof of guilt. Persons who died of suspected bewitchment were also autopsied in the backyard, since the *èvú* is known to devour the entrails of its victims. Fang women still hold their own secret societies devoted to protection against sterility or fecundity troubles. They perform rites in the backyard that are parallel and opposite to the male rite of passage and that unambiguously celebrate the female sex and force.

The bipolarity of courtyard and backyard among the Fang is a cultural attempt to regulate the risk of contracting a disease. Control of risks to health is certainly not the proximate reason for this binary opposition. Nevertheless, the many interactions between social, cultural, political, historical and ecological features combine to provide an evolutionary benefit in terms of disease regulation. This cultural management of the risk not only concerns the physical and functional layouts – which efficiently reduce the incidence of vector-borne and transmissible diseases – but also involves the symbolic control of supernatural forces, which are much less immediately tangible causes of sickness, pain, trouble, conflict and even death. In that respect, the women, who are in charge of the backyard, play a major role in maintaining a healthy environment for the entire community.

Source: Adapted from Dounias (2008)

PATHOCENOSIS, CULTURAL BEHAVIOUR AND SOCIAL CHANGE

Today, it is widely recognized that transmissible diseases have distinct expressions according to both their ecological context and the human societies affected. Grmek (1969) introduced the concept of pathocenosis, which suggests that diseases should not be considered isolated problems but instead need to be analysed as a set of diseases in interaction with a given human society and its environment. 'Human ecology' posits the same idea: that the response of an isolated individual to a single disease is no longer the central issue. Instead, one needs to address all the biological constraints that affect a group of humans who are marked by their own historical, religious, ideological and socioeconomic backgrounds.

The situations of social change that are occurring today among forest-dwellers are particularly revealing of the complexity of the interactions in play. Speedy, uncontrolled and forced social change can be harmful to people's health and may in turn be detrimental to their environments. Accordingly, such change puts added demands on the community, especially if it is neither desired nor anticipated. It may result in psychological loss, which if not managed properly may cause dysfunctional reactions.

Throughout their history, forest-dwellers have had to adapt to perennial change in forest ecosystems. However, the changes that they face today are much more brutal and radical than those experienced in the past. As deforestation, drastic modification of resource availability and the invasive influence of the cash economy occur more rapidly, local social, cultural, economic and political systems become increasingly difficult to accommodate (see Chapter 11, this volume). Past forager societies provide dramatic examples of populations that have been forced to make choices that are no longer validated by experience and have been revealed as costly in terms of ecological success (Dounias and Froment, 2006):

- Abandoning camp after a death to protect the living relatives from the wandering and harassing spirit of the dead was a common cultural practice among many foraging societies that efficiently reduced the community's exposure to infectious disease.
- Limited use of food storage among forest foraging peoples reduced the likelihood of proliferation of potential vectors of rodent-borne diseases. Similarly, some swidden agriculturalists cultivated crops that were propagated clonally and thus naturally stored in the field, not in the home.
- The nomadic Punan had little trouble with smallpox – a serious problem for Kenyah farmers – because the Punan practised silent barter, which served as a social fence protecting them from the epidemics that plagued their sedentary Kenyah neighbours (Knapen, 1998).
- In conditions of low pollution and high recycling by aquatic fauna, the Punan are less exposed to faecal pollution than some other forest-dwellers because of their use of the river for sanitary purposes. Wanting healthy rivers in which to collect clean water for domestic purposes explains why the Tubu Punan have refused to resettle downstream, closer to town (Levang et al, 2006).
- The native garb of the Punan was minimal. Today most Punan wear European clothes, urged on them by missionaries and local authorities, even though such clothes are not well suited for use in the forest. In the absence of soap, the same clothes are worn dirty until they wear out, creating a propitious ground for infectious skin diseases.
- Social regulations like mutual aid, collective activities and food sharing are still common among the Punan who live in the forest, but these customs are in constant decline among their suburban relatives. The increasingly individualistic behaviour jeopardizes high-risk persons like elderly widows who depend on the generosity of other members of the community for nutrition and health (Figure 13.6).

These examples reveal how social change can sometimes invalidate defence mechanisms and nutritional status. Deteriorating physical health can in turn compromise the social and cultural integrity of the society.

An important change in the dietary habits of many traditional societies throughout the past century is the decrease in plant-ash salt production and consumption. Salt was obtained by burning large quantities of plants to ash, then filtering the diluted ash through porous pottery. The filtrates – naturally occurring inorganic substances, including iodine, among others – were traditionally incorporated in dishes as condiments (Portères, 1957). Plants from an aquatic environment, which facilitates ionic uptake, were preferred (Lemonnier, 1984; Dounias, 1988). The development of goitre is fairly recent in some

Figure 13.6 *Meat sharing (here a barking deer) is still the rule among the remote Punan villagers (here in Long Pada, Tubu watershed, Eastern Kalimantan), but is no longer practised by their peri-urban and more individualistic relatives*

Source: Edmond Dounias, 2002

traditional societies – like the Azandé of the Democratic Republic of the Congo (Prinz, 1993) – and a few authors assume that this development may be the consequence of the decrease in ash salt consumption and its replacement by poorly iodized sodium chloride mineral salt. The situation in East Kalimantan differs markedly, however. Until quite recently, remote Bornean populations had very limited access to salt – indeed, it was highly prized, holding an important symbolic role among the Kenyah – whereas now people can usually buy inexpensive and iodized salt much more easily than in the past.

The Kenyah case reminds us that social change may not necessarily be accompanied by a less balanced biological optimum. There are also many examples revealing positive consequences of change, such as the neutralization of endemic warfare and related anthropophagous practices in Papua. Kuru disease, a progressive neurological disorder, used to be rampant among the Fore natives of the New Guinea highlands. The lethal degeneration was caused by a prion that was ingested during ritual acts of mortuary cannibalism. The disease declined with the progressive abandonment of the ritual cannibalistic practices (Gajdusek, 1996). Other examples include the renouncing of infanticide by Amerindian societies like the Yanomami of the Amazon (Early and Peters, 2000; see Chapter 11, this volume); the significant improvement in dental health from changes in food preparation, the abandonment of the use of teeth as tools, the renouncement of dental mutilation (pointed teeth, upper incisor pulling) for aesthetic purposes, and increased consumption of exotic foods that increase calcium in the diet (Walker et al, 1998); and the decrease in infant mortality among the Kenyah attributable to better nutrition and the presence of resident paramedics in the villages (Colfer, personal observation).

Conclusion

This chapter has provided case material primarily from long-term research conducted in Central Africa and Indonesia and also Amazonia about the kinds of ways that culture affects health. Although the particular examples we have provided are unique to these particular cultural contexts, important interactions between culture and forested environments are universal. One point we would like to emphasize is that each and every culture is unique, and thus it is impossible to develop standardized approaches to improving the health of people in forested areas: approaches that will work in all contexts are extremely rare. This also implies the necessity to study each group carefully or work closely with the people in each context to develop appropriate approaches.

Degrading diets and increasing illnesses are symptomatic warnings alerting us to ongoing ecological and sociocultural maladaptations. Yet few anthropologists have explored the cultural feeling of 'ill-being' expressed by forest peoples. There is a pressing need for further anthropological research, especially among the few remaining traditional societies subsisting in close interaction with their forest while confronted with drastic changes in their environment (Froment, 1997).

The complexity of the connections between the forest and human health is an increasing matter of debate on the international scene, but such debate remains rhetorical, for two reasons. First, the holistic approaches that would improve communication among conservationists and scientists in medicine, environmental health, ecology, anthropology and forestry – all dealing separately with similar issues – are rare, with a resulting failure to exchange views or share findings. The sociocultural factors that we have highlighted in this chapter, for instance, are generally neglected or, worse, considered anecdotal or retrograde expressions of folklore. Second, the systemic connections that everyone is debating remain poorly documented. In a recent review of roughly 650 documents exploring the state of knowledge concerning human health in forests, Colfer et al (2006) found few studies that looked at these issues in a systematic, comparative, interdisciplinary, longitudinal way. More long-term research studies devoted to these issues are in order, but time is short. Immediate action is required to protect the forest's fast-changing environments and human populations.

REFERENCES

Appell, G.N. (1986) 'The health consequences of development', *The Sarawak Museum Journal* 36(57): 43–74

Bahuchet, S. (1985) *Les Pygmées Aka et la forêt centrafricaine*, SELAF, Paris, France

Barfield, T. (ed) (1997) *The Dictionary of Anthropology*, Blackwell, Oxford

CIFOR (2006) Forest and Human Health Initiative website, www.cifor.cgiar.org/docs/_ref/research/livelihoods/forests_health/index.htm

Colfer, C.J.P., Sheil, D. and Kishi, M. (2006) 'Forests and human health: Assessing the evidence', CIFOR Occasional Paper 45, Bogor, Indonesia, available at www.cifor.cgiar.org/scripts/newscripts/publications/detail.asp?pid=2037

Conklin, H.C. (1980) *Ethnographic Atlas of Ifugao*, Yale University Press, New Haven, CT

Dounias, E. (1988) *Contribution à l'étude ethnoécologique et alimentaire des Koma Gimbe. Monts Alantika, Nord Cameroun*, Mémoire de maîtrise, ISTOM, Le Havre, France

Dounias, E. (1996) 'Agriculture des Mvae du sud Cameroun littoral forestier: Etude dynamique des composantes de l'agroécosystème et des plantes cultivées alimentaires', in A. Froment, I. de Garine and C. Binam Bikoï (eds) *Anthropologie alimentaire et développement en Afrique intertropicale: Du biologique au social*, ORSTOM-l'Harmattan, Paris, France, pp155–172

Dounias, E. (2001) 'The management of wild yam tubers by the Baka Pygmies in Southern Cameroon', *African Study Monographs* 26: 135–156

Dounias, E. (2005) 'Les "jardins" d'ignames sauvages des chasseurs-collecteurs Kubu des forêts de Sumatra', *Journal d'Agriculture Traditionnelle et de Botanique Appliquée* 42(1–2): 127–146

Dounias, E. (2008) 'Black and white: The ecological social, and symbolic opposition between frontyard and backyard in Fang homegardens in Southern Cameroon', in S. Heckler (ed) *Gardening and Dwelling: The Aesthetic and Pragmatic Value of Home Gardens*, Berghahn Books, Oxford

Dounias E. and Froment, A. (2006) 'Consequences of shift to sedentary lifestyle on the diet and health of former rainforest hunter-gatherers', *Unasylva* 57(2), Special Issue 'Forests and Human Health': 26–33

Dufour, D.L. (1993) 'The bitter is sweet: A case study of bitter cassava (*Manihot esculenta*) use in Amazonia', in C.M. Hladik, H. Pagezy, O.F. Linares, A. Hladik, A. Semple and M. Hadley (eds) *Tropical Forests, People and Food. Biocultural Interactions and Applications to Development*, UNESCO-Parthenon, Man and Biosphere Series, Paris, France, pp575–588

Durham, W.H. (1976) 'The adaptive significance of cultural behavior', *Human Ecology* 4: 89–121

Early, J.D. and Peters, J.F. (2000) *The Xilixana Yanomami of the Amazon: History, Social Culture, and Population Dynamics*, University Press of Florida, Gainesville, FL

Ellen, R.F. (1979) 'Sago subsistence and the trade in spices: A provisional model of ecological succession and imbalance in Moluccan history', in P.C. Burnham and R.F. Ellen (eds) *Social and Ecological Systems*, Academic Press, London, pp43–74

Etkin, N.L. (2001) 'Perspectives in ethnopharmacology: Forging a closer link between bioscience and traditional empirical knowledge', *Journal of Ethnopharmacology* 76: 177–182

Falconer, J. (1990) *The Major Significance of Minor Forest Products. The Local Use and Value of Forests in the West African Humid Forest Zone*, Community Forestry Note 6, FAO, Rome, Italy

Foley, W.H. (1985) 'Optimality theory in anthropology', *Man* 20: 222–242

Froment, A. (1997) 'Une approche écoanthropologique de la santé publique', *Natures, Sciences, Sociétés* 5(4): 5–11

Froment, A., Garine, I. de and Binam Bikoï, C. (eds) (1996) *Anthropologie alimentaire et développement en Afrique intertropicale: Du biologique au social*, ORSTOM-l'Harmattan, Paris, France

Gajdusek, D.C. (1996) 'Kuru: From the New Guinea field journal 1957–1962', *Grand Street* 15: 6–33

Garine, I. de (1972) 'The sociocultural aspect of nutrition', *Ecology of Food and Nutrition* 1(2): 143–164

Garine, E. de (1988) *Ngonje, note sur la possession chez les Yasa du Sud-Cameroun*, Mémoire de Maîtrise, Département d'Ethnologie et de Préhistoire, Université de Nanterre, Paris, France

Garine, E. de (1990) 'Food and traditional medicine among the Yassa of southern Cameroon', in C.M. Hladik, S. Bahuchet and I. de Garine (eds) *Food and Nutrition in the African Rain Forest*, UNESCO-MAB, Paris, France, pp83–86

Garine, I. de (1991) 'Ecological success in perspective', in G.A. Harrison (ed) 'Ecological success and its measurement', *Journal of Human Ecology* Special Issue 1: 55–72

Garine, I. de (1993) 'Food resources and preferences in the Cameroonian forest', in C.M. Hladik, H. Pagezy, O.F. Linares, A. Hladik, A. Semple and M. Hadley (eds) *Tropical Forests, People and Food: Biocultural Interactions and Applications to Development*, Man and Biosphere Series, UNESCO-Parthenon, Paris, France, pp561–574

Garine, I. de and Harrison, G.A. (eds) (1988) *Coping with Uncertainty in Food Supply*, Clarendon Press, Oxford

Grmek, M. (1969) 'Préliminaires d'une étude historique des maladies', *Annales ESC* 24: 1437–1483

Harrison, G.A. (1982) 'Life style, well-being and stress', *Human Biology* 54(2): 139–202

Hladik, C.M., Bahuchet, S. and Garine, I. de (eds) (1990) *Food and Nutrition in the African Rain Forest*, UNESCO-MAB, Paris, France

Hladik, C.M., Pagezy, H., Linares, O.F., Hladik, A., Semple, A. and Hadley, M. (eds) (1993) *Tropical Forests, People and Food: Biocultural Interactions and Applications to Development*, Man and Biosphere Series, UNESCO-Parthenon, Paris, France

Hugh-Jones, S. (1999) '"Foods" and "drugs" in north-west Amazonia', in D.A. Posey (ed) *Cultural and Spiritual Values of Biodiversity*, UNEP and Intermediate Technology Press, London and Nairobi, Kenya, pp278–280

Jackson, F.L.C. (1994) 'Bioanthropological impact of chronic exposure to sublethal cyanides from cassava in Africa', *Actae Horticulturae* 375: 295–309

Jelliffe, D.B. (1967) 'Parallel food classifications in developing and industrialized countries', *American Journal of Nutrition* 20: 279–281

Johns, T.A. (1990) *With Bitter Herbs They Shall Eat It*, University of Arizona Press, Tucson, AZ

Knapen, H. (1998) 'Lethal diseases in the history of Borneo: Mortality and the interplay between disease environment and human geography', in V.T. King (ed) *Environmental Challenges in South-East Asia*, Curzon Press, Richmond, Surrey, UK, pp69–94

Koppert, G.J.A., Dounias, E., Froment, A. and Pasquet, P. (1993) 'Food consumption in the forest populations of the southern coastal area of Cameroon', in C.M. Hladik, H. Pagezy, O.F. Linares, A. Hladik, A. Semple and M. Hadley (eds) *Tropical Forests, People and Food: Biocultural Interactions and Applications to Development*, Man and Biosphere Series, UNESCO-Parthenon, Paris, France, pp295–310

Lancaster, P.A., Ingram, J.S., Lim, M.Y. and Coursey, D.G. (1982) 'Traditional cassava-based food: Survey of processing techniques', *Economic Botany* 36: 12–45

Lathrap, D.W. (1973) 'The antiquity and importance of long-distance trade relationships in the moist tropics of Pre-Colombian South America', *World Archaeology* 5: 170–186

Lemonnier, P. (1984) 'La production de sel végétal chez les Anga (Papouasie-Nouvelle-Guinée)', *Journal d'Agriculture Traditionnelle et de Botanique Appliquée* 31(1–2): 71–126

Levang, P., Sitorus, S. and Dounias, E. (2006) 'City life in the middle of the forest: A Punan hunter-gatherer's vision of conservation and development', *Ecology and Society* 12(8): 18

Motte-Florac, E., Bahuchet, S. and Thomas, J.M.C. (1993) 'The role of food in the therapeutics of the Aka Pygmies of the Central African Republic', in C.M. Hladik, H. Pagezy, O.F. Linares, A. Hladik, A. Semple and M. Hadley (eds) *Tropical Forests, People and Food: Biocultural Interactions and Applications to Development,* Man and Biosphere Series, UNESCO-Parthenon, Paris, France, pp549–560

Pagezy, H. (1983) 'Attitude of the Ntomba society towards the primiparous woman and its biological effects', *Journal of Biosocial Science* 15: 421–431

Pagezy, H. (1988) 'Coping with uncertainty in food supply among the Oto and the Twa living in the equatorial forest near Lake Tumba', in I. de Garine and G.A. Harrison (eds) *Coping with Uncertainty in Food Supply,* Clarendon Press, Oxford, pp175–209

Pagezy, H. and Garine I. de (1990) 'Food and nutrition among "high-risk groups"', in C.M. Hladik, S. Bahuchet and I. de Garine (eds) *Food and Nutrition in the African Rain Forest,* UNESCO-MAB, Paris, France, pp73–76

Portères, R. (1957) 'Le sel culinaire et les cendres de plantes en dehors de l'Afrique', *Journal d'Agriculture Tropicale et de Botanique Appliquée* 4(3–4): 157–158

Prinz, A. (1993) 'Ash salt, cassava and goitre: Change in the diet and the development of endemic goitre among the Azandé in Central Africa', in C.M. Hladik, H. Pagezy, O.F. Linares, A. Hladik, A. Semple and M. Hadley (eds) *Tropical Forests, People and Food: Biocultural Interactions and Applications to Development,* Man and Biosphere Series, UNESCO-Parthenon, Paris, France, pp339–348

Shanley, P. and Luz, L. (2003) 'The impacts of forest degradations on medicinal plant use and implications for health care in Eastern Amazonia', *BioScience* 53(6): 573–584

Thomas, J.M.C. (1987) 'Des gouts et des dégouts chez les Aka, Ngbaka et autres (Centrafrique)', in B. Koechlin, F. Sigaut, J.F.C. Thomas and G. Toffin (eds) *De la voûte céleste au terroir, du jardin au foyer,* EHESS, Paris, pp489–504

Voeks, R.A. and Sercombe, P. (2000) 'The scope of hunter-gatherer ethnomedicine', *Social Science and Medicine* 51: 679–690

Walker, P.L., Sugiyama, L. and Chacon, R. (1998) 'Diet, dental health, and cultural change among recently contacted South American Indian hunter-horticulturalists', in J.R. Lukacs (ed) *Human Dental Development, Morphology, and Pathology: A Tribute to Albert A. Dahlberg,* Anthropological Paper 54, University of Oregon, Eugene, OR, pp355–386

PART III – HEALTH-CARE DELIVERY IN FORESTS

14

National Public Health Initiatives that Integrate Traditional Medicine

Cynthia Fowler

Integrative medicine = traditional healing systems + biomedicine

This chapter analyses the national health-care policies of countries in Africa, Southeast Asia and the western Pacific where there are active movements to integrate modern medicine with the traditional medicine of indigenous peoples living in forested ecosystems. It examines the character of national policies, using evidence from official international and national policies together with media reports from 2005 and 2006. The chapter assesses the motivations of policy-makers and the effects of their policies, and analyses what makes integrative policies successful or not. Four boxes, presenting ethnographic sketches of the health-care practices of forest-dwellers in Zimbabwe, Tanzania, Cameroon and East Kalimantan, provide examples of people who integrate traditional and modern medicine in their health-care practices.

Both 'traditional' and 'modern' medicines today are often syncretic or hybrid, combining elements from an array of knowledge systems and practices. Many of the countries that are developing integrative policies are post-colonial countries in the process of nation-building, which involves creating umbrella governments, regulating the activities of numerous ethnic groups, developing infrastructure and designing public services. Post-colonial nations often look outside their borders – to developed nations and international organizations – for public health models, advice and funding. Interesting global patterns are emerging as these nations attempt to develop biomedical industries, regulate traditional medicine and improve the health status of their citizens.

In some developing countries, traditional medicine is positively associated with cultural identity and post-colonial governance; modern or biological medicine is associated with the culture of colonizers or Westerners. Even in post-colonial nations with sizeable forest-dwelling indigenous populations, however, traditional medicine is not immune from being stigmatized. Conventional national health-care policies represent traditional medicine as primitive and ineffective, whereas modern medicine is positively associated with high-tech methods and other desirable aspects of modernity. Such policies express the generalized values, goals and priorities of politically powerful members of society.

Global efforts

A global movement to employ national governmental policies as a vehicle for integrating modern medicine and traditional medicine began in the 1970s. Seminal moments occurred in 1978 when the WHO released its Alma Ata Declaration (WHO, 1978) proposing a strategy for improving the health status of people around the world. The 1978 declaration followed a 1977 proposal to invest in research on traditional medicine (WHO, 1977). Alma Ata signifies the beginning of attempts to institutionalize traditional medicine globally (Kaboru et al, 2005). The aim of the Alma Ata Declaration is to make everyone in the world healthy using the primary health-care model (Pigg, 1995). The WHO aspires to prevent and treat diseases such as HIV/AIDS, malaria, sickle cell anaemia, tuberculosis and diabetes mellitus using the talents of indigenous healers to deliver biomedical information and modern pharmaceuticals. It also entertains the possibility of finding indigenous botanical therapies for these and other illnesses.

Alma Ata also encourages its member states to use indigenous medical practitioners in public health programmes (Green, 1988). International policy, particularly at the WHO, provides the pre-eminent template that developing nations are using to design public health policy for integrative medicine. Numerous international organizations have joined with the WHO to promote integrative medicine in the developing world, including the Association of South-East Asian Nations (ASEAN), the Pan-American Health Organization (Fink, 2002), the World Bank (DeJong, 1991; World Bank, 2001), the World Health Assembly and the United Nations Children's Fund (UNICEF). UNAIDS has a project in sub-Saharan Africa to train indigenous healers to treat HIV/AIDS with the objective of increasing access to anti-retroviral medications and treatment for opportunistic infections (UNAIDS, 2000). Doctors Without Borders operates the Access to Essential Medicines Campaign, based on a similar premise: it is easier for poor, rural people to get their medications from indigenous healers than from biomedical facilities (see also Chapter 3, this volume).

In 2003, the World Health Assembly adopted a resolution requesting that the WHO assist its member states with policy development by establishing international standards, hosting educational workshops on regulations, providing evidence from scientific trials on herbal medicines, and distributing information about their safety. The survey report implies that the most important aspect of traditional medicine policy is the regulation of herbal medicines by, for example, licensing herbalists, setting requirements for manufacture and sale, registering herbal medicines, evaluating the status of national pharmacopoeias, examining claims for the medicinal effects of botanicals, and conducting post-market surveillance and adverse-effect monitoring. Other major issues of concern are intellectual property rights (this is particularly important at the national level in India and other countries that are rationalizing ethno-medical systems) and sustainable harvesting of botanicals. The priorities in the public health policies of many African, Asian and Pacific nations, following the WHO, are to institutionalize traditional medicine, register indigenous healers and standardize herbal medications, but the policy details and success of the integrative medicine programmes vary.

45 of the WHO's 191 member states (or 32 per cent) have national policies governing the integration of biomedicine and traditional medicine (Table 14.1), and an additional 51 are developing policies (WHO, 2005a). 36 European countries and 35 African countries have policies and laws regarding traditional medicine, compared with 20 Western Pacific, 18 American, 16 Eastern Mediterranean and 10 Southeast Asian countries. In the WHO's assessment, 'Southeast Asia clearly leads the world in traditional medicine and herbal medicine research and policy development' (WHO, 2005a, p19).[1]

Table 14.1 *WHO member states with national policies on integrative medicine*

Africa	Republic of Côte d'Ivoire, Republic of Equatorial Guinea, Federal Democratic Republic of Ethiopia, Gabon, Ghana, Republic of Guinea, Republic of Mozambique, South Africa, Togo, Tanzania, Zambia
Americas	Guatemala, Mexico, Peru
Eastern Mediterranean	Egypt, Iran, Kuwait, Sudan, Syria
Europe	Belgium, Germany, Hungary, Norway, Russia, Ukraine, United Kingdom
Asia	Bangladesh, Korea, India, Indonesia, Maldives, Myanmar, Nepal, Sri Lanka, Thailand
Western Pacific	Australia, Cambodia, China, Laos, Malaysia, Mongolia, Philippines, Korea, Singapore, Solomon Islands

Many developing nations are enthusiastically pursuing the WHO's goals. Forty WHO member states have established national policies since 1990. Countries with exceptionally long legislative histories are India, China and Bhutan, which established national policies in 1940, 1949 and 1967, respectively. Adherence to the Alma Ata resolutions varies. The primary health-care systems of China, the Democratic People's Republic of Korea, Vietnam, Myanmar and Bhutan are integrated (WHO, 2005b). In the health-care systems of India and Thailand, allopathic and traditional medicine overlap and operate simultaneously but are less well integrated than in the countries listed above. Indonesia and East Timor have similar systems (WHO, 2004a).

In addition to enumerating the policies of member nations, the WHO's 2005 global survey report (2005a) summarizes its member states' laws and regulations governing traditional medicine and herbal medications, and the status of national pharmacopoeias. Currently, only 70 countries have laws that address the manufacture, sale and use of botanical remedies, but this number is increasing as more countries are developing legislation, such as Kenya's herbal medicine and medicinal plants policy, India's national design policy (which will protect intellectual property rights), and Thailand's initiative to ban smoking in traditional medicine establishments and other public spaces. In addition to the member states that have federal regulations, 58 have at least one national institute specializing in traditional medicine, 75 have national offices of traditional medicine and 61 have expert committees.

National public health policies

The integrative health-care policies of developing nations range from those that celebrate traditional medicine to those that merely acknowledge a need to pay more attention to it. An example of the latter is Myanmar, which does not have a separate policy for integrative medicine but repeatedly addresses the need to conduct research on and use traditional medicine in its national health policy (WHO, 2004a). An example of the former is China, where state policy emphatically promotes integrative medicine. In the 1940s, Mao Tse-tung proclaimed that China would integrate Western biomedicine and traditional Chinese medicine. He believed that Chinese medicine represented cultural pride and that an integrated system was more economical. Uganda's policy permits all therapeutic practices and requires practitioners to have adequate training and to be recognized in their community as authentic healers (Phillips, 1990). Bhutan's national policy is to 'preserve and promote traditional medicine through capacity building and establishing an effective system within the framework of the national health-care delivery system' (WHO, 2004a). Another variation is Thailand, whose policies reflect extraordinary enthusiasm for traditional medicine. In its 2002–2006 plan for national economic and social development, Thailand stipulates that the primary health-care system put traditional medicine into practice (WHO, 2004a). There is regional cooperation as well among developing countries. The health ministers of ASEAN are moving towards greater regional cooperation for handling health emergencies. One of their agenda items in 2006 was to incorporate traditional medicine into the public health systems of ASEAN's ten member nations, plus their dialogue partners China, Japan and South Korea.

Premise for integrative policies

Indigenous healers – and not biomedical physicians – are the primary health-care providers for many poor people in developing countries, and they are often more affordable (Tsey, 1997). In Zimbabwe, the Minister of Women's Affairs and Community Development publicly reinforced the trend among the nation's poor to seek care from traditional healers because of lower costs and easier access. In response to recent economic and political crises and corruption in public health bureaus, government officials in Zimbabwe have recommended that citizens seek care from *n'anga* (indigenous healers) and use botanical medicines (Chavanduka, 1994).

Box 14.1 Integrating modern and traditional medicine in Zimbabwe

Witness Kozanayi and Nontokozo Nemarundwe

With the HIV/AIDS epidemic and a crumbling health delivery service, Zimbabwe has seen an increase in the use of traditional medicine and the emergence of new herbalists, some of whom have been accused by registered herbalists of selling fake herbs to a desperate and unsuspecting public. Though some of the accusations by registered herbalists are efforts to protect their market, fake herbs are being sold (*Sunday News*, 2006) and have been traced to new herbalists who want to make quick money without adhering to the ethics of the herbalists' profession.

Because of the economic crisis in the country, which started around 2000, drugs are now scarce and unaffordable to many people. To most people, the only readily available medicine is traditional medicine. Even people who have access to biomedicine continue to use traditional medicines because they doubt the potency of some of the pharmaceuticals on the market. In addition, because most Africans have historically relied on traditional medicines, they believe that health care without traditional medicines is incomplete. For instance, the Shona believe that 'every illness or misfortune (ranging from AIDS to being knocked down by a drunken driver!) is the result of evil spells cast by the wicked or some enemy' (Bourdillon, 1998). Even if one receives modern medicine to treat, say, injuries from a car accident, traditional herbal remedies are needed to shoo away the evil spirit that cast the bad spell. It is believed that if one fails to protect oneself after a misfortune by using traditional herbs, the evil spirits will return to haunt the victim, in a more vicious manner.

It is common for medical doctors and nurses to prescribe moringa (*Moringa oleifera*, commonly referred to as 'the wonder tree') and muranga (*Warburgia salutaris*, 'the tree that disciplines all diseases') together with biochemical drugs. Moringa, a tree of Indian origin, boosts the immune system of HIV/AIDS and TB patients (www.treesforlife.org/project/moringa). Herbalists in Zimbabwe believe that every part of the muranga and moringa trees has healing powers. The roots, however, are believed to be the most potent, and the consequent demand for roots has dire consequences for regeneration (Mukamuri and Kozanayi, 1999). Some Zimbabweans who consider themselves more 'modern' condone the use of traditional herbs and are using moringa in the name of 'eating vegetables'.

The stigma associated with plant medicines (such as moringa) originally perceived as appropriate for HIV/AIDS patients only has decreased in recent years. Even modernists who want to be 'healthy' now casually mix traditional herbs with water, drinks, beer or porridge to strengthen the immune systems of adults and children. *Warburgia salutaris* (now a protected species) has been used in eastern Zimbabwe for a long time. Traditional healers say it is an ingredient of every concoction they give to their patients. Some NGOs and research institutes have made efforts to introduce commonly used plant medicines at the household, community and school levels. This has resulted in the demystification of some age-old traditions about *Warburgia salutaris* and other plant medicines, such as beliefs that a person must be in a trance or naked to harvest the tree bark and that storing too much muranga bark in one's house causes misfortune.

Another ethnographic reality is an impetus for creating pluralistic therapeutic environments (Good, 1977; Tovey et al, 2005). People often move fluidly between multiple modalities. A common health-seeking behaviour in Malaysia, for example, is to obtain care from biomedical providers as well as from traditional Chinese or Ayurvedic healers (Phillips, 1990). Parents in Kenya seek primary health-care services for their children's illnesses from both traditional and biomedical practitioners (Boerma and Salim, 1990).

The importance of ethnic identity in post-colonial cultural politics is evident in national efforts to institutionalize traditional medicine. India is particularly aware of its rich ethno-medical cultures. The national government invests in documenting, regulating and protecting traditional medicine. Twenty-one of India's state governments have departments of Indian medicine and homeopathy (WHO, 2004a). Numerous NGOs in India conduct research and operate projects related to traditional medicine. The Society for Research and Initiatives for Sustainable Technologies and Institutions, for example, is the research lab for the Honey Bee Network, which locates, tests and supports the development of innovations in traditional health practices. There are innumerable other examples of investments in traditional medicine by both governmental and non-governmental entities.

BOX 14.2 PREMISES OF INTEGRATIVE MEDICINE POLICIES

- The most effective way to manage major health problems is to recruit traditional practitioners.
- Traditional medicine is often more accessible (less expensive, closer to home) to poor people than biomedicine.
- Herbal medicines are more accessible (less expensive, easier to acquire) than allopathic pharmaceuticals.

BOX 14.3 TRADITIONAL MEDICINE GOES URBAN IN CAMEROON

Anne Marie Tiani

Bessomo started having headaches three months ago. She went to the health-care centre in her quarter in Yaounde. After three weeks, she was transferred to the central hospital, where she remained for four weeks. As the disease persisted, she was taken to a tradi-practitioner in a village 120km from the town.

The story of Bessomo is the usual itinerary for urban patients in Cameroon. For villagers, the order is reversed: one goes first to the tradi-practitioners, and then to the hospital as a last resort. Thus traditional medicine is very important for public health in town and in the country, and it is estimated that 80 per cent of Cameroonians partially or totally use traditional medicine.

Traditional medicine is different from phytotherapy practised by researchers, who extract active principles from plants, and stabilize, process and sell them in drugstores or to foreign pharmaceutical laboratories. There is also naturotherapy, using plant extracts generally imported from China or the USA. Both practices are flourishing in big cities like Douala and Yaounde, no longer just in rural areas. The generalization of poverty is one of the explanations for the pervasiveness of traditional medicine. Formerly discreet, the tradi-practitioners are increasingly visible. They display their commodities or advertise their prowess along the roads. More and more, they are also using modern communication channels such as radio, TV and newspapers.

However, the practice of traditional medicine is the subject of debate within Cameroonian civil society. Three attitudes clash. First, the illiterate tradi-practitioners consider their work to be of public utility, since they cure the poorest patients at low cost, but they do not get the recognition they deserve. They feel they are exploited by researchers and scientists who refuse to acknowledge them but seek their help when researching medicinal plants or send students to learn from them at no cost. They reproach the public authorities for not giving them support.

Second, modern doctors and the wealthier sectors of society recognize the validity of traditional medicine as an ancient sociocultural practice, used to restore the physical, mental, psychic or social balance of a community or individual. They also recognize the effectiveness of traditional medicine for treating mental illnesses, hepatitis, diabetes, hypertension and other diseases where modern medicine tends to perform poorly. However, in the absence of controls and regulations, many charlatans and swindlers have slipped into the trade, and consequently the practice has become dangerous for the public health.

According to a third point of view, traditional and conventional medicine have no common ground. The two approaches occupy different spheres and evolve according to different logics.

In fact, traditional medicine does not need modern medicine, and therefore it is neither necessary nor useful to further any integration or collaboration.

The debate shows, first, the resilience and even the rise of traditional medicine; and second, the complementarity of the two approaches. Today we see public authorities in Cameroon reinforcing the ability of tradi-practitioners to obtain value from their practice:

- A Department of Traditional Socio-Sanitary Services has been established in the Ministry of Public Health. However, it is chaired by modern doctors, who can create constraints to the promotion of this practice.
- With the assistance of the WHO, an ethical code worked out by the Cameroon government is being developed, to act as regulations for traditional medicine activities and their integration in systems of public health.
- Institutions such as OAPI (African Intellectual Property Organization for French-Speaking Africa) and the Ministry of Industry, Mines and Technological Development are setting up a framework to identify tradi-practitioners, protect trademarks, approve drugs and ensure their introduction into market. Thus far, however, only a few tradi-practitioners are involved.

Since the 1970s, the integration of traditional and modern medicines has suffered a setback. In that decade, the central hospital of Yaounde had a traditional medicine sector, directed by a modern doctor. But this experiment was not conclusive and the division was closed down.

PRINCIPAL COMPONENTS OF INTEGRATIVE POLICIES

One action item in the integrative medicine policies of several developing countries is to assemble traditional healers into professional associations. The Alma Ata Declaration suggests that traditional medical practitioners and traditional birth attendants will be in a better position to receive medical training if they are members of professional associations. Governmental and non-governmental organizations work through professional associations to train indigenous healers to deliver primary health care to people with HIV/AIDS, malaria, sickle cell anaemia, diabetes mellitus and other dangerous diseases (Green et al, 1995). There are many examples from Africa and Southeast Asia of public health projects that use indigenous healers as ambassadors to promote biomedicine and deliver information about family planning and other development projects (Green et al, 1995; Ndulo et al, 2001; Homsy et al, 2004; McMillen, 2004).

In Bangladesh, Concern Worldwide trains groups of traditional birth attendants to support maternal and child health (Fink, 2002). Numerous African countries have either created associations of birth attendants, herbalists, spiritual healers and other traditional practitioners, or established relations with existing grassroots alliances. Biomedical professionals in Tanzania have responded to the HIV/AIDS epidemic by recruiting the assistance of traditional medical practitioners (Green, 1997; King, 1999; 2002; McMillen, 2004). In South Africa, indigenous healers administer health care to AIDS/HIV patients (Green et al, 1995). In Uganda, indigenous healers use botanical remedies to treat the psychological and physiological symptoms of HIV/AIDS (Fink, 2002).

Box 14.4 Medicinal plant conservation for human health and biodiversity in Tanzania

Heather McMillen

Since 1990, Tanga AIDS Working Group (TAWG) programmes based on indigenous knowledge and medicinal plants have been addressing HIV/AIDS (McMillen, 2004). From 2003 to 2006, enrolment increased fivefold to 75–85 new patients each month. Annually, 2100 clients each receive about 23kg (a total of 48.3 tonnes) of plant medicines, derived from an estimated 190 tonnes of raw material.

Rising phytomedicine demand comes from several sources:

- the increased acceptability of seeking HIV testing and treatment;
- the unavailability of anti-retroviral therapies and a preference for traditional remedies; and
- the rising incidence of chronic diseases (AIDS, diabetes, high blood pressure) not easily treated by pharmaceuticals.

In 1990, healers and their trainees harvested plants around their villages to supply TAWG. Today they travel up to 40km. None of these plants are endemic or endangered, but harvesters report that stocks are decreasing because of expanding urban and agricultural areas, competing uses and a growing number of commercial harvesters. Although the plants are still available in outlying areas, local depletion is a critical issue that can affect human health, livelihoods, habitat conservation and genetic conservation.

In 2006, TAWG began to develop a forest farm. With guidance from the coastal forestry officer and the endorsement of the village that owned the land, TAWG acquired 100 acres for habitat conservation and the cultivation of medicinal and nutritious plants. An area of their parcel overlaps the east African coastal forest mosaic, internationally recognized for its high biodiversity and endemism, and will be conserved. The larger area will be developed as a nursery and education centre. The vision is that the forest farm will produce plants to support people living with HIV/AIDS, provide training (cultivation, conservation, plant uses, plant-harvesting techniques) and generate income.

Although still in development, this forest farm can serve as a model for innovative community-based conservation that benefits local livelihoods and makes direct contributions to human health. Primary stakeholders include the following:

- a healer and expert on plant sources, life cycles, uses and cultural significance;
- a forestry officer with experience in community-based conservation;
- an NGO with a focus on indigenous knowledge and a base of support from foundations, researchers, local healers and users of plant medicines; and
- villagers (administrators, traditional leaders, the environmental committee) with a vision to improve local livelihoods and health.

Those stakeholders share commitments to the following principles:

- a local base (the initiative is from local Tanzanian organizations);
- participation (planning has involved all stakeholders from the conception);
- shared goals (to address decreasing stocks and increased reliance on phytomedicines);

- transparency (open dealings with the forestry office and the healers who are sharing their knowledge);
- tiered results (short-term benefits for the villagers, with the recognition that the benefits from medicinal plant cultivation will come much later); and
- confidence (no waiting for clinical double-blind trials to validate the efficacy of plants that are already accepted and in demand).

A temporary nursery has been established. The major plants and stakeholders involved in medicinal plant trade and cultivation in the region have been identified. The forest and farm borders are being marked and measured. Water is being sourced, fire breaks have been established, and community outreach and education have begun. Next, the plan is to initiate income-generating activities (beekeeping, tree nurseries, poultry); training in medicinal plant propagation for home gardens (provided by healers and foresters) for other healers, community members and people living with HIV/AIDS; and HIV education and outreach for local healers, peer educators and foresters, provided by TAWG.

International and national development projects that utilize indigenous healers to improve public health have stimulated grassroots action by community-based healers. For example, AIDSCOM/AIDSCAP has inspired grassroots cooperatives in Capetown, South Africa (Green et al, 1995). Traditional and Modern Health Practitioners Together Against AIDS (THETA) is a grassroots organization in Uganda whose goals are to evaluate traditional remedies, standardize the processing of herbal medicines, assess spiritual therapies, prevent HIV, care for its victims and safeguard intellectual property rights (Homsy et al, 2004). This group also facilitates communication among health-care workers about preventing and treating HIV/AIDS. The Uganda Cooperative Society has organized the Herbalist and Farmers Development Initiative Cooperative Society, a group of people who produce, process and administer herbal medicines. Recent events demonstrate that grassroots organizations influence national policies. In June 2006, the Sierra Leone Traditional Healers Association drafted legislation to standardize traditional medicine that the Minister of the Department of Health and Sanitation is submitting to the parliament.

Regulation and systemization of herbal medicines is a common element of many integrative health policies. Several countries are developing compendia of traditional herbal remedies. In India, the Ministry of Health's Department of Ayurveda, Yoga, Unani, Siddha and Homeopathy (AYUSH) is managing the Traditional Knowledge Data Library, a catalogue of the Indian pharmacopoeia. This library is an inventory of 70,000 Ayurveda plant medicines, 65,000 Unani formulas, and 3000 Siddha mixtures. The database will eventually include 7000 more Unani and Siddha prescriptions and 1500 yoga postures.

The standardization of ethno-medicines is one of the most prominent issues in traditional medicine at the moment. Uganda, for example, is developing policies to standardize the packaging and dosage of herbal medicines. In Kenya, the Ministry for Planning and Development is developing a policy to regulate the use of herbal medicines and to coordinate scientists, medical doctors and traditional health practitioners in efforts to prevent HIV/AIDS.

Ensuring the safety and efficacy of herbal medicines is a priority for advocates of integrative medicine. The Centre for Traditional Medicine and Drug Research, a sector of Kenya Medical Research Institute, conducts research on the safety and effectiveness of herbal medicines with special interest in potential malaria and HIV/AIDS drugs. Regarding efficacy, the Zimbabwe National Traditional Healers Association has an HIV/AIDS policy governing prevention and treatment, stating that traditional healers should not claim they can cure AIDS because it may lead to mismanagement of disease symptoms. More generally, the association supports traditional values and discourages discrimination against HIV/AIDS patients (Xinhua, 2006).

The policies of several countries in Asia and Africa explicitly permit health-care workers to treat patients with traditional therapies. In Pakistan, medical clinics use both traditional and allopathic pharmaceuticals to treat cancer patients (Tovey et al, 2005). The Food and Drug Administration in Thailand recognizes the safety and effectiveness of 60 plants and recommends their use by primary health-care workers to treat 25 diseases (WHO, 2004a). The Thai Traditional Medical Development Centre hosts educational workshops on Thai herbal medications and massage and supports their use.

An increasing number of developing countries in Africa, Southeast Asia and the Southwest Pacific have government ministries that supervise traditional and integrative medicine. Sri Lanka has a Ministry of Health and Indigenous Medicine that oversees integrative medicine. Peru operates a National Program in Complementary Medicine (Fink, 2002). In Kenya, the Ministry for Planning and Development, the Ministry of Health and the Ministry of Gender, Sports, Culture, and Social Services each have some duties related to traditional medicine. Although Kenya does not yet have a formal national policy on traditional medicine, the Ministry of Culture is leading efforts to institutionalize it, which will be connected to conservation and education initiatives. In 2006, the President of Nigeria announced plans to establish the Traditional Herbal Institute to produce and market botanical remedies.

The protection of information, ideas and knowledge is an increasingly important legal issue that intersects with the growing popularity of traditional medicine. The UN Convention on Biological Diversity (Aguilar, 2001; Timmermans, 2003) and the Agreement on Trade-Related Aspects of Intellectual Property Rights (Timmermans, 2003) are two international policies that safeguard indigenous knowledge and healing techniques. India, China, Brazil and nine other nations are campaigning together for international protections against the patenting of herbal medicines by other countries. Indigenous peoples continue to struggle for intellectual property rights recognition and protection, but negotiating compensation for their knowledge or the commercialization of traditional medicine is 'complex, contentious, and politically nuanced' (Etkin and Elizabetsky, 2005, p26).

India's Ministry of Health is reluctant to make the Traditional Knowledge Data Library's contents publicly available to people in other countries, even though it is creating the database to prevent non-Indians from patenting Indian knowledge and materials. As a result of the reluctance to share its contents, about 18,000 plant medicines listed in the database have been patented in other countries since 2002 (Gupta, 2006). To avoid further redundancy, India's Union Cabinet recently authorized the National Institute of Science Communication and Information Resources to establish contracts with the patent offices of other nations and to allow access to the library. In exchange, other nations must agree not to issue any more patents on formulations it lists. India's parliament is currently considering a national design policy that addresses intellectual property rights and patenting of herbal medicines and other inventions.

Thailand's Public Health Ministry structures Thai traditional medicine with intellectual property rights regulations to protect indigenous medical knowledge, including the recipes for botanical medications. The Thai Ministry hopes to stimulate a 5 to 10 per cent increase in the proportion of the Thai population who use traditional medicine. They also hope to increase the number of commonly used plant medicines from 23 to 400. The Thai Traditional and Alternative Development Department has two health centres in each Thai province and offers traditional medical care in 117 hospitals around the country (MCOT News, 2006).

There are tremendous economic incentives for integrating traditional medicine into the global mainstream of health care. Equitable sharing of profits from the production of botanical supplements by pharmaceutical companies is one reason that India's government is working to institutionalize its medical traditions. In Kenya, leaders of national traditional medicine organizations emphasize the potential financial benefits from the global popularity of alternative health therapies. Malaysia's third industrial masterplan recommends the development of small and medium-sized enterprises based on traditional medicine. In 2006, Malaysia's Deputy Minister of International Trade and Industry stated that such enterprises should make the transition from basic operations of 'processing dried plant materials to herbal powders and fermentation to produce medicinal tonics' to a 'higher level of technology and make use of Malaysia's extensive biodiversity and local knowledge in traditional and alternative medicine' (Hamid, 2006).

Box 14.5 Major components of integrative medicine policies

- Assemble professional associations of indigenous healers.
- Train indigenous healers to deliver primary health care.
- Establish government ministries.
- Test and regulate traditional therapies, especially herbal medicines.
- Protect intellectual property rights.
- Produce financial profits.

Barriers to integration

The success of initiatives to integrate biomedicine and traditional medicine should be assessed in terms of broader sociocultural and politico-economic factors. The relationships that indigenous healers and consumers have with biomedical practitioners and consumers on the one hand, and with the staff and sponsors of integrative projects on the other hand, affect their willingness to comply with recommendations and directives. When the practitioners and consumers of traditional medicine 'fail' to comply with or participate in integrative projects, they may be expressing opposition to their living conditions or broader politico-economic issues. More directly, they may be resisting attempts to domesticate traditional medicine, co-opt the power of indigenous healers and reach the consumers of alternative therapies.

For many decades, Western health-care workers have been trying to figure out why some traditional communities are slow to adopt the biomedical modality. They have identified sociocultural differences, economic costs, accessibility of services and quality of care as major deterrents to adoption. Meanwhile, anthropologists have been documenting the ways traditional peoples select and adapt elements of biomedicine to manage their health needs. Ironically, advocates of alternative and complementary therapies wonder why biomedicine is slow to adopt traditional therapies.

BOX 14.6 POLYPHARMACY IN KALIMANTAN

Lisa X. Gollin

If rural health programmes are to succeed, the problems and promises of combining ethno-medical and biomedical modalities require special attention. The Indonesian Ministry of Health has established a system of village clinics and cadres – the *Puskesmas-Posyandu* system (see Chapter 15, this volume) – and promotes traditional medicine. Most of Kalimantan, the Indonesian-held portion of Borneo, is rural. It is in these areas that health services are most deficient. Government efforts do not fully meet the needs of Indonesia's diverse and dispersed populations, or take advantage of local biocultural resources. In East Kalimantan, where the author conducted ethno-botanical research among the Kenyah Leppo` Ke (Gollin, 2001), a combination of medical options is available in the villages: primarily, plant remedies gathered or grown in the forest and field; secondarily, popular Indonesian botanicals and ointments; and occasionally, access to biomedical care and pharmaceuticals from government clinics and apothecaries downriver (see also Chapter 16, this volume).

As is true elsewhere, polypharmacy – use of local therapies in conjunction with pharmaceuticals – is common in Kalimantan. Biomedicines are incorporated and used according to pre-existing notions of healing. Bitter pills (e.g., quinine) are used in much the same way as bitter plants: to expel fever from the body or a foetus from the womb (oxytocic to hasten labour). Iron tablets, injections of the red serum cobalamin (vitamin B12), and a red fern (*Stenochlaena palustris*) all possess a red sign of their efficacy as blood fortifiers. The equation of biomedicines with indigenous medicines and the synergistic or antagonistic effects of combining pharmaceuticals with bioactive plants can be serendipitously beneficial, deleterious or benign.

Health-care providers prescribe multi-drug cocktails and often do not label pills or provide adequate instructions for use. A woman diagnosed with ulcers is given antibiotics to address the infection, iron to address the anaemia associated with women, an analgesic to relieve pain, and an antihistamine prescribed for its side-effect as a sleeping aid.

Limited consultation time, lab facilities and diagnostic capabilities make it easy for villagers to present faux maladies. Villagers, particularly the semi-nomadic Punan, present anticipated ailments during consultations so that they can stockpile medicines for forest sojourns. Households have candy jars full of unlabelled pills.

Villagers obtain syringes and ampoules from downriver apothecaries. One village 'shot expert' injects metamizole because it makes one sweat in conjunction with plant medicines, encouraging disease egress in patients with fever convulsions. A powerful analgesic and fever reducer, metamizole has been banned in the USA and Europe because of the risk of a potentially fatal condition, agranulocytosis.

Women have negative experiences with the biomedical establishment where reproductive health is concerned (see Chapter 7, this volume). Government clinicians promote the subcutaneous prophylactic implant Norplant to village women, many of whom have had reactions such as fever, violent headaches and chronic fatigue. Women employ a variety of botanical therapies to mitigate Norplant side-effects.

The Kenyah pharmacopoeia includes birth control plants. A few are harmful. The antidote vine (*Aristolochia* sp), used to treat hangovers, can also be used to prevent pregnancy. Aristolic acid has demonstrated implantation-inhibiting properties but is also associated with kidney failure and carcinogenicity.

Acknowledging that plant medicines are a readily accessible form of primary health care, *Puskesmas-Posyandu* workers promote village medicinal gardens. Some are designed without regard to the extensive local pharmacopoeia, featuring Javanese flora (many health workers are from Java). Such gardens are ignored and even disdained by villagers, who regard them as a form of cultural-botanical imperialism.

The safety, efficacy and availability of ethno-botanical therapies have yet to be fully investigated, and the information needs to be disseminated in a locally meaningful way (e.g., via ethno- and biomedical practitioners, learning gardens, guidebooks and more).

The globalization of traditional medicine reveals difficulties in applying generic concepts to local situations (Pigg, 1995). Development agencies (e.g., the World Bank, USAID) frequently use a generic version of 'traditional' or 'indigenous' medicine in their plans to integrate non-biomedical systems into health improvement programmes. In effect, integration fails because the medical system adopted resembles the agencies' definition of 'traditional' more than the local community's (Pigg, 1995). Traditional medicine includes the health management practices of an incredibly diverse collection of complex communities, some of which are rooted in specific locations and others that are dispersed. Traditional healing systems are temporally dynamic and spatially variable and operate within broader sociocultural configurations.

'Traditional medical practitioners' is a broad category that masks considerable diversity. Like attempts to standardize traditional medicine, efforts to institutionalize traditional medical practitioners ignore the wide variety of healing paradigms and methods.

CONCEPTUAL DISJUNCTIONS

Several profound conceptual disjunctions complicate the movement to rationalize traditional medicine and produce national and international standards. The WHO (2005b) believes that the main barriers to effective national legislation are 'lack of research data, lack of appropriate control mechanisms, lack of education and training and lack of expertise'. Some national-level as well as some grassroots organizations that promote integrative medicine have objectives and strategies that are very similar to the WHO's, so the cultural differences are not always present or obvious. Among the goals of India's Department of AYUSH, for example, are the 'standardization and quality control of traditional herbal drugs, documentation, dissemination of knowledge and to explore its potential in improvement of health status of people in the country' (Pune Newsline, 2006).

Epistemological differences – including explanations of illness and knowledge transmission – and the underlying objectives of participants are major barriers to integrating indigenous medicine and biomedicine. The different health paradigms can complicate communication about disease (e.g., contagions) and illness (e.g., culture-bound syndromes). A material effect is that donors resist funding health-care practices where knowledge resides in oral history (Fink, 2002). It is important to continue investing in social research and in medical schools with traditional medicine programmes to guide governmental and non-governmental organizations in developing integrative medicine policies (Tsey, 1997). Ghana and Hawaii, for example, offer instruction in traditional medicine to their students. The Institute of Traditional Medicine in Burma operates a five-year degree in traditional medicine and has 128 instructors who teach classes in plant and animal medications, panchakarma, massage, acupuncture and other topics (PTI, 2006).

Attempts to submit traditional medicine to scientific scrutiny (Tsey, 1997; Alter, 2004) present barriers to integrative medicine. In the biomedical framework, safety and efficacy are determined by scientific methodologies and sanctioned by governmental bureaucrats. Many biomedical practitioners are not willing to use or recommend an alternative therapy unless it has been proven safe and effective in multiple randomized controlled trials (Nahin et al, 2005; see Chapter 5, this volume). The WHO's policies reveal ambivalence about traditional medicine. On the one hand, they express the sentiment that traditional medicine is beneficial because poor people have better access to indigenous practitioners than to biomedical doctors. On the other hand, policy documents from the WHO question the safety of traditional medicine and the sustainable production of herbal medicines. The WHO's approach to providing health care to the rural poor in the developing world is not to integrate traditional perceptions and methods into public health programmes. Instead, it employs the personnel of traditional health care – native practitioners – to deliver modern medicine without necessarily altering the premises and techniques of biomedicine. One solution to these problems is to select healers from existing community-based organizations (Green et al, 1995).

Western scientists have difficulty locating evidence of efficacy in traditional medicine (Vickers, 2001), and attempts to medicalize it can cause problems. The commercialization of plant medicines has the potential to change their effects because it involves standardizing dosages, which are easily miscalculated. Western scientists also have difficulty recording and measuring plant consumption in all of its contexts because biochemically active plants are consumed in different contexts and various dosages (Etkin and Elizabetsky, 2005). In indigenous contexts, the effectiveness of healing practices is determined in the long-term, dynamic negotiations of interpersonal relationships in conjunction with the changing conditions of patients and the treatments they receive, along with many other contextual factors (Waldram, 2000).

Conventional medicine continues to exclude ethno-nosologies (classifications) and minimize ethno-etiologies (causes). There are many indigenous cures for unhealthy social conditions, for example, yet the biomedical establishment has difficulty addressing the diagnosis that psychosocial issues cause illnesses. A related disconnect is that biomedical practitioners do not understand spiritual healing in indigenous religions (Tsey, 1997). Spiritual healers in some cultures believe that they must be called to the profession by supernatural forces or beings and cannot be 'taught' to heal (Tsey, 1997). Indigenous healers may be reluctant to participate in projects if they will be 'educated' to treat their patients. Spiritual healing is a valued type of healing in Ghana's Ewe communities, but, like preventive medicine, it has not yet been acknowledged by public health workers (Tsey, 1997).

Efforts to integrate medical systems have aggravated the difficult power relations that exist between traditional healers and biomedical doctors – even, for example, in Ghana's monoracial society (Tsey, 1997), where the 'ways of the ancestors' have been stigmatized by some elite members of society (Concord Times, 2006). International aid organizations (e.g., USAID) and NGOs can potentially exacerbate racial inequalities when selecting healers to participate in public health programmes. Although these organizations often produce tremendous public health benefits, they can perpetuate dominant–subordinate relationships between Western and indigenous cultures if they do not respect indigenous healers and recognize the value of the therapies they use.

COMPLEX MEANINGS

Contemporary empirical assessments of traditional medicine contextualize the health-related beliefs and practices of local communities in history, culture, politics, economics and local ecologies. Recent studies situate indigenous healing systems within the context of creolized scientific-indigenous health-care formations and within the context of rapid flows of information between indigenous groups and other communities around the world. Traditional medicines are syncretic, pluralistic and have a diversity of complex meanings.

Traditional medicine can have different meanings within a single society. Among some groups in Madagascar, indigenous healers and their clientele are ashamed to reveal their involvement in traditional medicine and therefore practise it as an underground activity (Fink, 2002). In other segments of the population, where traditional medicine symbolizes ethnic pride, healers and clientele are proud to associate with it. Native Hawaiians are negotiating tenuous boundaries between the celebration of traditional medicine as anti-colonial sentiment and the standardization of traditional medicine by local, state and federal governments. Some segments of the native Hawaiian population celebrate indigenous knowledge, elevate the status of indigenous healers and promote the use of traditional herbal medications. At the same time, local communities are proud of the native Hawaiian medical practitioners who are working within biomedical systems. Some health-care providers in Hawaii integrate native Hawaiian and biomedical systems, such as the Waimanalo Health Clinic, which offers both allopathic and ethno-medical treatments.

To further illustrate the notion that traditional medicines have complex meanings, we can consider the fact that some indigenous healers resist government intervention (Fink, 2002) while others welcome and even encourage attention to their healing traditions. An example of the former is Don Warne, a Lakota (Native American) healer and medical doctor with a master's in public health who works for the National Institutes of Health. Warne says, 'We don't need the government coming in and saying whether or not practices are useful, safe or effective because we've done it for thousands of years' (Fink, 2002, p739). An example of the latter is Kassomo, a traditional healer in Tanga town, who has been spotlighted in the media (a UNAIDS publication, World Bank reports, a BBC broadcast and anthropological literature) because of his collaborations with biomedical institutions (McMillen, 2004). Another example is in India, where native practitioners are very proud of their medical traditions and are leading the movement to 'unify the entire system and codify [ethno-medicines]' (Mukherjee and Wahile, 2006).

FACTORS THAT FACILITATE INTEGRATION

What are the conditions that will enable integrative public health policies to succeed? Policy-makers should use empirical evidence to design legislation that reflects the ethnographic data on health-seeking behaviours and perceptions of health. Social scientists have documented many cases where people use both traditional and modern medicine to manage health and illness. These researchers can advise governmental and non-governmental health projects on how to translate information between biomedical practitioners and indigenous healers (e.g., Green et al, 1993) and help overcome the lack of coordination between public health officials and traditional healers (Fink, 2002).

Health policy-makers ought to consider the opinions of people who use traditional medicine or practise medical pluralism. In rural Zambia, local people support collaboration between traditional and modern health practitioners in caring for HIV/AIDS patients. According to Zambians, the following conditions are necessary for the success of integrative initiatives (Kaboru et al, 2005):

- traditional healers should be paid for their work;
- the secret techniques and materials that traditional healers use should not be revealed;
- traditional healers should be provided with training;
- traditional and modern doctors should be taught how to cooperate; and
- local communities should be allowed to participate.

Community participation is essential for the success of formal attempts to integrate traditional and modern medicine (Kaboru et al, 2005). It is profitable to work through existing community-based organizations in addition to collaborating with larger NGOs. Research shows that the traditional medical practitioners who are the most effective collaborators with HIV/AIDS prevention programmes are motivated to learn about biomedicine and are respected by other members of their communities, not those who belong to official healer organizations or are formally educated (Green et al, 1995).

Traditional healers who receive attention from government health officials and NGOs appreciate the 'recognition and encouragement' (Fink, 2002, p1734). Respect for traditional medicine and indigenous healers is crucial in integrative medicine projects that aim to improve the health of poor people in the developing world (WHO, 2004b), particularly for people living in forests, where diseases are common and conventional formal health care is rare.

Pluralistic medical environments can yield numerous benefits. For instance, integrating traditional medicine into the medical establishment has the potential to empower native forest communities in addition to improving healer–patient relations and health-care delivery. National governments that adopt elements of traditional medicine are better equipped to cure and heal patients. Curing, the goal of globalized medicine, is the successful treatment of the biomedical and biochemical causes of disease. Many traditional medical systems heal patients by treating the whole illness experience (Strathern and Steward, 1999). Pluralistic medical systems that combine curing with healing medicine are more holistic and potentially more effective for restoring patients' wellbeing.

National public health policies affect the lives of real people even though individuals do not necessarily consult official policies when deciding how to manage their health.

Box 14.7 Potential impacts of institutionalization of traditional medicine

- status of participating individuals;
- technical management of natural products;
- preparation and dosage of traditional medicines;
- social contexts of healing;
- conceptualizations of healing and illness;
- reproduction of indigenous knowledge; and
- cost, availability and access to traditional medicines.

Globalization, post-colonialism and integrative medicine

The institutionalization of traditional medicine is linked to global governance and post-colonialism in the late 20th and early 21st centuries. Globalization facilitates the integration of traditional and allopathic medicine. Globalized culture in the 21st century has a positive opinion of indigenous knowledge. The popularity of traditional medicine has caused the leaders of globalized medicine to acknowledge its value. Policy-makers at international development and aid organizations that are operating in forested countries and encouraging national governments to adopt integrative medicine are representatives of globalized culture. The global movement towards integrative medicine may confer higher status to traditional medicine and indigenous healers, boosting their profile in national forums and perhaps in local communities. It is unlikely, however, that the employment of indigenous healers and the regulation of traditional medicine will confer financial and political power that is equal to the power of government leaders or biomedical practitioners.

Post-colonialism is good for integrative medicine. Self-consciousness of cultural identity is an after-effect of colonialism. People living in many formerly colonized nations are refashioning materials, behaviours and ideas that they associate with their identity. One of the ways they express ethnic pride is by celebrating indigenous knowledge and natural resource management, which are frequently represented by native healing systems and herbal remedies. The growing popularity of traditional plant medicines may have the unintended or intended consequence of preserving indigenous knowledge (Mukherjee and Wahile, 2006). This particular condition of post-colonialism contributes to the growth of integrative medicine even though there may be other aspects that hinder or negatively affect indigenous healing systems.

Globalization and post-colonialism operate in different ways. The flow of ideas and materials is typically top–down in globalization but bottom–up in post-colonialism. Traditional medicine is thus being reinforced by the dominant forces and classes in global society and by subaltern forces and classes in local communities. Those national governments that already have or are currently developing relevant policies provide additional substantiation. Traditional medicine is thriving from the attention it is receiving – in 2006 it was worth US$60 billion per year – and it is predicted that the industry will continue to grow by 7 per cent annually (Etkin and Elizabetsky, 2005).

Traditional medicine is extraordinarily dynamic. It is likely to assume multiple new forms where it continues to merge with biomedicine. Perhaps integrative medicine will replace Western allopathy as the dominant globalized medicine. The financial and health incentives supporting integrative medicine are substantial. Will the rising number of developing countries whose national public health policies support traditional medicine yield a political economic realignment, such that forest-dwelling communities have more equitable shares of power and money?

NOTE

1 *National Policy on Traditional Medicine and Regulation of Herbal Medicines: Report of a WHO Global Survey* (WHO 2005a), is a valuable resource for information on the WHO's interactions with their member states regarding traditional medicine, particularly Chapter 5 ('Country Summaries'), which has an overview of the status of policy and herbal medicine regulations in 134 countries around the world.

REFERENCES

Aguilar, G. (2001) 'Access to traditional resources and protection of traditional knowledge in the territories of indigenous peoples', *Environmental Science and Policy* 4(4–5): 241–256

Alter, J.S. (2004) *Yoga in Modern India: The Body between Science and Philosophy*, Princeton University Press, Princeton, NJ

Boerma, J.T. and Salim, B.M. (1990) 'Maternal and child health in an ethnomedical perspective: Traditional and modern medicine in coastal Kenya', *Health Policy Planning* 5(4): 347–357

Bourdillon, M.F.C. (1998) *The Shona Peoples*, Mambo Press, Gweru, Zimbabwe

Chavanduka, G.L. (1994) *Traditional Medicine in Modern Zimbabwe*, University of Zimbabwe, Harare, Zimbabwe

Concord Times (2006) 'The elites and traditional medicine', www.concordtimess1.com/biz.htm

DeJong, J. (1991) 'Traditional medicine in sub-Saharan Africa: Its importance and potential policy options', Working Paper, World Bank, Washington, DC

Etkin, N. and Elizabetsky, E. (2005) 'Seeking a transdisciplinary and culturally germane science: The future of ethnopharmacology', *Journal of Ethnopharmacology* 100(1–2): 23–26

Fink, S. (2002) 'International efforts spotlight traditional, complementary, and alternative medicine', *American Journal of Public Health* 92(11): 1734–1740

Gollin, L.X. (2001) 'The taste and smell of taban Kenyah (Kenyah medicine): An exploration of chemosensory selection criteria for medicinal plants among the Kenyah Leppo` Ke of East Kalimantan, Borneo, Indonesia', PhD dissertation, Anthropology, University of Hawaii, Honolulu, HI

Good, C.M. (1977) 'Traditional medicine: An agenda for medical geography', *Social Science and Medicine* 11: 705–713

Green, E. (1988) 'Can collaborative programs between biomedical and African indigenous health practitioners succeed?' *Social Science and Medicine* 27(11): 1125–1130

Green, E. (1997) 'The participation of African traditional healers in AIDS/STD prevention programmes', *Tropical Doctor* 27 (Supplement 1): 56–59

Green, E., Jurg, A. and Dgedge, A. (1993) 'Sexually transmitted diseases, AIDS and traditional healers in Mozambique', *Medical Anthropology* 15: 261–281

Green, E.C., Zokwe, B. and Dupree, J.D. (1995) 'The experience of an AIDS prevention program focused on South African traditional healers', *Social Science and Medicine* 40(4): 503–515

Gupta, M. (2006) 'Patents slip out as India nurses data', available at www.business-standard.com/common/storypage.php?storyflag=y&leftnm=lmnu2&leftindx=2&lselect=1&chk login=N&autono=212625 (accessed 22 January 2006)

Hamid, H. (2006) 'Traditional medicine, herbal products a growth area', www.herbamalaysia.net/index.php?option=com_content&task=view&id=111&Itemid=45

Homsy, J., King, R., Tenywa, J., Kyeyune, P., Opio, A. and Balaba, D. (2004) 'Defining minimum standards of practice for incorporating African traditional medicine into HIV/AIDS prevention, care, and support: A regional initiative in eastern and southern Africa', *Journal of Alternative and Complementary Medicine* 10(5): 905–910

Kaboru, B.B., Falkenburg, T., Ndulo, J., Muchimba, M., Solo, K., Faxelid, E. and The Bridging Gap's Project's Research Team (2005) 'Communities' views on prerequisites for collaboration between modern and traditional health sectors in relation to STI/HIV/AIDS care in Zambia: health Policy', www.ncbi.nlm.nih.gov/sites/entrez?db=pubmed&uid=16290128&cmd=show detailview&indexed=google

King, R. (1999) *Collaboration with Traditional Healers in AIDS Prevention and Care in Sub-Saharan Africa: A Comparative Case Study Using UNAIDS Best Practice Criteria*, UNAIDS, Geneva, Switzerland

King, R. (2002) 'Ancient remedies, new disease: Involving traditional healers in increasing access to AIDS care and prevention in East Africa', UNAIDS Case Study, UNAIDS Best Practice Collection, http://unaids.org/publications/IRC-pub02/jc761-ancientremedies_en.pdf

McMillen, H. (2004) 'The adapting healer: Pioneering through shifting epidemiological and sociocultural landscapes', *Social Science and Medicine* 59(5): 889–902

MCOT News (2006) 'Public health ministry encourages Thai traditional medicine', MCOT English News, Thai News Agency

Mukamuri, B. and Kozanayi, W. (1999) 'Institutions surrounding the use of marketed bark products: The case of *Berchemia discolors*, *Adansonia digitata*, and *Warburgia salutaris*', IES working paper No. 17, Institute of Environmental Studies (IES), University of Zimbabwe, Harare, Zimbabwe

Mukherjee, P.K. and Wahile, A. (2006) 'Integrated approaches toward drug development in Ayurveda and other Indian systems of medicines', *Journal of Ethnopharmacology* 103(1): 25–35

Nahin, R.L., Pontzer, C.H. and Chesney, M.A. (2005) 'Racing toward the integration of complementary and alternative medicine: A marathon or a sprint?' *Health Affairs* 24(4): 991–993

Ndulo, J., Faxelid, E. and Krantz, I. (2001) 'Traditional healers in Zambia and their care for patients with urethral/vaginal discharge', *The Journal of Alternative and Complementary Medicine* 7(5): 529–536.

Phillips, D.R.F. (1990) 'Traditional and modern health care in the Third World', in *Health and Health Care in the Third World*, Longman Scientific and Technical, Harlow, Essex, UK, pp63–102

Pigg, S.L. (1995) 'Acronyms and effacement: Traditional medical practitioners (TMPs) in international health development', *Social Science and Medicine* 41(1): 47–68

PTI (2006) 'Indian president continues Burma visit', http://0-web.lexis-nexis.com.library.wofford.edu/universe/document?_m=35a86a4a85fac85ce6ba9f649f31ca86&_docnum=121&wchp=dGLbVzb-zSkVA&_md5=6ece6743c943eaf92bfa381c928f6d18 (accessed 31 July 2006)

Pune Newsline (2006) 'Bushan Patwardan on AYUSH Panel', http://cities.expressindia.com/fullstory.php?newsid=188899, (accessed 18 July 2006)

Strathern, A. and Stewart, P. (1999) *Curing and Healing: Medical Anthropology in Global Perspective*, Carolina Academic Press, Durham, NC

Sunday News (2006) 'Fake ARVs on sale', 23 April, Zimbabwe

Timmermans, K. (2003) 'Intellectual property rights and traditional medicine: Policy dilemmas at the interface', *Social Science and Medicine* 57(4): 745–756

Tovey, B.A., Chatwin, J., Hafeez, M. and Ahmad, S. (2005) 'Patient assessment of effectiveness and satisfaction with traditional medicine, globalized complementary and alternative medicines, and allopathic medicines for cancer in Pakistan', *Integrative Cancer Therapies* 4(3): 242–248

Tsey, K. (1997) 'Traditional medicine in contemporary Ghana: A public policy analysis', *Social Science and Medicine* 45(7): 1065–1074

UNAIDS (2000) 'Collaboration with traditional healers in HIV/AIDS prevention and care in sub-Saharan Africa: A literature review', UNAIDS Best Practice Collection, Joint United Nations Programme on AIDS, Geneva, Switzerland

Vickers, A.J. (2001) 'Message to complementary and alternative medicine: Evidence is a better friend than power', *BMC Complementary and Alternative Medicine* 1(1): 1–3

Waldram, J.B. (2000) 'The efficacy of traditional medicine: Current theoretical and methodological issues', *Medical Anthropology Quarterly* 14(4): 603–625

WHO (1977) 'Promotion and development of training and research in traditional medicine', Thirtieth World Health Assembly, Document No WHO 30.49, World Health Organization, Geneva

WHO (1978) 'Declaration of the Alma-Ata International Conference on Primary Health Care', Alma-Ata, USSR, 6–12 September, www.who.int/hpr/NPH/docs/declaration_almaata.html (accessed 2 March 2006)

WHO (2004a) 'Review of traditional medicine in the South-East Asia region', Report of the Regional Working Group New Delhi, World Health Organization, Geneva, Switzerland

WHO (2004b) 'Commission on intellectual property rights, innovation and public health', Framework Paper, World Health Organization, Geneva, Switzerland

WHO (2005a) 'National policy on traditional medicine and regulation of herbal medicines', Report of WHO Global Survey, www.correofarmaceutico.com/documentos/190905_plantas.pdf (accessed 20 July 2006)

WHO (2005b) 'Traditional medicine fact sheet', www.who.int/mediacentre/factsheets/fs134/en/ (accessed 11 January 2006)

World Bank (2001) 'Learning from local farmers and healers: World Bank discuss validation of herbal treatments', http://web.worldbank.org/WBSITE/EXTERNAL/COUNTRIES/ECAEXT/ALBANIAEXTN/0,,contentMDK:20020456~menuPK:34460~pagePK:34370~piPK:34424~theSitePK:301412,00.html

Xinhua (2006) 'Zimbabwe to launch anti-AIDS policy framework', http://english.people.com.cn/200601/19/eng20060119_236638.html (accessed 22 January 2006)

15
Approaching Conservation through Health

Robbie Ali

Access can be a double-edged sword. Remote areas of developing countries are often home to people who lack access to even the most basic health care. Some of these areas, such as rainforests, may also have high value in terms of biodiversity. In fact, remoteness and lack of access in these locations explain why these areas have remained ecologically intact and species-rich. In Kalimantan, for example, most logging occurs along roads and rivers; forested areas farther from roads and rivers are much more likely to have remained intact. The building of a new road may provide much-needed access for isolated local peoples, but it may also be the first step in a process that ends in the clear-cutting of the surrounding forest.

In recent years, conservation groups working at many sites in developing countries have made attempts to work more closely with local communities to help preserve biodiversity. The basic premise of such community-based conservation is that the decisions and activities of local communities can influence conservation outcomes.

In many projects, notably those known as integrated conservation and development projects (ICDPs), activities more directly related to conservation are conducted along with those addressing local economic development, social services, physical infrastructure and civic capacity. Such projects, like other conservation efforts, have yielded some successes but also failures, due to mistakes in project design or implementation or to factors outside the project managers' control.

Because health services for people living in or adjacent to the rainforest are often lacking, and because health issues often rank high among local perceived needs (Puchong, 1999; Taylor and Taylor, 2002), the potential would seem to exist for health-related activities to lead to better community-based conservation. Health and conservation initiatives have indeed been linked in at least 40 programmes in developing countries during the past decade. Melnyk (2001) and Margoluis et al (2001) reviewed recent projects that linked health activities to biodiversity conservation. Engleman (1998) and Vogel and Engleman (1999) reviewed projects that specifically linked reproductive health and family planning activities to biodiversity conservation. Mogelgaard (2003) has reviewed ICDPs in developing countries in Latin America, Africa and Asia, undertaken by conservation organizations and international and local NGOs, often as collaborative efforts. Many projects reported positive impacts of health activities on conservation outcomes, but reports were typically anecdotal, since most programmes were not designed as research with hypothesis testing or attribution of particular outcomes to particular activities.

Margoluis et al (2001) distinguish between two kinds of health–conservation linkages. Community health activities that are *conceptually linked* to conservation are 'directly related to the maintenance of intact biodiversity in the local environment'. Examples include projects that focus on:

- health education linked to environmental education promoting more sustainable use of natural resources;
- diarrhoeal disease and erosion or watershed management;
- malarial mosquitoes as related to ecosystem disruption (see Chapter 9, this volume);
- medicinal plants (see Chapter 3, this volume);
- respiratory health as related to fuelwood and forest fires (see Chapter 5, this volume);
- nutrition and food security (see Chapter 4, this volume); and
- eased long-term demands on forests through better family planning and slower population growth (see Chapter 6, this volume).

Health activities *operationally linked* to conservation, on the other hand, may address health concerns not directly related to intact biodiversity. The keys to operational linkages are 'the *ways* in which project managers utilize health activities to achieve conservation outcomes, i.e., *how* conservation practitioners functionally connect health interventions and conservation outcomes in project design and implementation' (Margoluis et al, 2001, emphasis added). For example, a mobile clinic providing monthly childhood immunizations would not be a health activity conceptually linked to conservation, but it could be linked operationally if clinic sessions also included an environmental education component, or if the mobile clinic was developed as a response to community need and so led to increased goodwill towards the conservation organization.

Margoluis et al (2001) provide a useful outline of four strategies commonly utilized for linking health and conservation: bartering, creating an entry point, building a bridge and finding symbiosis. I propose a modification to this list of strategies to both clarify and amplify the ways in which health activities may benefit conservation: adding a fifth strategy for linking health activities with conservation – empowerment (Table 15.1).

Note that the barter strategy may involve contingency. That is, a health service provided as an incentive can potentially be withdrawn if a community does not fulfil its side of a 'conservation contract'. This clearly has ethical implications that should be carefully considered. Note also that the primary difference between the symbiotic strategy and the bridge strategy is whether community residents are aware of the conceptual linkage between the health activity and conservation. This strategy has overlap with the entry-point strategy, but here I recommend that the definition of the latter be narrowed to read, 'health projects or services used to build strong relationships with community residents'. The empowerment strategy can then be defined as 'health interventions used to increase community capacity for conservation'.

The empowerment strategy is particularly valuable when working with communities that tend to be natural allies for conservation, such as indigenous groups or workers in conservation-related fields. Examples of the empowerment strategy include:

1 improving health care to improve forest workers' physical wellbeing and, thereby, conservation, such as creating an employee health programme for wildlife trackers and guides (Ali et al, 2004);

Table 15.1 *Strategies for linking health activities and conservation*

Strategy	*Definition*	*Example*
Barter	Quid pro quo: health services provided to a community in exchange for behaviour that sustains biodiversity conservation.	Establishment of a community clinic in exchange for an agreement not to cut forest in a nearby protected area.
Entry point	Health projects or services used to build strong relationships with community residents.	Establishment of a community clinic that eventually leads to environmental education activities and community land-use planning.
Empowerment	Health interventions used to increase community capacity for conservation.	Training local health-care workers in record-keeping and other skills useful in forest monitoring.
Bridge	Health interventions undertaken with the intention of linking them conceptually to conservation activities.	Installation of a village water line to promote future watershed protection.
Symbiotic	Health project interventions developed based on known common ground between the health needs of a population and its conservation goals.	Installation of a village water line when community members are already aware of the importance of watershed protection.

Source: Adapted from Margoluis et al, 2001.

2 providing health-care training for local people that develops transferable skills (e.g., reading, record-keeping) that become useful in conservation-based employment or entrepreneurship;
3 using health as a focal point for community learning about collective management that leads to increased social capital (e.g., the capacity for collective decision-making and community mobilization);
4 establishing or strengthening community structures or organizations for health that also come to be used for natural resource governance; and
5 employing local people in health-related fields as an economic alternative to employment in environmentally destructive logging.

COMBINING HEALTH CARE AND CONSERVATION AGENDAS

But how can the health-care agenda and the conservation agenda be best combined? An obvious goal of physicians and other health practitioners is to improve health and health care, particularly for those in greatest need. Clearly, most remote areas of developing countries are inhabited by groups of people who are underserved in health care and especially lack access to well-trained physicians.

Health-care services are typically concentrated in urban areas of high population density, but the activities of conservation agencies (whether national or international, governmental or non-governmental) are, conversely, usually in less populated rural and remote areas. Because resources are limited, areas chosen as conservation priorities must first be identified as having high conservation value (e.g., in biodiversity and ecotourism potential). Next, somewhat analogously to what happens in medical disasters, when mass casualties have overwhelmed responders and not all can be saved, the task becomes triage: the identification of those areas already lost, those in peril and those still viable.

Most conservation activities thus have the goal of protecting areas of high conservation value that are under threat and yet still viable. For the conservation of a given eco-region, initial decision-making steps may, broadly speaking, include site triage, threat analysis, stakeholder analysis, and planning and implementation of conservation strategies (Box 15.1).

Just as it is true that not all sites can be protected through conservation efforts, it is also true that not all conservation can be helped through health and health care. From the conservationist's perspective, in considering whether health and health care ought to be part of conservation activities at a given site, the question is not: Is there a need for better health care here? (in a developing country the answer will nearly always be yes), but rather: Will responding to the need for better health and health care here (along with responding to other needs, such as economic development and education) help with conservation? To answer this question for a given place, we revisit Box 15.1 and add the questions in Box 15.2.

An ethical argument can be made that any major conservation activity in a developing country ought to include a health-care component whether or not it promotes conservation, since it is unjust to spend conservation dollars without addressing the health needs of impoverished local people. This obligation may extend to helping improve rural health care in the entire region, not just in villages considered critical for conservation. On the other hand, conservation organizations have limited resources and are accountable to donors, who expect that their gifts will be devoted to conservation work and to helping local people in ways that also benefit conservation. It is thus incumbent upon conservation

BOX 15.1 DECISION-MAKING STEPS FOR CONSERVATION PLANNING

- Site triage: What sites in the region have the highest conservation priority because they possess high conservation value, are in peril and yet are still viable?
- Threat analysis: What are the threats to the site, and at what level do they exist? Are there opportunities for cost-effective interventions that will reduce these threats?
- Stakeholder analysis: What groups have a stake in the conservation or other use of the site? What role does each group play in the threats to the site? What are the costs and benefits of conservation to each group?
- Planning and implementation of conservation strategies: How can conservation agencies engage the stakeholders to reduce threats to the site (e.g., encouraging sustainable economic development, addressing needs and interests of local people, providing political and financial opportunities for land protection)?

BOX 15.2 DECIDING WHETHER HEALTH CAN BE LINKED TO CONSERVATION

Key question: Will meeting an unmet need for better health and health care help with conservation?

- From which stakeholder groups do the identified threats to conservation arise?
- Are there stakeholder groups whose health and wellbeing can be linked to conservation?
- Are health and health care among the priority needs and interests of local people?
- Can health and health care be linked to sustainable economic development or to political or financial opportunities for land protection?

groups to strive to meet their responsibilities both to donors and to the communities where they work. Assisting with health care, especially if it can further conservation, is sometimes a way to do this. Moreover, if designed with generalizability in mind (i.e., working with the existing local health system and not being overly resource-intensive), a well-run health-care system in a conservation area can indeed serve as a regional or national model for good rural health care.

CONTEXT FOR COMBINED CONSERVATION AND HEALTH INTERVENTION

The Kelay Conservation Health Program (KCHP) has been in operation for three years. This programme brings together The Nature Conservancy (TNC), Community Outreach Initiatives (CORI, an Indonesian NGO specializing in health education), and the Berau District Health Department. Cooperation between a conservation agency and a government health department, though rare, is an example of medicine in the service of conservation. The KCHP operates in Indonesian Borneo (Kalimantan) at a site noted for its exceptionally high biodiversity value, with a human population whose health and wellbeing depend on a rainforest ecosystem (see Chapter 13, this volume). It has a dual agenda: to improve health for local people, and to allow more effective conservation of critical rainforest habitat.

Conservation concern

The number of orangutans left in the world is estimated to have decreased from 200,000 or more a century ago to about 20,000 today. Their range, once including much of Southeast Asia, is now restricted to lowland tropical rainforests in Borneo and Sumatra. Unless the rampant logging of these forests can be stopped or slowed, orangutans face extinction in the wild within the next 20 years (Van Schaik et al, 2001), and all but a few fragments of the rainforests they inhabit are likely to be lost by 2010 (Jepson et al, 2001). Efforts to save Indonesia's rainforests have had little success so far against the onslaught of legal and illegal logging. A 1999 World Bank review of ICDPs in Indonesia (none of which included a health component) reported that, despite well over US$100 million in external financing over the past decade, 'most of the attempts to enhance biodiversity conservation in Indonesia through ICDPs are unconvincing and unlikely to be successful under current conditions' (Wells et al, 1999). Challenges for protecting what remains of Indonesia's forests include financial constraints, rapid encroachment by logging, local poverty, widespread corruption and political instability. Only major financial commitments by developed nations or private donors may hope to avert a catastrophic loss of forests and biodiversity in Indonesia, and only if they are made soon and serve to finance well-planned and implemented conservation measures (Brack and Hayman, 2001).

Health concern

In post-decentralization Indonesia, newly powerful local governments now plan land use and seek to meet the needs of their constituencies (Barr et al, 2002; Casson, 2002). Poverty is the norm in rural areas, where annual per capita incomes are typically lower than the Indonesian national average of US$680 (World Bank, 2002). Roads, electricity, water and sanitation are often lacking: only 65 per cent of Indonesia's rural population has access to safe water, and only 52 per cent has access to adequate sanitation (WHO, 2000). Kalimantan's population is 70 per cent rural (Brookfield et al, 1995). The forests of Kalimantan are the traditional home of about 3 million indigenous peoples (Dayaks), out of a total Kalimantan population of 9 million (WHO, 1999). As in many parts of the world, indigenous peoples' wellbeing has typically been neglected by both government and industries in favour of the interests of people from more developed parts of the country, who in Kalimantan are typically transmigrants from Java and other islands (Fearnside, 1997).

The Indonesian Ministry of Health has for more than two decades implemented a rural health system along the Alma Ata model for villages throughout the country (Peterson, 2000). Doctors and hospitals are heavily concentrated in cities, with rural areas, including the Kelay River villages, notoriously underserved. *Posyandus* (village health posts) were originally supposed to be in all villages, but although some villages have a resident nurse or midwife, most *Posyandus* are only staffed by minimally trained village volunteers, and in many villages there is in fact no *Posyandu* at all (Robinson et al, 2001). The level of training of medical personnel in rural areas is also typically low; equipment and supplies are minimal, with vital items often lacking. Maternal, infant and under-five mortality rates in rural areas are all higher than national averages. Less than half of rural deliveries are attended by trained personnel. As in most developing countries, health-care costs in government facilities in Indonesia are exceedingly low compared with the USA, even after adjusting for income differences. Annual Indonesian government expenditures on health care in 2001 amounted to roughly US$6 per person.

KCHP: HISTORY AND APPROACH

A surprising number of orangutans (some 1500 to 2000, perhaps 10 per cent of the total world population) reside in one of the last large tracts of lowland rainforest in Indonesia, a remote part of the Berau District of East Kalimantan. Seeking an appropriate model of sustainable natural resource management suitable for replication in other areas of Indonesia, TNC made this area a conservation priority and in 2001 began developing a programme to address the needs of Punan Dayak villages along the Kelay River adjoining these forests as part of a broad-based multi-stakeholder conservation approach. Three major stakeholders were identified: local communities, local government and logging concessions. Conservation goals included reducing orangutan hunting by local people, creating and maintaining agreements with logging concessions not to operate in critical habitat areas, preventing illegal logging by outside groups, and improving governance at the village, sub-district and district level regarding natural resources management. In programme villages, conservation strategies included strengthening traditional land-use *(adat)* laws and their enforcement, developing economic alternatives to employment in logging, improving agricultural practices and market access, strengthening community structures and capacity, and providing meaningful incentives for conservation.

I began discussions in April 2002 with USAID and TNC, and we decided that because health care was an unmet need among the subsistence farmers and hunter-gatherers of the Punan villages, this site might be appropriate for a conservation health programme to complement and augment TNC's other conservation work along the Kelay River. Human population density in the district was low and, notably, logging had not yet expanded to levels seen in other parts of the island. The local Punan villagers, before being moved to their present location 20 years ago through government and church settlement schemes, had lived farther upriver in the rainforest, where they led a semi-nomadic hunter-gatherer existence (Abe et al, 1995). These Punan thus represented an indigenous group with close economic and cultural ties to the rainforest and an intrinsic interest in conservation, and in many ways were themselves a part of the ecosystem TNC was trying to protect. Local health care was reported by TNC staff to be rudimentary, and both local communities and local government emphasized the need for better health services. After consulting with local community leaders and government officials, TNC agreed to add a health component to its conservation effort with the idea of generating goodwill in the villages and the government, and linking health care both operationally and conceptually to conservation. This was an entry-point strategy; the barter approach was discussed but abandoned on the grounds that it would not be ethical to make health care a bargaining point. With initial funding of US$50,000 for the first year, TNC staff and I carried out a conservation and health assessment in January and February 2003, which led to the formulation of a conservation and health agenda for the programme. In July and August 2003 we conducted a baseline survey of health and conservation in seven villages along the Kelay River and in three comparison villages along the nearby Segah River.

The KCHP, which was planned in close collaboration with the Berau Department of Health, aims for cost-effective and sustainable improvements in both conservation outcomes, and human health and wellbeing in project villages. The intention was to improve the health care of the target population (especially mothers and children) until the Department of Health could upgrade its own services to this remote area through new nurses and facilities, which was to happen within a year. The KCHP was designed as a nurse-led community programme that would augment and support the local government health system at the *Puskesmas* (health centre) and *Posyandu* (community-sponsored village health-service post) levels, and train volunteer village *Kaders* (health workers) and villagers in disease prevention and primary care measures.

Posyandus represent the first level in the Indonesian government health-care system, providing basic elements of maternal and child health care at the village level, including family planning, immunizations, nutrition and diarrhoeal disease control. *Posyandus* are typically staffed by *Kaders*, volunteers who have received three to six days of initial training from *Puskesmas* nurses or nurse-midwives. They are supervised by *Puskesmas* staff, who visit monthly. Prenatal and well-child clinics are also held during these monthly visits (Peterson, 2000). As in much of rural Kalimantan, the *Puskesmas-Posyandu* system in the Kelay area had suffered from shortages of funding, staffing, transportation, and facility infrastructure. The system was rudimentary in the two downriver villages within the project area, and practically non-existent in the five upriver villages: the *Puskesmas* immunization team had visited the upriver villages only twice in the previous year.

In the first half of 2003, planning meetings for project implementation were held with district Ministry of Health staff and sub-district *Puskesmas* staff. The most critical needs in project area villages were for additional nursing staff and the establishment of monthly *Posyandus*. Beginning in June 2003, TNC hired one nurse to work temporarily in the most upriver village (Long Sului) until a permanent government nurse was placed there approximately six months later. Over the past two years, TNC and the Kelay Health Department have provided additional help:

- Physical infrastructure. A pipe system providing clean water from a mountain spring was installed in Long Sului, with the assistance of village residents. A few months later, a similar piped-water system was installed at Long Boi. Health posts were planned and constructed and demonstration garden plots growing forest medicinal plants were started in programme villages.
- Health worker and *Kader* training. Initial training by Community Outreach Initiative staff took place in three modules, each lasting approximately one week:

    1 training of trainers (including TNC and government health centre staff) in how to be effective mobilizing agents for community health;
    2 training for nurses in the diagnosis and management of common clinical conditions; and
    3 village-level *Posyandu Kader* training. *Posyandu Kaders* were selected by local government and project staff along with local village heads, using the criteria that they be literate, enthusiastic villagers suitable for the role.

- Mobile health clinic. A monthly clinic (*Posyandu/Puskesmas Keliling*) using boats, staffed by both TNC and Kelay *Puskesmas* staff, began operating in five programme villages and within a month was visiting all seven programme villages. Services include immunization of children and pregnant mothers, weighing of children, provision of supplemental food for children under five years who are below normal weight, and education of both *Kaders* and mothers. In 2004, a doctor joined the *Puskesmas Keliling* visits to programme villages every other month. *Posyandu/Puskesmas Keliling* visits continue up to the present time.

- Liaison between villages and district government. TNC has been working to coordinate programme activities with government agencies, including the Family Welfare Bureau, a politically influential governmental organization that includes many wives of high-ranking officials, such as the district chief (*Bupati*). This coordination has led to good support from these sectors for the KCHP's activities, collaboration between TNC and these sectors for certain programmes relevant to health and conservation, and visits to the project villages and KCHP by government officials. TNC also supported a meeting of *Kaders* from all programme villages with district government officials in the nearby town of Tanjung Redeb. The relationships of these aspects and components of the Kelay Conservation Health Program health-conservation linkages are shown in Table 15.2.

Table 15.2 *KCHP and health–conservation linkages*

Component	*Strategy*	*Linkage*
Conservation health assessment Participatory planning, implementation and management of programme with local communities and government Baseline assessments Village health-post construction and staffing Training of trainers, health workers and *Kaders* Monthly mobile health clinic via boat	Entry point, empowerment	Conceptual Operational
Water line Medicinal plant garden Diarrhoeal disease prevention Nutrition from sustainable agriculture and forest products	Bridge or symbiotic (depending on existing community awareness)	Conceptual
Literacy training Community–government liaisons Participatory evaluation	Empowerment	Operational
Community health education	Entry point or bridge (depending on topic)	Either (depending on topic)
Family planning	Symbiotic	Conceptual
Involvement of medical school and students	Bridge	Operational

Notes: The KCHP generally avoids the barter strategy on ethical grounds. Empowerment for conservation in the KCHP is through increased health of conservation allies, transferable skills, employment, or community capacity.

Programme planners noticed that village men tended to participate more in decision-making and high-impact uses of natural resources such as logging and hunting, while nearly all programme *Kaders* were women. Yet many *Kaders* had been informal leaders in their villages prior to the programme, and the programme works to further empower them so that they can better take part in natural resource decision-making. In addition, many health workers are men, as are all the clinic builders and boat drivers. The KCHP water projects benefit all village residents, and the overall programme is appreciated by both women and men as especially benefiting the children of their community.

Environmental monitoring for the programme area includes data on water quality, tree cover and orangutan populations. Community surveys of conservation knowledge, attitudes and practices are also done periodically. Health data, including nutritional status, on pregnant women and children under five are collected monthly.

ACCOMPLISHMENTS AND LESSONS LEARNED

An evaluation team, comprising representatives from TNC, the Berau Health Department, the Medical College at Mulawarman University and the Graduate School of Public Health at the University of Pittsburgh, conducted a comprehensive evaluation of the KCHP in 2005. Programme costs were determined from TNC records. Processes and activities were evaluated through interviews with programme staff and a review of programme records. The evaluation of outcomes involved qualitative components, such as interviews, focus groups and direct observations, as well as a quantitative assessment via baseline and follow-up household surveys, including a quasi-experimental study by which programme impacts could be gauged. Evaluation methods were designed and implemented with the idea that they were in themselves an intervention in a setting of scarce health resources, and that their utility to the conservation health programme should therefore be maximized. Thus, for example, interviews with officials were also a chance for TNC staff to strengthen relationships with them; visits to the villages to conduct the survey gave TNC staff opportunities to conduct *Posyandu* clinics as well as educate village nurses; and the medical students involved in the evaluation received an educational experience and also strengthened relationships between TNC and the local medical college, and had opportunities to influence the local medical profession.

Health impacts

After 20 months of operation and expenditures of about US$62,500, the KCHP is delivering basic health services to a remote population of 1100 people. It has had an especially dramatic impact on antenatal and well-child care, with immunization rates for children going from zero in the baseline survey to 85 per cent in the follow-up survey. There were smaller effects on community health knowledge. The tendency of upriver families to spend months at a time in forests or in farm fields away from their home villages presents an obvious challenge. Nevertheless, the programme has made a positive impression on stakeholders at all levels by providing basic health care. Its strong community base, solid documentation and support systems – seldom available in other remote areas in Indonesia – mean that the programme has the potential to serve as a model for rural and remote health-services delivery in other locations. The involvement of the Medical College at Mulawarman University is bringing increasing attention to the links between conservation and health, and the importance of rural health services in Kalimantan. Mulawarman students were enthusiastic and hard-working volunteer partners in both the baseline and the follow-up survey, and the medical college is now working towards making it a formal clinical training site for its students. These spillover effects are broadening the programme's impact beyond the programme villages themselves, across Kalimantan and potentially to other parts of the country.

Conservation impacts

Whether the KCHP has benefited conservation is critical to assessing its value to TNC as an organization. In addition, the costs of the programme need to be justified in terms of conservation benefits. Conservation impacts are typically much more difficult to detect than health impacts, especially within a short timeframe. It is also difficult to attribute a given outcome to a single intervention. TNC's conservation effort in Berau is both a political and a social process involving multiple strategies and actions. This process may in turn be influenced by political, economic, organizational and individual factors, many of which are simply beyond project managers' control.

That being said, TNC's conservation efforts in East Kalimantan over the past three years have been remarkably successful. During this time, TNC has facilitated the establishment of two locally managed protected areas, the Wahea Conservation Area (38,000ha) in the East Kutai District, and the Lesan River area (a core area of 12,000ha) in the Berau District. In addition, TNC is negotiating with the shareholders of the Gunung Gajah timber concession to set aside approximately 20,000 hectares from further logging. In the Upper Kelay communities specifically, TNC conservation successes include the following:

- No-hunting agreements. TNC negotiated and signed conservation agreements with the villages that include a commitment not to hunt orangutans.
- Local participation. TNC facilitated the formation of an inter-village forum that meets monthly and is now divided into four sections: health and clean water, timber company relations, alternative livelihoods and development. Forum members have been active in most TNC activities on the Kelay, including participatory mapping of village borders and village land-use planning.
- Local empowerment. TNC has trained the forum in communications, financial record-keeping, small grants proposal writing, and drafting of village legislation, and TNC and forum members are assisting four villages in creating laws related to natural resource use and the prohibition of logging without permits. The forum, if strengthened, will also become an effective negotiating body with the concessionaires and local government.

TNC's conservation strategy in East Kalimantan is to engage communities and local government in collaborative management. The KCHP is one component of this strategy for achieving TNC's overall project goal (as stated to USAID) of 'creating an enabling environment at the village level'. The present evaluation confirms that the programme did indeed meet this objective and, moreover, that it did so in a cost-effective way. Scott Stanley, TNC programme director for East Kalimantan, confirms that it has also helped to create an 'enabling environment' at the level of the local government. He credits the programme with helping to establish TNC's reputation among government leaders as an agency that assists and works well with local communities. He also states that TNC's health programme was one of the main reasons why the Berau District Regent has twice refused requests from small timber concessions to shut TNC down in the district. All government officials interviewed in this evaluation (including the head of the Kelay sub-district, Berau Health Department staff, other district and sub-district staff, and village heads) were enthusiastic about how much the KCHP is helping people in Kelay, and most of these people also emphasized the need for TNC's continued involvement.

Because the KCHP is structured around the community *Kaders* and focuses on community education and empowerment, it may also be helping TNC meet another goal: building community organizations. For example, the liaison component of the KCHP brought a group of 32 village *Kaders* to the district capital to meet with government officials for the first time. Using health as an entry point for community capacity-building could potentially dovetail with TNC's development of the inter-village forum (one section of which already addresses health and water issues).

The KCHP has thus made a valuable and cost-effective contribution to conservation both at the community level in the Upper Kelay and at the political level in the district of Berau and the province of East Kalimantan. The challenge now will be to strengthen conservation–health synergies and community empowerment strategies for conservation, and simultaneously transfer the commitment for health care to the government. TNC's village water and medicinal plant garden projects entail solid health–conservation links, but there are additional opportunities for links to conservation. For example, TNC's relationship with the village *Kaders* and monthly *Posyandu* puts the organization in a rare position for:

- developing a regular education and action forum for health, environment and conservation;
- transforming the *Kaders* into active community health leaders, mobilizers and educators through training and guided activities; and
- linking the *Kader* group to an inter-village forum that can better negotiate with local government and timber concessions.

These ideas are further discussed below. In addition, better 'marketing' of the health programme at the village level could better link it conceptually and operationally to conservation goals.

SPECIFIC PROJECT ELEMENTS

The *Posyandu* system is the main means by which KCHP promotes community-managed health in programme villages. Great strides have been made in establishing the *Posyandu* in the villages and instituting a monthly mobile clinic for Kelay. Rates of immunization and vitamin A administration increased dramatically in just 20 months of activity. User satisfaction with the *Posyandu* and health services provided by TNC is very high. In the 2005 survey, satisfaction with village health care in Kelay was 92 per cent among recipients, and satisfaction with the *Posyandu* in Kelay was 100 per cent. The perceived value of the health programme among residents of the villages is also high, with 63 per cent of respondents in Kelay who had heard of TNC affirming that it was the most useful thing that TNC was doing in their village.

Areas for possible improvement include the following:

- The involvement of the two partners, the Berau Health Department and Family Welfare Bureau, could be strengthened, particularly in the monthly clinics.
- Having the same doctor present at the monthly clinics and coordinating better with the village nurses could improve the tracking and care of the chronically ill, clinical training, and liaison work with the hospital in Tanjung Redeb.
- New government nurses, fresh from high school with no previous work experience, need additional supervision and training as clinicians, public health practitioners and educators. Ideally the project should continue to work with the Berau Health Department to accomplish this training. The Kelay nurses should then prove important village-level allies in future community work.
- New outreach and tracking strategies can raise the already-high rates of immunization and vitamin A administration closer to 100 per cent.
- The village *Kaders* could be even more active in the community. To become empowered as health and environmental advocates and as community teachers and leaders, *Kaders* require continued support as part of a team, with modest incentives, refresher training courses and morale-building activities. A recruitment mechanism is also needed to replace *Kaders* who become inactive or move on.
- Educational components of the monthly clinics could be strengthened, including training and support for community teachers, through more participatory learning methods.
- Monthly village clinics present excellent opportunities for integration with community building, education and other conservation work, such as coordinating with teaching forums for environmental issues, organizing forums for economic initiatives, and holding political forums to discuss issues such as land tenure.
- More active promotion of family planning has become important, given increased child survival. In the Kelay follow-up survey, 46 per cent of respondents stated that they were using contraceptives, but 67 per cent said they *would* use contraceptives if available free of charge.

There is high demand at both the village and government levels for TNC to stay involved in the monthly clinics. Staffing and boat transport will be needed for this, but coordination with conservation activities can minimize extra costs. Ideal roles include TNC as facilitator and educator in the clinic, the Family Welfare Bureau as supporter of the supplemental food programme, and the Berau Health Department to maintain responsibility for clinic functions, such as immunizations and child-growth monitoring.

The water projects in Long Sului and Long Boi are good examples of strong health–conservation links and of the programme's participatory style of management. In the 2005 survey, 58 of 61 respondents (95 per cent) reported getting their water from the new supply system at Long Sului; in 2003, all respondents reported getting their water from the river. Moreover, the residents of Long Sului have a strong sense of ownership of their water system, evident from the regulations that the village has already drafted to maintain and manage it, including a two-week rotation duty, shared among all residents, to keep the water basin clean and the establishment of a village fund for system repairs. The health–conservation link is also evident in these regulations, which explicitly protect the surrounding forest in perpetuity as a community watershed, prohibiting logging or cutting wood in the area for any purpose under penalty of law.

Encouraging the cultivation and use of medicinal plants is an excellent conceptual link between health and conservation. Suggested strategies for strengthening this early initiative in other villages include designating one or more garden caretakers in each village, providing community-wide education about growing and using medicinal plants, and developing markets.

TNC's support seems to have helped the Berau Health Department decide to build three new health sub-posts in the programme villages, increase staffing with six new nurses, and supply medicines and equipment. TNC's liaison work was also successful in establishing forums to bring *Kaders* and other Kelay residents together with district government leadership. These forums should increase the political capacity of Kelay communities over time and allow their fuller participation in the political decision-making process, while making TNC a critical intermediary between communities and district government that may help serve conservation goals.

The KCHP manager's excellent monthly and quarterly reports are under-utilized. Such information can inform ongoing decisions related to *Posyandu* activities and provide feedback to village health staff. With a little additional initial investment, given TNC's strength in GIS mapping and informatics, an exemplary community information system could be developed linking health, demographic, environmental, economic and other data. Such a system could serve many uses in monitoring and improving both health and conservation.

CONCLUSION

The design and implementation of the KCHP, the close collaboration with the local health department and the strong participation from the communities it serves have been important keys to its success. Initial meetings and training were critical in terms of establishing and demonstrating coordination, team spirit and government support. Joint planning with the Berau Health Department also allowed TNC to avoid duplication of services and to coordinate implementation with the department's own plan to increase health facilities and staffing in the upriver Kelay villages. It should be noted, however, that for a variety of complex reasons, the Health Department did not fully meet its agreed-upon responsibilities in the partnership on time, and that TNC has sometimes had to fill in unexpected gaps. Although the Health Department is committed to continued funding for health services in Kelay, ensuring that programme quality does not deteriorate will probably be a challenge. Continued health-system needs in Kelay will include staffing (such as regular visits by a physician to programme villages, and a nurse-midwife to train village midwives), basic equipment, a regular supply of medications, and ongoing training for *Kaders* and nurses. These needs are ultimately the responsibility of the Berau Health Department, though an organization like TNC can play a valuable role by continuing to argue for them and providing capacity-building and training.

REFERENCES

Abe, T., Ohtsuka, R., Watanabe, M., Yoshida, M. and Futatsuka, M. (1995) 'Adaptation of the resettled Kenyah Dayak villagers to riverine environment in East Kalimantan: A preliminary report', *Journal of Human Ergology* (Tokyo) 24(1): 33–36

Ali, R., Cranfield, M., Gaffikin, L., Mudakikwa, T., Ngeruka, L. and Whittier, C. (2004) 'Occupational health and gorilla conservation in Rwanda', *International Journal of Occupational and Environmental Health* 10(3): 319–325

Barr, C., Wollenberg, E., Limberg, G., Anau, N., Iwan, R., Sudana, I.M., Moeliono, M. and Djogo, T. (2002) *The Impacts of Decentralisation on Forests and Forest-Dependent Communities in Malinau District, East Kalimantan (Case Study 3)*, Center for International Forestry Research (CIFOR), Bogor, Indonesia

Brack, D. and Hayman, G. (2001) *Intergovernmental Actions on Illegal Logging: Options for Intergovernmental Action to Help Combat Illegal Logging and Illegal Trade in Timber and Forest Products*, The Royal Institute of International Affairs, prepared for UK Department for International Development, London

Brookfield, H., Potter, L. and Byron, Y. (1995) *In Place of the Forest: Environmental and Socio-economic Transformation in Borneo and the Eastern Malay Peninsula (Part 2: Issues of Endangerment and Criticality; Urban Development and Social Welfare; Levels of Urbanization)*, United Nations University Press, New York, NY

Casson, A. (2002) *Decentralisation of Policies Affecting Forests and Estate Crops in Kutai Barat District, East Kalimantan (Case Study 4)*, Center for International Forestry Research (CIFOR), Bogor, Indonesia

Engelman, R. (1998) *Plan and Conserve: A Source Book on Linking Population and Environmental Services in Communities*, Population Action International, Washington, DC

Fearnside, P.M. (1997) 'Transmigration in Indonesia: Lessons from its environmental and social impacts', *Environmental Management* 21(4): 553–570

Jepson, P., Jarvie, J., MacKinnon, K. and Monk, K. (2001) 'The end of Indonesia's lowland forests?' *Science* 292(5518): 859

Margoluis, R., Myers, S., Allen, J., Roca, J., Melnyk, M. and Swanson, J. (2001) *An Ounce of Prevention: Making the Link between Health and Conservation*, Biodiversity Support Program, WWF, Washington, DC

Melnyk, M. (2001) *Community Health and Conservation: A Review of Projects,* Biodiversity Support Program, Washington, DC

Mogelgaard, K. (2003) *Helping People, Saving Biodiversity: An Overview of Integrated Approaches to Conservation and Development,* Population Action International, Washington, DC

Peterson, C.E. (2000) *The 1993 Indonesian Family Life Survey Data, Appendix E: User's Guide; Appendix C: Organization of the Indonesian Health Sector,* National Institute for Child Health and Human Development (NICHD) and Agency for International Development (AID), Washington, DC, pp75–81

Puchong (1999) 'Pendeba action in the Qomolangma (Mt Everest) National Nature Preserve', in *Sustainable Development International,* ICG Publishing, London, pp169–174

Robinson, J.S., Burkhalter, B.R., Rasmussen, B. and Sugiono, R. (2001) 'Low-cost on-the-job peer training of nurses improved immunization coverage in Indonesia', *Bulletin of the World Health Organization* 79(2): 150–158

Taylor, D. and Taylor, C.E. (2002) *Just and Lasting Change: When Communities Own Their Futures,* Johns Hopkins University Press, Baltimore, MD

Van Schaik, C.P., Monk, K. and Robertson, J.M. (2001) 'Dramatic decline in orang-utan numbers in the Leuser Ecosystem, northern Sumatra', *Oryx 35: 14–25*

Vogel, C.G. and Engleman, R. (1999) *Forging the Link: Emerging Accounts of Population and Environment Work in Communities,* Population Action International, Washington, DC

Wells, M., Guggenheim, S., Khan, A., Wardogo, W. and Jepson, P. (1999) *Investing in Biodiversity: A Review of Indonesia's Integrated Conservation and Development Projects,* World Bank, Washington, DC

WHO (1999) Director-General Dr Gro Harlem Brundtland: speech to the International Consultation on the Health of Indigenous Peoples, 23 November 1999, World Health Organization, Geneva, Switzerland

WHO (2000) *Global Water Supply and Sanitation Assessment 2000 Report,* World Health Organization, Geneva, Switzerland

World Bank (2002) *Operational Manual: Operational Policies 3.10 – Annex C,* World Bank, Washington, DC

16

Hidden Suffering on the Island of Siberut, West Sumatra

Gerard A. Persoon

This chapter describes the past and present health-care conditions on the island of Siberut (Kabupaten Mentawai archipelago), in West Sumatra, Indonesia, and recent efforts to overcome the prevailing health problems of the local population.

The island of Siberut (see Figures 16.1 and 16.2) is inhabited by about 30,000 people. Most of them are ethnic Mentawaians (see Persoon, 1994; Persoon et al, 2002, for more about the Mentawaian Islands). Small groups of migrants have come from other parts of Sumatra (Minangkabau, Batak and Nias) or Java. The traditional population of Siberut consisted of patrilineal groups living in isolated settlements, or *uma*-communities, on the banks of rivers that cut through the tropical rainforest (Box 16.1). There was little division of labour on the island; every man and woman could basically perform all tasks. The medicine men, however, were specialized; they were responsible for maintaining a balance within the community, as well as with the environment, through extensive rituals (see Schefold, 1988). These medicine men based their healing practices on extensive knowledge of medicinal plants and curing rituals passed on to them by the previous generations.

Box 16.1 Traditional life on Siberut

Siberut is the largest of the Mentawaian Islands off the west coast of Sumatra. The population is predominantly Mentawaian (about 30,000 people). Relatively small groups of people originate from other ethnic groups such as the Minangkabau, Javanese and Batak. Traditionally, the people lived in small settlements along the banks of the rivers that cut through the thick forest. People lived by hunting, gathering, and sago and taro cultivation in the swampy areas along the riverbanks, and also practised animal husbandry with semi-domesticated pigs and chickens. The rivers and coastal areas provided ample opportunities for catching fish and crabs, and occasionally turtles. Their agroforestry practices in the swidden fields provided a wide range of food products, like bananas, cassava and fruits.

Shifting cultivation on Siberut is characterized by perennial crops, and its most remarkable aspect is the lack of fire. Vegetation is not burned after it has been cut; instead, trees and branches are left in the field. Seeds of fruit trees are planted before the trees are cut, and gradually the seedlings find their way through the withering vegetation. In this way the soil is never directly exposed to wind, sun and rain, thereby reducing erosion, and the nutrients are slowly released from the vegetation. Sago is the staple food. Annual crops like upland rice or corn were unknown until recently.

Exchange with traders from the mainland has ancient roots. Turtle shells, fish, copra and various kinds of forest products such as rattan and resins were offered in exchange for ironware, salt, cloth, tobacco and glass beads. In more recent times, numerous other products have been added to this list. Resource extraction from the natural environment offered products that were in substantial demand by the traders from Sumatra.

Politically, the people of Siberut were organized in small autonomous settlements consisting of one extended patrilineal family of up to 50 or 60 people. Friendly relations were maintained with other settlements within the same watershed area, but people never organized themselves at this level. The Dutch colonial administration imposed a system of village heads on the traditional situation. The estimated 250 to 300 *uma*-communities were forced into a smaller number of *kampung* in the early 1950s, each headed by a *kepala kampung*. The implementation of a 1979 law on village administration reduced the number of *kampung* to about 20 *desa*. Since 1999, however, the Mentawaian Islands have become a *kabupaten*, or district, of which the administrative centre is situated on the island of Sipora.

The people of Siberut have always been considered primitive and backward by outsiders. They were supposed to be in urgent need of development. Missionaries brought Christianity and specially designed development programmes for the isolated people, to bring them into the mainstream of Indonesian social and economic life. In the past few decades, logging operations and ecotourism have fundamentally changed conditions on the island. The state has claimed the primary forest resources on the island for granting logging concessions since the early 1970s, even though there is no 'empty' land on the island, according to the local people. The patrilineal groups consider even the closed-canopy forest as their property.

The weak resistance of the local Mentawaian people to outside interventions can be attributed to the lack of internal organization and lack of experience in dealing with more powerful outside forces. The autonomous *uma*-communities simply could not resist external powers; there were never any serious attempts to unite with other *uma*-communities for the sake of resistance. Instead, the people have sought to avoid conflict by retreating, giving in or fighting with the 'weapons of the weak': unobtrusive non-compliance with the regulations or temporary obedience. Intimidation or threat of force has usually been sufficient to impose policies upon the local people or their environment.

Considerable external attention has been aimed at 'saving Siberut', the largest island with a good deal of rainforest and with the most traditional population. The combination of a relatively small island with traditional people in their intact natural environment has a strong attraction for Western tourists as well as for donor organizations working in nature conservation. International films on the 'traditional' conditions on the island have been instrumental in attracting this kind of attention. Many of these films were broadcast in Western countries, but also in Japan and Singapore. All of them portray the harmonious life of isolated islanders in an abundant environment facing an aggressive outside world in search of adherents of new religions, timber, or with a mission to 'civilize' these 'primitive' indigenous people.

Though the people of Siberut always appeared healthy to outsiders and had a diversified diet, they suffered from a wide range of diseases and physical ailments. Malaria, rheumatism, tuberculosis, high child mortality, cholera and numerous other diseases, in combination with certain cultural practices, kept the total population relatively low. Since Indonesian independence, numerous efforts have been undertaken to 'civilize' the island.

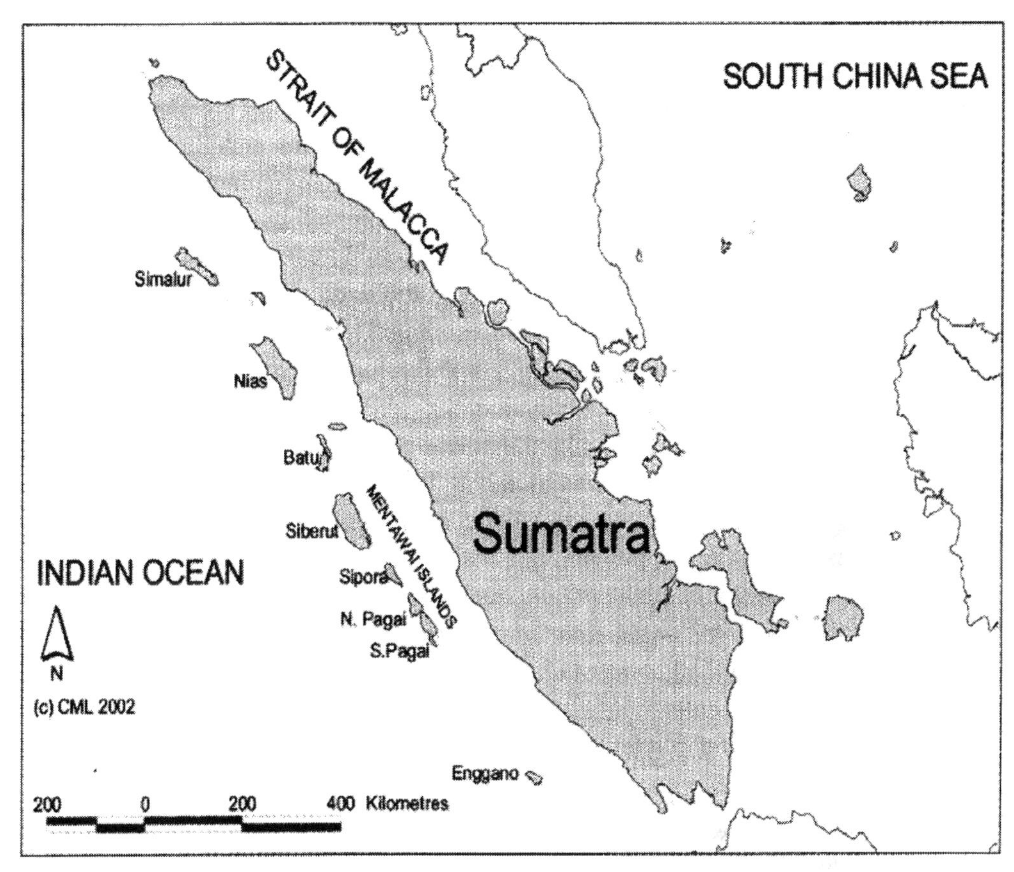

Figure 16.1 *Islands off the west coast of Sumatra*

Forced settlement in larger villages, a shift towards permanent agriculture and outlawing of the 'pagan' aspects of their activities, including those of the medicine men, were among the most prominent policy measures.

Despite the numerous development activities, many of which were combined with efforts to protect the rich biodiversity on the island, general health conditions barely improved. Some reasons involve the way modern health care was implemented on the island; other factors include the cultural perceptions and habits of the islanders. Modern health-care facilities, like those provided by the *puskesmas* (community health centres) and the clinics of the Catholic and Protestant churches, are concentrated in the main harbour villages and do not reach the isolated river valleys. Modern health-care workers have neither cooperated nor exchanged information. Over the years, they have consistently refused to recognize the roles of local healers and accept their assistance in improving general health conditions.

The islanders' own cultural practices have contributed to the persistence of health problems and much hidden suffering on Siberut. Perceptions about life, illness, cures and death are closely connected to religious beliefs and cannot easily be replaced by modern

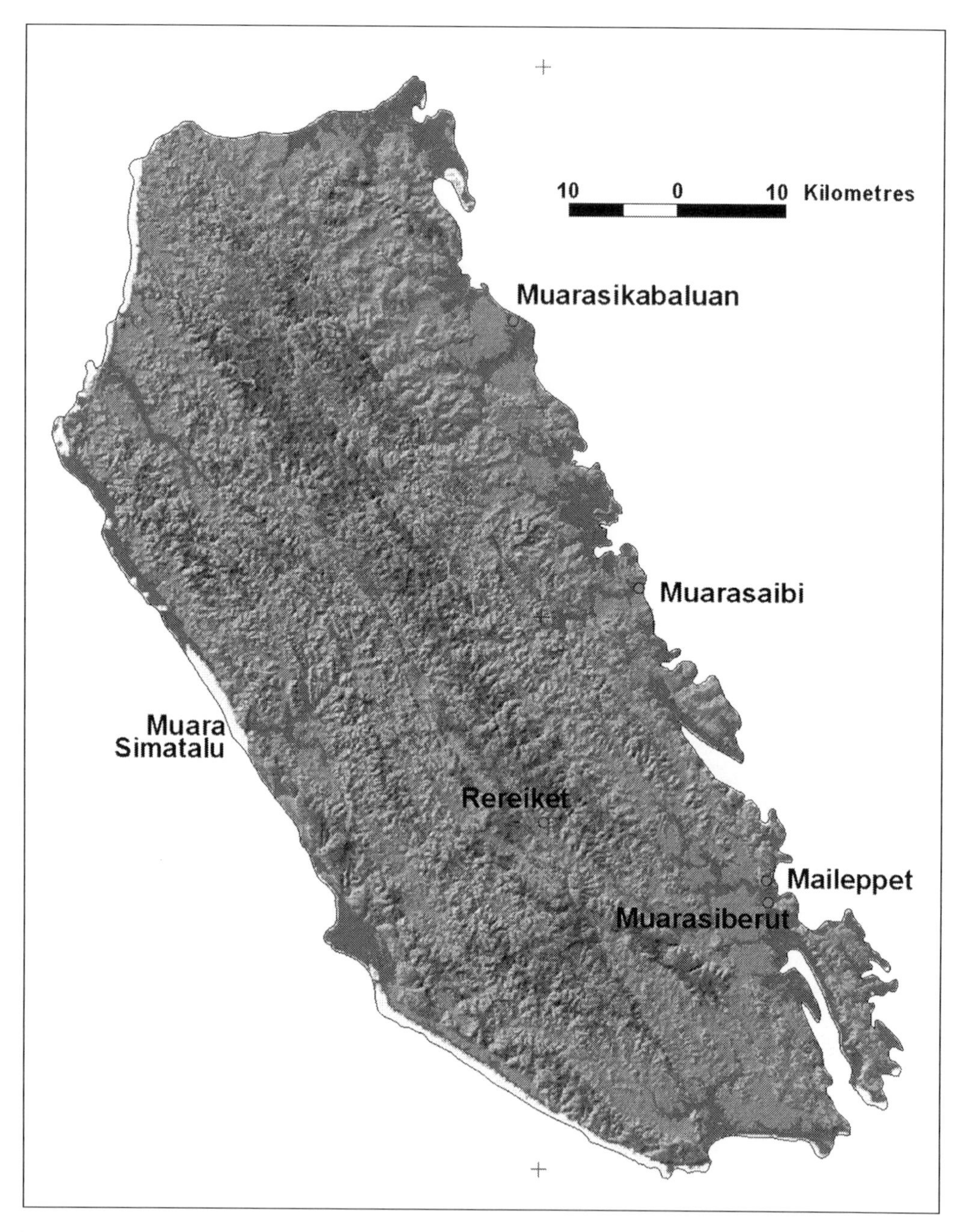

Figure 16.2 *The island of Siberut*

health care, which focuses on physical condition with little attention to the cultural perception of causes of illness.

A new, more culturally sensitive programme for community health is now being implemented. It aims to encourage cooperation among all parties involved in modern health care, while bridging the gap with traditional healers and involving the local communities – in particular the women – in improving health-care conditions.

TRADITIONAL HEALTH AND HEALTH CARE

Travellers to Siberut in the 19th and early 20th centuries usually portrayed the local population in very positive terms: the men and women were well fed, they looked strong, and they possessed an impressive material and religious culture (see Suzuki, 1958, for a bibliography of these accounts). Given the abundance of food in the forest and the relatively low population density, hunger was unknown on the island. The combination of wild and domesticated foods provided the inhabitants with sufficient nutrition throughout the year. People did not seem to suffer from major health problems. Many of the early visitors had no medical training, however, and they were perhaps impressed by the colourful rituals, the fully tattooed bodies and the communal housing. Most likely they did not have a sharp eye for health conditions.

Conditions in a moist tropical lowland and swamp forest are not always favourable to health. Problems are compounded by nutritional habits, sanitation and sleeping arrangements. Among the diseases and causes of death that have been reported for Siberut are typhoid, cholera, infertility (particularly of males), rheumatism, malaria, tuberculosis and tropical ulcers. The infant mortality rate exceeds 25 per cent, and other health problems exist (Höfner and Pech, 1978; see also Ciba Foundation, 1977, for general health and disease in tribal societies, and Baer, 1999, for similar health conditions among the Orang Asli of Malaysia).

The *kerei*, or medicine man, was called if people fell ill or if the sense of harmony within the *uma*-community was seriously disturbed. The medicine man was not a healer only in the physical sense; health problems were also seen as related to spiritual wellbeing. In many cases illness was thought to be caused by misbehaviour toward other people or the environment. Because Mentawaians believe that everything in the environment has a spirit, maintaining a respectful relationship with that environment is essential for societal and individual wellbeing. Healing for that reason implied also the restoration of the disturbed relationship with the environment. The environment in which people live is also thought to be the world of their ancestors, who need to be respected (Schefold, 1988; 2002). Thus, healing illness and physical discomfort was a complicated issue. The tasks of the medicine man were manifold, and long years of learning and experience were required. In addition to knowing the medicinal properties of hundreds of plants, the medicine man must learn songs and prayers to communicate with the spirits (Ave and Sunito, 1991). Healing ceremonies are therefore complicated events that combine physical and spiritual elements, and often involve many members of the *uma*-community.

Certain cultural practices and habits among the Mentawaians do not always favour good health: small infants receive mouth-to-mouth feeding and undergo extensive ritual bathing in cold water. The high infant mortality is also related to the lack of sufficient vitamin-rich foods after breastfeeding. A relatively large number of children die between their first and fourth year as a result of malnutrition. There are also no specialized midwives in the communities to assist in complicated deliveries. Another aspect of community life that facilitates the rapid spread of diseases is the practice of attending illness and healing rituals. As a result of these community meetings, some diseases spread rapidly, including cholera, typhoid, tuberculosis and typical childhood diseases.

MODERN HEALTH CARE

During the colonial period, from about 1910 to 1942, modern health-care facilities on the island were largely limited to the capital, the harbour village of Muara Siberut, where a medical doctor was stationed. There were some efforts to keep pigs out of the settlements. But in general, the Dutch did not have a large impact on the health conditions on the island. Interestingly, one of the last medical doctors who worked on the island wrote his memoirs about his experiences on Siberut and in particular about his 'colleagues', the medicine men, for whom he expressed his respect (Van Beukering, 1947; 1978).

In the years after Indonesian independence, health facilities gradually improved. In addition to clinics (*puskesmas*) run by the government in the two main harbour villages, small clinics were established by Italian Catholic and German Protestant missionaries. The clinics were usually staffed by trained nurses; the government *puskesmas* were led by a doctor and a midwife. Because the foreign Catholics and Protestants and the predominantly Minangkabau government officials from West Sumatra (who are all Muslims) had little contact with one another, there was little cooperation among the health-care workers of these organizations. All were established in the two main harbour villages of the sub-districts (Muara Siberut in southern Siberut and Muara Sikabaluan in the north). There were no modern facilities in other parts of the island. Occasionally, nurses from the Protestant and Catholic clinics in particular visited isolated villages on a monthly basis, but these visits were discontinued because of transportation problems or staff turnover.

Dental care is completely absent. The health facilities are generally limited to relatively simple treatments of diseases and injuries. Patients with complications have to be taken to Padang, the capital city of the province of West Sumatra, which is a difficult 10- to 12-hour boat trip and a major journey for people who have never left the island. It may also be financially prohibitive, since Siberut is a self-sufficient economy with limited exchange of imported and exported goods and little wage labour, and thus cash is limited. The cost of medical care in modern hospitals is beyond the reach of most islanders.

In the numerous resettlement villages that have been built over the past few decades, scores of buildings have been constructed as health-care centres. Most of them, however, were never used for that purpose because staff and medicines were lacking. These resettlement villages were part of the 'development and civilization' programme of the Indonesian government, which wanted to bring the scattered *uma*-communities of Siberut into the mainstream of Indonesian society. People still identify with their *uma*-community, however, and solidarity and community spirit in these villages are therefore limited. Feelings of jealousy and mistrust dominate when it comes to distribution of benefits from the government or other organizations.

In general, health-care conditions failed to improve because of the large concentration of people, the absence of good sanitation and an impoverishment of their diet. The change in diet came with emphasis on the cultivation of cash crops like cloves and *nilam*, which caused people to neglect their more diversified sources of food production and gathering, as practised in the old settlements. People were also forced to cultivate rice instead of sago, the latter being widely considered a low-status and 'lazy man's' food. Some shops in the resettlement villages sell a small range of medicines for common health problems like malaria, coughs, skin infections and wounds.

As part of the push toward modernization, traditional health care and the medicine men in particular became the object of severe attacks. In the early 1980s, traditional health care was once more condemned as 'primitive superstition' and 'paganism'. Under the threat of police force, the medicine men gave up their healing rituals and abandoned the glass beads, bells and other paraphernalia they used during these rituals. Many of the medicine men gave up their specialization, at least for some time.

When foreign tourists started to come to Siberut in larger numbers in the late 1980s, the main attraction of the island was the colourful population and, in particular, the medicine men and their elaborate rituals. The government started to take a somewhat more relaxed attitude to the medicine men. Some picked up their old profession again, and new medicine men were inaugurated. This trend has largely continued until the present day, even though tourism has suffered dramatically from the political unrest following the end of the Suharto regime in 1998 (see Persoon and Heuveling van Beek, 1998; Bakker, 1999).

For the villagers, there remains a clear differentiation between traditional health care and modern facilities. People in the village generally turn to the medicine man for advice first. Other people have now obtained knowledge about medicinal plants but are not inaugurated in the magico-religious world of the medicine men. *Siagai laggek*, 'those who know about medicine', can help people who suffer from health problems that can be cured relatively easily by certain medicinal plants. The medicine man would look more deeply into the problem, find the appropriate medicinal plants, and then perform one or more healing rituals, which in most cases combine physical treatment with medicinal plants and a curing ritual to please the soul of the patient. If necessary, such a performance might be repeated. The medicine man might receive a payment in kind for his services. If the healing ritual requires a larger ceremony, with many people, a communal ritual involving the killing of pigs might be prepared for the entire *uma*-community of the patient.

If the patient does not improve or if the problem is beyond the capacities of the medicine man, he or she can go to one of the clinics in the harbour villages. Moving sick people over large distances downriver or along the coast in small boats is a problem, however, both because of the distance and also because of the people's perception of disease. There is a strong belief that sick people should not be taken away because if the soul leaves the body of the ill person, it might not be able to return.

People who do go to the clinics are not accustomed to being away from their *uma*-community for a long time in an unfamiliar environment, paying cash for medical treatment (a situation that causes frustration and irritation between the patients and the doctors or nurses), or undergoing prolonged medical treatment. If the medicine is good, patients assume, it should work almost immediately. Instructions to take medicine for many days or even weeks are met with distrust and often not obeyed. As a result, patients often 'shop' for health care.

Development and conservation projects

Over the past few decades, numerous development and conservation projects have been implemented on the island. Most have been initiated by the government, in particular the Department of Social Affairs (see Departemen Sosial, 1997), but some have been financed by international institutions, such as UNESCO (the United Nations Educational, Social and Cultural Organization), the Asian Development Bank (ADB), or the World Wide Fund for Nature (ADB, 1995; WWF, 1980; 1996). Recently, several small NGOs have started to work on the island (SKEPHI, 1992; Yayasan Citra Mandiri, 2002). Health care has never been a major component of these projects, however; instead, the focus is on general development, construction of new settlements and schools, introduction of new food crops, infrastructure (roads, bridges and electricity) and conversion to one of the officially accepted religions. Nature conservation in particular has been the dominant concern of interventions by foreign agencies. The largest effort so far has been the ADB-funded Indonesia Biodiversity Conservation project, which was implemented on Siberut and Flores. In reviewing all these activities over the past 25 years, it is somewhat surprising to realize how little attention has been paid to primary health care and the hidden suffering caused by conditions in the tropical lowland forest and poor health-care facilities, including poor sanitation. There are no toilets on the island, and people are used to relieving themselves in the forest. Though all these projects mentioned improving local livelihoods and welfare of the population, the amount of serious attention was strikingly small. Trained staff was not hired, and efforts to improve the situation were not undertaken wholeheartedly. The focus of attention was always on other topics. Once in a while, a major outbreak of cholera with numerous casualties prompted quick action by officials in Padang or even Jakarta, but attention quickly faded away once the immediate problem was solved.

As a result, the island is still regularly struck by epidemics, infant mortality remains unacceptably high, and the hidden suffering of the population continues. Diseases like malaria and typhoid continue to affect numerous victims on Siberut. Eyesight problems and rheumatism are widespread. Dental problems are frequent, now that the people consume large quantities of sugar and sweets but still have not become familiar with dental care. In addition, simple infections and injuries cause unnecessary suffering, pain and discomfort. Nevertheless, in the regular development programmes of the government, this aspect of life on the island receives relatively little attention compared with infrastructural improvements.

When in the mid-1990s Siberut was discovered as a surfers' paradise and began to attract large numbers of Australian surfers, the health conditions on the island became a concern. Widespread malaria and other diseases discouraged the surfers from spending time on the island; instead, they took fully-equipped boats from Padang to the island and remained on board. Interaction with the local population was very limited. At one point, however, a medical doctor and avid surfer saw the conditions on the island and took action. He established a foundation called Surf Aid, which is aimed at providing medical care to the local people and takes voluntary contributions from the surfers, who are generally well off. The money is used to provide mosquito nets to the people in the villages close to surfing areas, and a fully trained Australian doctor was stationed on the island for a time. But because of the political unrest in the country and in particular the October

2002 bombing in Bali, which killed more than a hundred Australians, the number of surfers has declined dramatically, and the financial resources of the foundation have diminished accordingly.

NEW COMMUNITY HEALTH-CARE PROJECT

Largely thanks to the initiative of a man from Siberut who was trained as a medical anthropologist in The Netherlands, a new proposal to improve the local health facilities was developed. Kirekat, an NGO, submitted it to a Dutch co-funding agency, Cordaid, and with substantial help from a Jakarta-based Catholic medical organization, Perdhaki, the proposal was accepted. An Indonesian branch of the NGO, Yayasan Kirekat Indonesia, began implementation in the second half of 2003. Funds raised by Memisa, an organization affiliated with Cordaid, will support this project for a number of years.

The Kirekat project aims to promote integrated community health care by paying more attention to the needs and capacities of the local population. Project organizers hope to overcome some of the problems that stymied previous projects, such as inadequate staff, lack of local cooperation, and ignorance of local conditions, culture and social relations. An inventory of the prevailing health-care problems is part of the project's activities. The project will gauge the willingness of the population to help improve general health conditions and provide education about the causes of diseases and epidemics, how they can be treated, and the health risks associated with some cultural practices. The project also makes a sincere effort to involve the local medicine men in the promotion of better health care. Instead of condemning their skills and experience as superstition or silly forms of paganism, organizers propose to involve the medicine men in an improved system of health care and raise awareness among them about which diseases and injuries are beyond their treatments, apart from their role in healing the patients spiritually.

A crucial element in the Kirekat project is stimulating cooperation between the *puskesmas* run by the government, the district's health-care officials, and the various health-care activities of the Catholic and Protestant organizations. Because of a history of mistrust, this is a major challenge. To create a common understanding of the health-care problems and share experiences, workshops have been organized with the nurses and doctors of the clinics as well as with officials from the district health department. At one of these meetings, it was surprising to see how much people shared in terms of knowledge, experience and concerns, yet they had never joined forces. With the new project in a facilitating role, these kinds of problems may now be overcome.

The Kirekat project has limited funds, but the basic idea is less to provide the local people with new facilities and medicine than to mobilize the local capacities to engage in self-help activities, change health-related behaviours and promote cooperation among the existing institutions. But the last aspect should be the final stage in a community health-care process. Training of local community members in primary health-care activities will be part of the programme. Special attention will be paid to women, and to infant and child care. In the past, health care was organized largely through the typical official administrative structures, in which males dominate and women are hardly ever heard (see Chapter 6, this volume).

Improvements cannot be made overnight. Decades of mistrust, condemnation of the medicine men, and competition instead of cooperation between the religious and governmental health-care organizations are not easy to overcome. Mutual trust can develop only over time and on the basis of new and positive experiences. That is why all parties involved in improving the general health conditions on the island need to make long-term commitments. Traditional formal medical approaches will need to incorporate the role of the medicine men and local healing practices. A learning and listening attitude will need to replace the culturally insensitive ways of modern health-care organizations. Because nearly all health-care workers are of a non-Mentawaian ethnic origin and have little or no training in the local cultural traditions, they have much to learn. The government health-care workers sent to the island from Sumatra must now take a sincere interest in a culture that they have previously considered backward and primitive.

PRELIMINARY RESULTS AND CHALLENGES AHEAD

In its first two years, the Kirekat project has organized workshops with institutions at the regional and district level, including representatives of the Padang-based Catholic hospital. The aim of the workshops was to discuss the health conditions on the island and the contributions that each of the parties could make to solving the problems. Creating an atmosphere of collaboration instead of competition is one of the main goals of such meetings.

At the local level, health records and the health-seeking behaviour of the villagers were studied. Intensive discussions were held with the villagers about their health problems and experiences in seeking improvement in their physical and mental wellbeing. Based on these inventories and consultative meetings with experts, an integrated plan for improving primary health care was designed for two watershed areas, Rereiket and Simatalu, that had not previously received help.

The medical staff of the project have worked on training health cadres recruited from among the villagers, establishing a small health-care centre, and interacting with the traditional healers, the medicine men as well as *siagai laggek* ('those who know about medicine'). Simatalu is by far the most isolated area on Siberut, and local health facilities are very poor. The area was seldom visited by health-care workers in the past. The Kirekat project has now established two permanently staffed posts with an adequate stock of medicines, including vaccines. Radio communication has also been established to make travel to the main harbour villages less necessary. Additional supplies of medicines or assistance in case of emergencies can now be requested immediately through the head office of Kirekat near Muara Siberut.

Moreover, in both areas, Kirekat has selected a few villagers for basic training in medical care, medicine and sanitation. Safe drinking water, in particular during the dry season, is a major concern in most settlements. Sanitary facilities are also absent in most villages, and cause pollution problems with the increased density of people in larger settlements. The amounts of household waste and human excrement become more and more problematic and require serious attention. Meetings with the villagers increase their awareness about the problems arising from these conditions (Yayasan Kirekat Indonesia, 2005; 2006).

One of the important achievements of the Kirekat project is a more culturally sensitive approach to health care. Involvement of the medicine men, attention to the women, and an *uma*-focused approach are some of the elements. This approach (see Figure 16.3) allows every *uma*-community to become involved in the project activities by sending a representative to the cadre groups in every village. This helps overcome people's deep-seated loyalty to their own *uma*-community in the village. Instead of ignoring this loyalty, as many previous projects have done, here it is a starting point for any activity.

It is not easy to determine the improvement that has been achieved over the past few years because of the absence of reliable health records. But based on the promising results of the first phase, Kirekat has submitted a proposal for continuation and extension of the project for another three years. This proposal has been received favourably.

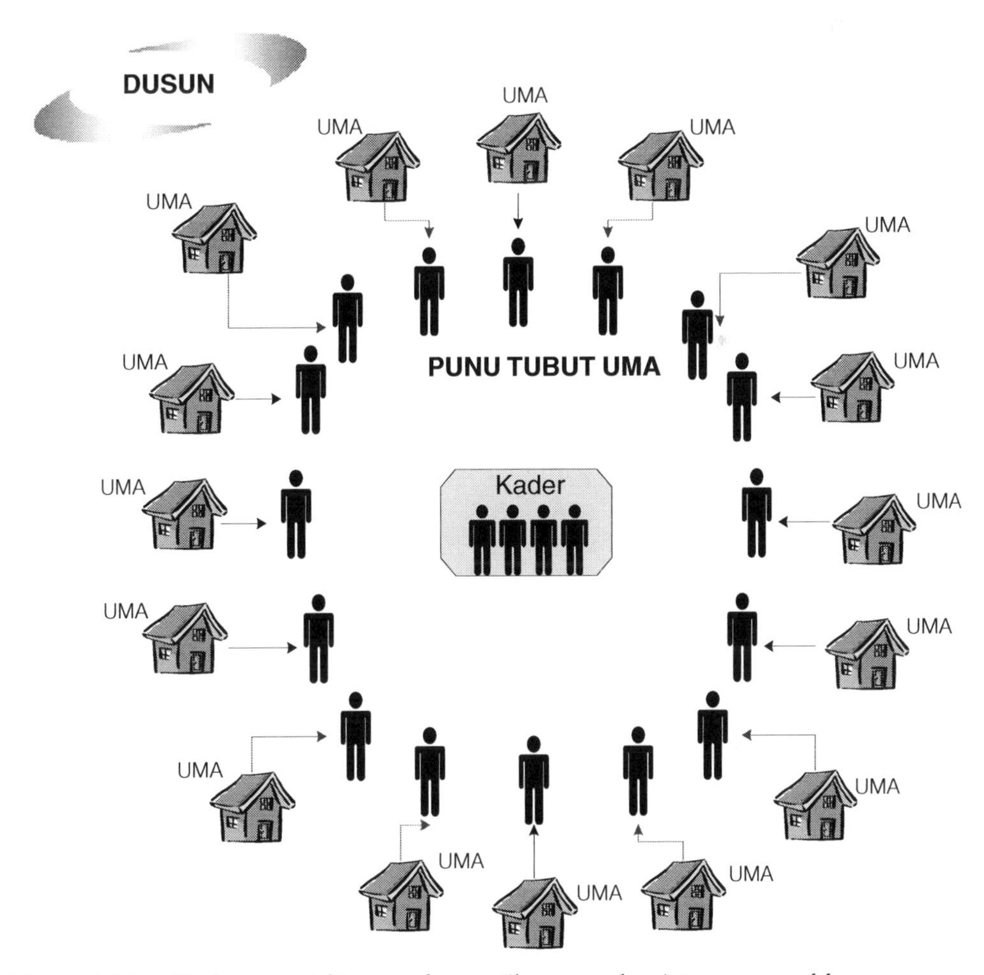

Figure 16.3 *Each* uma *within a settlement (*dusun *or* desa*) is represented by one person to jointly form the cadre for implementing the health-care programmes*

CONCLUSION

Health-care problems on the island of Siberut were ignored and even denied for a very long time. Other priorities in development planning and biodiversity conservation have led to persistent problems in health conditions and considerable hidden suffering by the local people.

The limited modern health-care facilities provided by the government and religious organizations have not been culturally sensitive. They provide health care from the point of view of modern Western medical science alone, ignoring traditional health care or condemning it as irrelevant and primitive.

A new project is integrating available facilities and involving the local communities in improving general health conditions and making health care more efficient. Though first results are promising, it will be a matter of continued hard work to arrive at truly integrated community health care that will improve the lives of the people of Siberut.

REFERENCES

ADB (Asian Development Bank) (1995) *Siberut National Park Integrated Conservation and Development Management Plan (Volume I/III)*, Ministry of Forestry, Jakarta, Indonesia

Ave, W. and Sunito, S. (1991) *Medicinal Plants of Siberut: A World Wide Fund for Nature Report*, WWF, Zeist, The Netherlands

Baer, A. (1999) *Health, Disease and Survival: A Biomedical and Genetic Analysis of the Orang Asli of Malaysia*, Centre for Orang Asli Concerns (COAC), Kuala Lumpur, Malaysia

Bakker, L. (1999) 'Tiele! Turis! The social and ethnic impact of tourism in Siberut (Mentawai)', MA thesis, Leiden, The Netherlands

Ciba Foundation (1977) *Health and Disease in Tribal Societies*, Symposium 49 (new series), Elsevier, Amsterdam, The Netherlands

Departemen Sosial (1997) *Pembinaan kesejahteraan masyarakat terasing*, Departemen Sosial, Jakarta, Indonesia

Höfner, W. and Pech, H. (1978) 'Ernährungs- und Gesundheitszustand von unter Fünfjährigen der Mentawai-Insel Sipora/Indonesien', *Tropenmedizin und Parasitologie* 29: 497–508

Persoon, G.A. (1994) *Vluchten of veranderen. Processen van verandering en ontwikkeling bij tribale groepen in Indonesië*, FSW, Leiden, The Netherlands

Persoon, G.A. and Heuveling van Beek, H. (1998) 'Uninvited guests: Tourism and environment on Siberut', in V. King (ed) *Environmental Challenges in Southeast Asia*, Curzon, London, pp317–341

Persoon, G.A., Schefold, R., de Roos, E. and Marschall, W. (2002) 'Bibliography on the islands off the west coast of Sumatra (1984–2002)', in R. Schefold and G.A. Persoon (eds) *Special Issue, Indonesia and the Malay World* 30(88): 368–378

Schefold, R. (1988) *Lia, das grosse Ritual auf den Mentawai-Inseln*, Dietrich Reimer Verlag, Berlin, Germany

Schefold, R. (2002) 'Visions of the wilderness on Siberut (Mentawai) in a comparative Southeast Asian perspective', in G. Benjamin and C. Chou (eds) *Tribal Communities in the Malay World: Historical, Cultural and Social Perspectives*, Institute of Southeast Asian Studies (ISEAS), Singapore, pp422–438

SKEPHI (1992) *Destruction of the World's Heritage: Siberut Vanishing Forest, People and Culture*, SKEPHI, Jakarta, Indonesia

Suzuki, P. (1958) *Critical Survey of Studies on the Anthropology of Nias, Mentawei and Enggano*, Koninklijk Instituut voor Taal-, Land- en Volkenkunde Bibliographical Series 3, Martinus Nijhoff, 's-Gravenhage [The Hague], The Netherlands

Van Beukering, J.A. (1947) 'Bijdrage tot de anthropologie der Mentaweiers', PhD dissertation, Kemink en Zoon N.V., Utrecht, The Netherlands

Van Beukering, J.A. (1978) *Arts in de tropen. Dokter in 't oude 'Indië' en 't nieuwe Afrika*, Ad. Donker, Rotterdam, The Netherlands

WWF (1980) *Saving Siberut: A Conversation Master Plan*, WWF, Bogor, Indonesia

WWF (1996) *Indigenous Peoples and Conservation: WWF Statement of Principles*, WWF, Gland, Switzerland

Yayasan Citra Mandiri (2002) *Implementasi KAM (LGC UNAND) di Siberut* (3 CD-ROM), Padang, Indonesia

Yayasan Kirekat Indonesia (2005) *Improving Health Care Projects in Isolated Areas on Siberut, Annual Report*, Yayasan Kirekat Indonesia, Padang, Indonesia

Yayasan Kirekat Indonesia (2006) *Improving Health Care Projects in Isolated Areas on Siberut, Annual Report*, Yayasan Kirekat Indonesia, Padang, Indonesia

17

Conclusions and Ways Forward[1]

Carol J. Pierce Colfer

Our perusal of the literature on human health and forests (Colfer et al, 2006) convinced us of the need for more focused attention to this topic, and this book represents an early contribution to this effort. In the previous study, we focused specifically on forest foods, forest diseases, traditional medicine and health care, and forest cultures. This collection consists of 15 essays on more specific topics central to the study of human health and forests, each written by one or more authors with expertise in that field. One goal for this book has been to look at the central issues in more detail.

Another goal has been to pull together into one volume the diverse approaches to this field of study. A need for improved interdisciplinary communication among experts dealing with health and forests was identified in our earlier work. The existing information on this topic is scattered throughout the literatures of many disciplines, making it difficult for a researcher or student to find all the relevant material on any one aspect of health and forests. The diversity of approaches represented in this book is a strength in addressing this problem and alerting researchers to other bodies of pertinent knowledge. However, it also means that any given reader will find some chapters – those closer to his or her own discipline – easier to read and understand than others. Part of the difficulty of interdisciplinary communication derives from the different intellectual traditions and norms that characterize the different disciplines. Some of these differences are clear in this collection. Although the review process, which included people from other disciplines, helped smooth over these interdisciplinary communication difficulties, some inevitably remain.

Finally, there are different academic traditions in anglophone versus francophone countries, in America versus Europe, and in the North versus the South. This book reflects these differences. Francophone authors have written Chapters 8, 12, and 13. Chapters 4, 6, 10, and 16 have non-francophone European lead authors. And Chapters 2, 3, 4, 6, and 9 were written or co-authored by experts from Australia, Ghana, India, Chile and Japan. Excluding myself, North Americans were lead authors on Chapters 3, 5, 11, 14, and 15. Box authors come from Germany, the Netherlands, the Philippines, US and Zimbabwe.

Cross-cutting features and issues

Precisely because this book highlights the multitudinous ways in which forests and health concerns intersect, a succinct summary of the contents is impossible. Table 17.1 tries to capture a bit of this diversity by clustering some of the conclusions into six topics: foods,

Table 17.1 *Six topics linking forests and human health*

		Chapter
Foods		
	Forest people's health can be improved with better nutrition.	2
	An enormous diversity of forest foods provide important nutrients, especially for vulnerable groups.	4, 12, 13
	Forest foods have a safety-net function (seasonality, famine, war, disease).	4, 10
	Forest products are important parts of household economies.	4, 10
Medicines		
	Individual women benefit in various ways from the ability to control their fertility.	7
	Forest foods and medicines are disappearing with habitat shrinkage and over-use.	3, 10, 12
	There is increasing interest in integrative medicine that brings together formal Western and traditional medical systems.	14
	Declining wild stocks of medicinal plants are accompanied by adulteration and species substitutions, which in turn reduce efficacy, quality and safety.	3
Diseases		
	Exposure to smoke from fuelwood adversely affects health (causing chronic obstructive pulmonary disease, acute lower respiratory disease and other problems).	5
	There are significant gender differences related to diseases affecting forest peoples.	7
	Working conditions make forest managers highly vulnerable to HIV/AIDS.	10
	HIV-affected households are more dependent on forests.	10
	Many diseases can be shared between humans and other primates.	12
	External exploitation of natural resources can introduce diseases.	11
	New pathologies for forest-dwellers are emerging (alcoholism, violence, poverty).	11, 12, 13
	Bats and other forest creatures serve as reservoirs and vectors of disease.	7, 8, 9, 12
	Combating malaria will require more attention to human behaviours.	9
Governance and institutions		
	Improved governance is an important component of health in forests.	2
	Political and corporate actors often do not recognize the value of high biodiversity, which may lead to further extinctions.	11
	Significant efforts have been made to marry traditional and 'modern' health-care systems.	14

Table 17.1 *(cont'd)*

	Ensuring effective 'handovers' of health projects to local government partners can be a challenge.	15
	Outsiders can play important facilitation and training roles in collaboration with local health departments.	15
Environment–health links		
	Use of fuelwood is apparently rising because of an energy crisis.	5
	Population pressures have impacts on the environment.	7
	Population, childbearing, family health and women's economic, educational and sociopolitical opportunities are intimately linked.	7
	Environmental change alters vectors' habitats, bringing them into increasing contact with humans and thus causing more disease.	8, 9
	The speed and frequency of global transport have allowed diseases to spread much further and more rapidly than in the past.	8, 12
	For many forest peoples, there is a close relationship between their wellbeing and the forest itself.	11, 12, 13, 16
	External exploitation of natural resources can cause environmental and social problems.	9, 11
Culture and way of life		
	Cultural patterns (e.g., silent trade) can have protective health functions.	11, 12, 13
	Cultural patterns can exacerbate health problems.	5, 16
	Interaction among peoples with differing ways of life can adversely affect health.	10, 11, 12,
	Marriages of different health-care systems can be beneficial.	14, 15
	Health-care systems can be culturally insensitive, with adverse effects on people's health.	7, 16
	Cultural images affect our attitudes towards disease vectors.	8

medicines, diseases, governance and institutions, direct environment–health links, and culture and way of life.

I have somewhat arbitrarily divided the topics I would like to address further here into 'features' and 'issues'. Features are the characteristics that we have found in forested areas that need to be taken into account when planning health and forestry research and action. Issues are the topics that have emerged repeatedly in the analyses in the book, and these also need to be considered in such plans. I then provide some comments on possible ways forward in addressing the problems identified.

Features of human health and forests

The first feature is health differences among groups of people – both differences among contemporaries and changes over time. Froment (Chapter 12) takes an evolutionary view (see also Butler, Chapter 2), examining the cultural patterns that have both protected local people and encouraged disease transmission among hunter-gatherers. Froment and Dounias with Colfer (Chapter 13), look at health differences between hunter-gatherers and sedentary peoples, as well as change over time in Central Africa. Dounias and Colfer provide examples of cultural differences among forest-dwellers that have implications for health status, and they describe specific customs of the Punan of Borneo that have protected their health in the past, compared with nearby settled populations. They also indicate some of the recent behavioural changes with both adverse and positive health effects (discussed by Ali in Chapter 15 in the context of health delivery). Pattanayak and Yasuoka (Chapter 9) emphasize the enormous variability in different peoples' behaviour, in interaction with vectors, diseases and environments, and the implications of such complexity for malaria incidence and control. Persoon (Chapter 16), in describing efforts to improve health care on the Indonesian island of Siberut, stresses the changing external perceptions about local people's health and the disconnect between the traditional health system and modern health care.

Of particular significance to this feature are the differences between the susceptibility of people long resident in forests and recent in-migrants. Gómez (Chapter 11) describes the shocking increase in malaria incidence among the Yanomami after exposure to a flood of infected gold miners from other areas. Besides the urgent need for improved detection and treatment, these findings clearly suggest tailoring health-care interventions carefully to the context and to the populations at risk – in this case, both the Yanomami and the gold miners (also stressed in Chapter 9).

The second feature, introduced in Chapter 2, is the important role insects and forest animals play in disease transmission. Many forest animals can serve as disease reservoirs, and some can serve as vectors when they come into direct contact with human beings. This issue was most thoroughly addressed by Gonzalez et al (Chapter 8), regarding the role of bats in Nipah virus, Ebola virus and SARS. Lahm (Box 12.1) describes how gorillas and other primates have transmitted Ebola. There are many other examples of forest animals' involvement in human disease in the literature. The mosquito's role in malaria is highlighted by Pattanayak and Yasuoka (Chapter 9), who emphasize how certain species have adapted to both environmental and behavioural changes, including attempts at vector control. Froment (Chapter 12) calls germs the 'most dangerous predators of humankind' and frankly discusses the disease-friendly context humid tropical rainforests provide; see also Allotey et al (Chapter 7), who explain the vectors involved in eight diseases. The roles of forest animals, insects and bacteria in human disease stimulate considerable debate, with some researchers (though not the authors in this book) even calling for forests – as reservoirs for disease – to be cut down. Others argue that human health problems are exacerbated when forests are degraded or removed (discussed further below). The complexity and dynamism of the relationships among reservoirs, vectors and human disease are striking, and again we are reminded of the importance of attending to contextual variables – including human contextual variables – in trying to understand diseases and their transmission and persistence.

A third feature pertains to the value of medicinal forest plants in maintaining and improving forest peoples' health. In many remote forested areas, traditional medicines are the only option for the people, a fact amply demonstrated by Cunningham et al (Chapter 3). The same authors stress the importance of forest medicines in more urban environments as well. Lopez (Chapter 10) describes the value of medicinal plants in alleviating HIV/AIDS symptoms in southern Africa (also stressed by McMillen, Box 14.3, and Fowler, Chapter 14). There is a huge literature on medicinal plants, related issues of intellectual property rights and experimental efforts to work with local communities (such as Shaman Pharmaceuticals), and there is some recognition of the value of forests as reservoirs of medicinal substances not yet discovered (at least partially reviewed in Colfer et al, 2006).

The fourth feature is foods from the forest (see Colfer et al, 2006). This topic is dealt with exhaustively by Vinceti et al (Chapter 4), but it also emerges in the discussion of HIV/AIDS by Lopez (Chapter 10). These authors find an increased susceptibility of poorly nourished peoples to disease, as well as the importance of forests as sources of food (and income) for the poor. Dounias with Colfer (Chapter 13) stress the cultural meanings of food and the importance of looking at forests, food and health in a holistic way. As with medicinal plants, the literature on forest foods is extensive but seems to have been largely ignored in efforts to manage forests and improve the lives of people in or near forests.

Issues for health-care planning

The first issue I would like to emphasize pertains to gender, health and forests. The fact that I (a gender specialist) was surprised to see the centrality of gender in some of the analyses highlights the degree to which gender issues have been underplayed in this field. Smith and Allotey et al (Chapters 5 and 7, respectively), while noting data inadequacies, see dramatic differences between men and women in *exposure* to certain diseases and pollutants. Differences in *susceptibility* to certain diseases also appear as an under-researched issue among forest peoples, with Allotey et al providing some intriguing bits of evidence on such differences. Gender differences in *presentation* – whether a man or a woman is willing or able to recognize a disease and come forward for treatment – have also been identified. Variations in presentation can relate to differences in men's, women's and health workers' understanding of the causes of the health problem, recognition of disease symptoms, taboos and stigma relating to specific diseases, and norms about modesty and anticipated treatment at health clinics, among other things. Finally, differences in *treatment* have been identified. Women in many areas have complained that they are treated with disdain, ignored or kept ignorant, particularly by the formal health-care establishment (compare Dounias's observations on the excellent treatment of first-time mothers among the Ntomba of the Democratic Republic of Congo, Chapter 13). Women often have less access to funds to pay for treatment, they may be subject to cultural restrictions on their mobility, or they may have household obligations – child care, elder care, domestic duties – that make trips to clinics much more difficult than for men. Colfer et al (Chapter 6) delineate in broad strokes the interactions among women's reproductive and work roles, family health, women's educational and income-generating opportunities, and their status. The chapter highlights the implications of excessive fertility for individual women's lives and the dangers of human population growth for forests. This book points to a need for greater attention to gender differences among forest peoples with regard to their health-care needs.

The second issue relates to emerging diseases (including HIV/AIDS, SARS and Ebola), the spread of which has sparked global concern. Though some recently recognized diseases have captured more attention than long-standing, underfunded and deadly diseases like malaria in Africa (which some argue are unaddressed because they do not threaten people in the North), one cannot deny that these new diseases represent global threats. Rapid transportation and the resulting increased mobility of humankind enables diseases to spread around the world in hours. Whereas in the past, efforts to address forest health problems had to rely on the 'carrot' of a mild global concern for justice or the environment, there is now a potentially effective 'stick' that must be recognized (cf. Garrett, 1994). Diseases that affect forest peoples or emerge in forests are highlighted in Chapters 2, 7, 8, 9 and 10 and considered indirectly in Chapters 11 and 12.

The third issue involves deforestation or land-use change and human health. This is well introduced by Butler (Chapter 2), who touches on the scientific controversies over these interactions (specifically, the idea that intact forests serve a regulating function with regard to disease). Pattanayak and Yasuoka (Chapter 9) provide quantitative, correlational evidence at macro-, meso- and micro-scales showing positive health effects linked to intact forests and forest protection. Gómez (Chapter 11) provides dramatic, long-term ethnographic evidence showing the adverse effects of forest disturbance on the health of the Yanomami of Brazil (see also the introduction to Chapter 1 and Dounias with Colfer, Chapter 13). The continuing controversies over human welfare and nature conservation are introduced by Butler (Chapter 2) and expanded on by Ali (Chapter 15), who describes an effort to provide health care to forest peoples in an area in Indonesian Borneo that an NGO wants to protect. Persoon (Chapter 16) makes similar points about the people of Siberut, also in Indonesia. Smith (Chapter 5) takes a different but related stance: his presentation of the research on the adverse effects of using firewood has not yet been taken up by those who study or address the role of fuelwood collection in deforestation in many areas.

The final theme pertains to existing efforts to improve health in forested areas and the different conceptual frameworks that characterize 'modern' and 'traditional' health care. Fowler (Chapter 14) describes global and national policies that encourage linking traditional medical systems with modern medicine, but she also describes the constraints to such integration (a topic further pursued by Persoon, Chapter 16, and Dounias with Colfer, Chapter 13).

Ways forward

As with any attempt to summarize the contents of this book, the suggested ways forward are multiple and complex. Table 17.2 provides some of the specific suggestions from different chapters. These focus on the needs for research, for improved interactions in various spheres and for improved practices on the ground.

Here I focus only on two particularly important avenues for improving the lives of forest dwellers and maintaining the forests themselves: interdisciplinary work, and participatory approaches that include women and youth. Attention to these two issues

Table 17.2 *Three categories of recommendations for improving human health and forest linkages*

Research on	Chapter
Various aspects of forest foods	4
Smoke inhalation	5
Forest-dwellers, their ill-health and their relationship to their environment	13
Systematic, comparative, longitudinal, holistic, interdisciplinary studies on health and forests	9, 13
Safety, efficacy and quality of medicinal plants	3
Gender differences in health	7
Improved interactions	
Multi-sectoral recognition of interactions among environment, population, health, income generation, education and women's status	6
Multi-sectoral approaches to solving the problem of HIV/AIDS in southern Africa	10
Greater interaction/cooperation between environment and health sectors	10, 13, 15,16
Interdisciplinary cooperation in health and forest interactions	2, 13
Better integration between traditional and modern health sectors	3, 14
Improved practices	
Better education/information relating to health, and use of traditional forest medicines	2, 3
Policy changes that recognize the value of medicinal plants and integrate them with formal health-care systems	3
Investigation and selection of appropriate certification and marketing networks for medicinal plants and producers	3
Improvements in combined treatment and prevention	2, 8
Conservation and sustainable use of forest foods and medicines	3, 4
New social understanding, new technology, new organizational approaches to prevent illness connected with smoke inhalation	5
Greater accessibility to family planning in forested areas, for both human and forest wellbeing	6
Reduction in human contact with vectors, improved disease recognition, epidemiology and biosecurity	8, 9
Greater involvement of forest sector in sustainable forest management to benefit human health	10
Closer, more effective partnerships between conservation and health professionals	15, 16

would seem to allow us to address the research, coordination and improved practices suggested by this book's contributors. Implicit in these suggestions is greater global attention to human health and forests more generally.

Interdisciplinary communication (also mentioned in Table 17.2) has been called for by many, yet examples of collaboration between foresters and medical people, for instance, or between ecologists and demographers remain too few. We should neither underestimate the difficulties of accomplishing such collaboration nor reject the possibility out of hand.

Any given reader of this book has undoubtedly found some chapters more compatible with his or her own thinking than others. The reader may even have skipped some chapters because the perspective or methodology seemed too foreign. Different fields use different vocabulary and have different conceptual frameworks and analytical conventions; there may be errors of fact in areas peripheral to an author's expertise; the conclusions may pose dilemmas for readers in other fields. These kinds of substantive barrier to better cooperation are important and well recognized.

But a bigger barrier is differences in implicit norms of behaviour. These are rarely discussed, but in my view, they block constructive interactions more than do scientific differences. We are all trained to read, study and learn, and most of us can, with effort, figure out what people in other disciplines are talking about.

In Box 17.1, drawing on my long years of interdisciplinary experience, my anthropological training and experience in analysing patterns of social interaction, I venture to offer some suggestions for 'best practices' in interactions among people from different disciplines.

Box 17.1 'Best practices' for interdisciplinary teams

- Recognize the importance of each discipline (including community members as experts on their own systems) in understanding the whole picture and your mutual interdependence. Avoid hierarchy and guard against a sense that your own discipline is superior.
- Recognize that all individuals and disciplines have strengths and weaknesses. Play to the strengths of each, compensating for the weaknesses.
- When a statement by someone from another discipline appears not to make sense, assume a difficulty in communication rather than incompetence.
- Keep communication channels open so that you can address misunderstandings and conflicts before they turn into major problems.
- Ask and answer questions freely and honestly.
- Minimize jargon.
- Be generous with praise and appreciation – strong motivators and trust-builders within the team – and take care to give credit to all who have participated. Sharing too generously is better than failing to acknowledge a contribution.
- Recognize the different incentive structures (such as criteria for advancement and sources of prestige) in different fields.
- Promise only what you can deliver. If you can't deliver on time, explain and apologize. Remember that others are relying on you.

The second important way forward follows from the global neglect of people living in forests. As noted earlier, the comparatively small populations living in forests, and the logistical difficulties reaching them, have resulted in political and financial decisions to allocate scarce health resources elsewhere. There is an almost total absence of formal health services and facilities for many forest peoples. Such problems are often compounded by changing cultural and ecological systems, where traditional safety nets (like foods and medicines from forests, networks of mutual aid within communities, belief systems that have contributed to mental health and conflict resolution) have been seriously compromised. Additionally, there is a growing awareness of both our global need for forests (e.g., for climate and bio- and cultural diversity, for sources of medicines now and in the future) and their potential as a source of serious, even deadly diseases. These facts demand attention by the health-care community, by governments and by donors.

There is a crying need for many kinds of research: from high-tech to low-tech, from conventional to experimental to participatory action research, in the whole range of forested countries, using the variety of disciplines already looking at these issues (usually individually). But as we respond, we must also link our research efforts with delivery mechanisms, whether traditional or modern, informal or formal, community-based or clinic-based – ideally, all of the above.

This collection has raised many questions pertaining to health and forests, and disciplinary specialists will see different critical issues that need addressing. Given the small populations and logistical constraints to providing conventional health care to forest peoples, I see the most potential for improving health in forested areas to be in working collaboratively with local people. Field experience in the 1990s brought to light the general agreement that good forest management ultimately depends on improving the health of people living in forests (CIFOR, 1999). Nearly ten years ago, a CIFOR team began experimenting with adaptive collaborative management (ACM) of forests, initially with emphasis on community forest management (e.g., Hartanto et al, 2003; Colfer, 2005; Kusumanto et al, 2005). Our experience also brought to light the strong demand for health services, including a willingness to work on health issues by the communities themselves (Hartanto et al, 2003). And we again found that women were most consistently the informal health-care providers within families.

The timing is right for taking an approach that involves women. Whereas graduate students and others have been consistently interested in such issues, the funding and institutional support from universities, governments and donors seemed steadily to decline over the past 30 years. Now, however, the Millennium Development Goals have emphasized health-related issues that governments and donors have begun to buy into. Western philanthropists like Bill and Lucinda Gates and some governments are allocating more funds for health in developing countries (though often still neglecting the critical details of people's behaviours and beliefs that affect health). And researchers in various fields have been fine-tuning methodological approaches to working with communities and governments to help address the continuing shortage of both funds and health professionals willing to work in remote areas.

With our partners, we at CIFOR are now trying to expand our ACM approach more explicitly into the health sphere. Part of our motivation comes from our recognition of the potential contribution of women in both environmental protection and health care. Women represent more than half the world's population, they have well-documented knowledge and experience with both family health and the natural resources in their areas, and their roles in reproduction and health care are widely acknowledged. Yet few formal programmes either tap their potential or contribute directly or effectively to women's wellbeing, even though there is ample evidence that improvement in women's wellbeing has spillover effects on the rest of the family. Figure 17.1 illustrates one hypothesis we think worth investigating.

A related group that has equivalent needs (and perhaps also potential) is children and youth. The plight of children in AIDS-afflicted areas of Africa is well known, and the prospects for these children, growing up without the benefits of parental guidance or education, are frightening. Again, we have a carrot (protecting children in need) and a stick (a continent – Africa – where conflict and ignorance may grow over time). The situation of children in most forested areas is less dire than that of AIDS orphans, but forest-dwelling children still suffer the health problems of forest populations in general.

Another potentially worrying dilemma is the conflict between the maintenance of indigenous knowledge systems – which can serve as a cultural 'insurance policy' for humanity, comparable to biodiversity in the biological world – and formal education. Many, such as Butler (Chapter 2), see great promise in educating the world's youth. Nevertheless, great care will need to be taken to ensure that the treasures that inhere in indigenous knowledge are not lost as we provide formal education to marginalized groups.

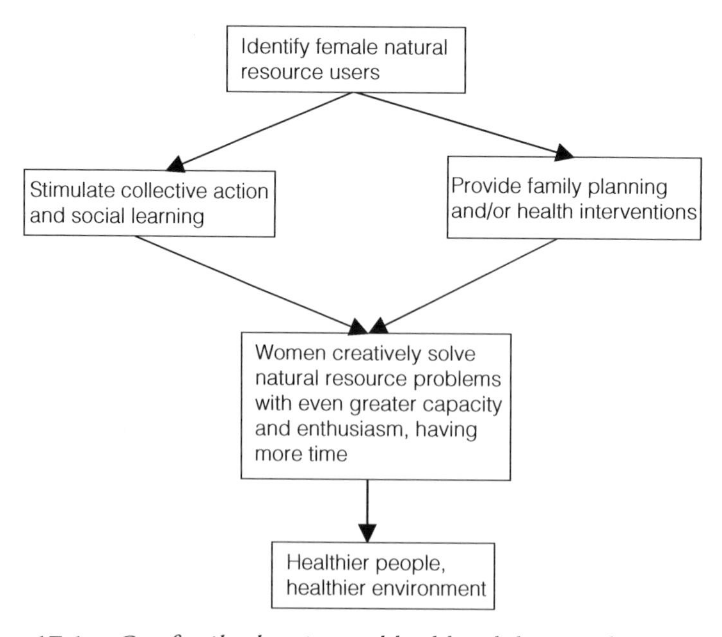

Figure 17.1 *Can family planning and health, while strengthening women's collective action, improve environmental conditions?*

Whereas many tropical forest women are illiterate, unable to speak national languages and over-burdened with work (three barriers to working with them), young people are likely to have some education, energy and free time. CIFOR and Riak Bumi, an NGO in West Kalimantan, are testing an unusual approach: pairing youths with women in a potentially powerful partnership for improving health in forested areas. The intent is to bring literacy and youthful energy together with the wisdom of their elders and with women's motivations to improve their families' health. This approach could also increase the sustainability of health efforts by involving youth from an early age in health measures, and strengthening both women's and youth's capacity for collective action in all spheres (including environmental protection).

This book has brought together analyses from a wide variety of disciplinary perspectives, all focused on the relationships among people, forests and human health. We have seen broad diversity in approaches and kinds of data, and we have identified issues that call out for further study and action. After working together on this collection, we have been even more struck by the dynamism and complexity that characterize forest and human systems. The importance of taking up the challenges presented in this book is clearer than ever to us. We hope that readers will renew their interest in and commitment to improving both human and forest health, as a contribution to improving life on Earth for all living beings.

NOTE

1 This chapter has benefited from useful and constructive comments on earlier drafts by Peter Kunstadter and Mike Arnold, though they bear no responsibility for any remaining misinterpretations or errors. An observation from Kirk Smith prompted a recommendation in Box 17.1. The work of Linda Yuliani and Yayan Indriatmoko (of CIFOR) and Seselia Ernawati (of the NGO Riak Bumi) in bringing together women and youth has been enthusiastically received and emboldens me to recommend this approach.

REFERENCES

CIFOR (1999) *C&I Toolbox*, CIFOR, Bogor, Indonesia

Colfer, C.J.P. (2005) *The Complex Forest: Community, Uncertainty and Adaptive Collaborative Management*, CIFOR and Resources for the Future, Washington, DC

Colfer, C.J.P., Sheil, D. and Kishi, M. (2006) 'Forests and human health: Assessing the evidence', Occasional Paper 45, CIFOR, Bogor, Indonesia

Garrett, L. (1994) *The Coming Plague: Newly Emerging Diseases in a World Out of Balance*, Penguin Books, New York, NY

Hartanto, H., Lorenzo, C., Valmores, C., Arda-Minas, L. and Burton, E.M. (2003) *Adaptive Collaborative Management: Enhancing Community Forestry in the Philippines*, CIFOR, Bogor, Indonesia

Kusumanto, T., Yuliani, L. and Macoun, P. (2005) *Learning to Adapt: Managing Forests Together in Indonesia*, CIFOR, Bogor, Indonesia

Afterword

Jonathan A. Patz

During the 1990s, approximately 9 million hectares of tropical forests were cleared and the annual rate of deforestation for Brazil alone reached 2.6 million hectares in 2004; at this rate, the world's rainforests could disappear before this century ends. Irreversible feedbacks could destroy our forests even sooner if one considers biophysical processes (e.g., when forest patches become so small that changes in evapotranspiration disrupt local climate conditions – culminating in drying and potentially self-combustion of the remaining stands of forest). While the value of forests is well recognized for the preservation of biodiversity across species, this book by Colfer and colleagues brings forth the broader array of 'ecosystem services' forests provide for our human species.

Human Health and Forests serves as a valuable update and companion text to a large international effort to measure the value of preserving ecosystems around the world. That report, the *Millennium Ecosystem Assessment*, asked the broad question: 'how does ecosystem sustainability relate to human wellbeing across many sectors and regions?' In the area of human health, for example, deforestation affects: the risk of injury and psychological distress from increased run-off and landslides in disturbed areas; livelihoods and mental health of indigenous forest-dwellers when their settlements are displaced or their sustaining natural resources taken away; and a number of emerging and/or resurging infectious diseases (see box below).

Partial list of conclusions of the *Millenium Ecosystem Assessment* (2005) on the linkages between ecosystem degradation and infectious diseases

1 Anthropogenic drivers that especially affect infectious disease risk include destruction or encroachment into wildlife habitat, particularly through logging and road building; changes in the distribution and availability of surface waters, such as through dam construction, irrigation, or stream diversion; agricultural land-use changes, including proliferation of both livestock and crops; deposition of chemical pollutants, including nutrients, fertilizers and pesticides; uncontrolled urbanization or urban sprawl; climate variability and change; migration and international travel and trade; and either accidental or intentional human introduction of pathogens (*medium certainty*).

2 There are inherent trade-offs in many types of ecosystem changes associated with economic development, where the costs of disease emergence or resurgence must be weighed against a project's benefits to health and wellbeing. Such trade-offs particularly

exist between infectious disease risk and development projects geared to food production, electrical power and economic gain. To the extent that many of the risk mechanisms are understood, disease prevention or risk reduction can be achieved through strategic environmental management or measures of individual and group protection (*high certainty*).

3 Intact ecosystems play an important role in regulating the transmission of many infectious diseases. The reasons for the emergence or re-emergence of some diseases are unknown, but the main biological mechanisms that have altered the incidence of many infectious diseases include altered habitat, leading to changes in the number of vector breeding sites or reservoir host distribution; niche invasions or interspecies host transfers; changes in biodiversity (including loss of predator species and changes in host population density); human-induced genetic changes of disease vectors or pathogens (such as mosquito resistance to pesticides or the emergence of antibiotic-resistant bacteria); and environmental contamination of infectious disease agents (*high certainty*).

Source: Patz et al (2005)

Malaria, for example, has been associated with land-use changes and human resettlement throughout the tropics. The clearance of forest has facilitated the colonization of humans into previously unsettled areas, as well as created forest gaps, cleared lands and culverts, in which anophelene mosquito species thrive. In Africa, deforestation is known to favourably promote a major disease vector, *Anopheles gambiae*, and the expansion of malaria is also occurring in Amazonia where deforestation has been shown to provide suitable breeding sites for *Anopheles darlingi*. In deforested areas breeding sites yield over a 100-fold increase in *An. darlingi* biting rates, even after controlling for human population density.

As forests decline, therefore, so can our health – particularly the health of communities dependent on the supporting services that intact forests offer. This book covers well the human dimension of deforestation, and realistically blurs the distinction between human and natural systems, which are inextricably coupled together. Trade-offs obviously exist between extraction of materials for near-term profit versus long-term sustainability of communities that rely on the preservation of forests for future generations.

The challenge then is to identify and promote optimal situations whereby the maximum number of people – not forgetting future generations (as well as number of species) – can benefit from informed forest management policies. Research and education are essential in this regard. If, for example, wood products are decided to be the highest valued resource of a forest, would this decision have been made if all services of the forest are counted (such as CO_2 sequestration, plant pharmaceutical or local food production, local climate stabilization, biodiversity maintenance, forest-related cultural values, or infectious disease suppression)? These key services (which are beautifully laid out in this book) should be balanced with other competing interests as we make decisions on the fate of the world's forests.

Yet even with eyes wide open to the diverse societal benefits afforded by intact forest ecosystems, we may be on the verge of inadvertently wiping out our remaining forests. In 2007, when the Fourth Assessment Report of the United Nations Intergovernmental Panel on Climate Change (IPCC) concluded, with '90 per cent certainty', that human activity (primarily burning fossil fuels) is causing global warming, the political will to reduce greenhouse gases finally arrived: good news! While reducing our reliance on fossil fuels is immediately and obviously necessary, an unchecked development of biofuels as the potential mainstay source for mobile energy could threaten our world's food supply – the analogy being that instead of feeding people, we take our grains to feed combustion engines! But most related to this book, as pressure builds to expand croplands to meet energy demand, our remaining forests will be (and in some cases already are) threatened by oil-palm, soybean and sugar-cane production.

Growing awareness of the linkages between human wellbeing and forest ecosystems is encouraging and a necessary reality as we attempt to achieve both human and environmental sustainability. There are myriad stakeholders with an interest in the fate of tropical forests. Clearly, the issue of equity pervades as one weighs the interests of local communities and transnational corporations, livelihoods across generations, and survival and health across species. Education and communication to arrive at fully informed decisions on natural resource management and conservation policy are requirements as we consider the best path forward. *Human Health and Forests* is a marvellous step towards fulfilling this requirement.

REFERENCE

Patz, J.A. and Confalonieri, U.E.C. (convening lead authors), Amerasinghe, F., Chua, K.B., Daszak, P., Hyatt, A.D., Molyneux, D., Thomson, M., Yameogo, L., Malecela-Lazaro, M., Vasconcelos, P. and Rubio-Palis, Y. (2005) 'Health health: Ecosystem regulation of infectious diseases', in Millennium Ecosystem Assessment (eds) *Ecosystems and Human Well-Being: Current State and Trends. Findings of the Condition and Trends Working Group, Millennium Ecosystem Assessment Series*, Island Press, Washington, DC

Index